Global Perspectives on Ecology

Thomas C. Emmel

with Lieselotte Hofmann

Mayfield Publishing Company

First edition 1977

Library of Congress Catalog Card Number: 76-56504
International Standard Book Number: 0-87484-338-3

Manufactured in the United States of America
Mayfield Publishing Company
285 Hamilton Avenue, Palo Alto, California 94301

This book was set in Elegante by Applied Typographic Systems and was printed and bound by the George Banta Company. Sponsoring editor was Alden C. Paine and Carole Norton supervised editing. The text and cover were designed by Nancy Sears, and Michelle Hogan supervised production. Inside cover photo by Kent Reno/Jeroboam and part title page artwork adapted by permission from W. D. Billings (1952).

Contents

Preface

There is an old saying that it is always light at the end of the tunnel. But, as one wag recently suggested, that light at the end of the tunnel may be the headlight of an oncoming train. If you agree, the readings in this book may tend to confirm your view. If, on the other hand, you are an optimist who regards the glimmer of light as the beacon of a new dawn, you may find support for your view in these readings, too.

While a great many books of readings on ecology—and, specifically, global ecology—exist, this one is unique in what it offers: a globally balanced approach to the pressing environmental problems and the potential solutions that deserve to be considered. Authors of general biology texts do not devote sufficient space to exploring these problems, yet case histories and the general ecological principles involved are frequently the most interesting and worthwhile parts of biology to the lay citizen.

This book can readily serve as a main reading text in ecology and environmental courses and as a supplementary text to any of the recent general biology texts (one, for example, is my *Worlds within Worlds: An Introduction to Biology*, 1977) to be used in ecology sections of introductory biology courses for majors and nonmajors alike. The articles have been chosen for their readability, currency, and biological emphasis as well as for their global balance. Because of their cultural, economic, and political import, they might well be used as source material in courses in, for instance, environmental politics, law, economics, and geography. Throughout, the case for a new way of looking at the earth—not as a fragmented world but as an interdependent one—is presented forthrightly but without strident doomsaying or undocumented assertions.

Too many of us are blind to our inheritance, indifferent to the terrible injuries sustained by our plucky planet. That we treat the earth as though it were a machine is reflected in the misleading term Spaceship Earth, which suggests that this planet is something man-made, a hapless, helpless object to do with as we will. We have closed our minds not only to many of our desecrations but to our own capabilities for changing the status quo. Fortunately for mankind and our earth, the collapse of human hopes is not imminent: there are a great many people who are trying to reverse the depressing trends and re-establish ecological sanity and stability. As Albert Mayer has noted, "Trend is not destiny." All the groups that are fighting daily to protect resources and wild places around the world depend upon the grassroots support of citizens in every nation, and therein lies the key to solving environmental problems. Enough people must care enough to try to change our present course.

We must ask ourselves: Is continual growth worthwhile? Is it not vital to save a part of the remaining natural world so that people everywhere can regain perspective on their essential and intimate relationships with that natural world? Out of a maelstrom of despair and poverty and inequity, amidst highly diverse societies and political orders, can an ecological consciousness be gained in time? Or is it now too late?

We shall soon be finding out the answers and, as in a marriage, will be faced with the situation "for better or worse." Unlike marriage, however, there is only one choice of a world for us to live with, and it depends on our actions within the next two decades as to the future character, indeed even the existence, of that partnership.

I thank the many authors and publishers who kindly granted permission to reprint the included papers, and numerous colleagues, especially Jonathan Reiskind of the University of Florida and Charles F. Leck of Rutgers University, who have given me the benefit of their sage advice and comments on the selections. Lieselotte Hofmann has been an invaluable editorial collaborator in preparing the introductory essays and the final selection material. This book had its inception in the efforts of Alden C. Paine to get me to commit in written form a series of ideas on global ecology that were presented in the four McElderry Fund Seminars on global ecology and environmental problems at the Idyllwild campus of the University of Southern California. Carole Norton of Mayfield has ably followed through on the editorial job of producing the final book and I am most grateful for her unceasing interest in this project.

Thomas C. Emmel

University of Florida
Gainesville

Introduction

A strange thing happened on the way to the "energy crisis" of 1973. In the late 1960s people around the world, but especially those in the industrialized nations, were suddenly caught up in a surge of concern over the headlong pace of environmental degradation.

Nowhere was this concern more evident than in the United States, where the mounting distress culminated in Earth Day on April 22, 1970. Temporarily casting aside their differences, mavericks and Establishmentarians uneasily joined hands and gave the earth their blessing. No longer was the environment merely the darling of do-gooders, bird watchers, nature nuts, and little old ladies in tennis shoes.

To sustain the euphoria of Earth Day, however, was obviously impossible, and when the momentum started faltering, there was relief in some quarters. As long as the embryonic environmental movement comforted itself with reams of rhetoric and piles of platitudes (which it did), the status quo was secure. Besides, said the debunkers, the movement was just a fad. In that case, retorted its supporters, it was the last fad.

From among the tens of thousands who jumped on or drove the bandwagon (see Selection 9), not a few jumped off when the movement showed signs of muscle, when rhetoric was translated into action. Despite intermittent and sometimes gloating reports to the contrary, the movement has not lost its punch. As the complexities of the political and socioeconomic ramifications have become apparent, it has become less frenzied, and at the same time more organized, more institutionalized, more dedicated; and its numbers have multiplied, though not by legions.

One of the main problems that environmentalists face is that their crusade involves threats to the earth's life-support system that are long-term and gradual. But policymakers, in business and in government, are, on the whole, myopic; they are concerned with short-term problems and short-term goals. In the environmentalists' battlecry to avert the threats to the earth or at least dull their impact, they see a threat of repression. And they are not far wrong. For salvaging the earth does mean clapping restraints on those who encourage mindless and profligate despoliation of our planet.

So an ecological backlash has been inevitable. "Anti-ecologists" charge that the movement, which may have started out with good intentions, has degenerated into an eco-cult replete with technophobes, prophets of doom, "elitists," and extremists. As one detractor puts it, "No reasonable person wants foul water or polluted air; but not everybody who wants clear water and clean air is reasonable."[1] Evangelical environmentalists now have to hold the fort against big business, labor unions, technocrats, and the federal, state, and local governments. Environmental legislation enacted during the first, heady days of the movement has occasionally been tinkered with to quiet the protests of those whose interests are not compatible with the best interests of a vulnerable biosphere. Worried ecologists and others in the life sciences find themselves trying to convince less-troubled scientists and laymen alike that the resources of the earth are finite, that the biosphere is fragile, and that someday all hell is going to break loose.

Why did all this trepidation over mankind's onslaught on the world's land, air, water, and wildlife become a torment only in the last decade or so? After all, even paleolithic man* was exhausting his environment and fouling his nest. He, however, could always move on to exhaust a new environment and foul another nest. The world was an open place, unexplored, teeming with riches to be used and squandered. And since man had a superior capacity to learn, he was able to expand over the entire earth, become the dominant species, and begin serious plundering of the treasures of "the only decent piece of real estate in an enormous volume of space."[2] Through unsound agricultural practices, deforestation and overgrazing, great civilizations—in Africa, Asia, Europe, and Latin America—devastated their surroundings and thereby destroyed themselves, leaving littered ruins and bleak landscapes as testimony to how nature fought back.† The countries where these civilizations once

---

*The word "man" to indicate both man and woman is used, however reluctantly, in this and other introductory essays to avoid complex, awkward syntax.

†The dichotomy of nature and man is, of course, a false one and is used here only as a literary convenience. Man is part of, not apart from, nature; he acts upon and is in turn acted upon by the rest of creation.

flourished are today among the poorest in the world. Aldo Leopold explains it this way:

> Civilization is not, as [historians] often assume, the enslavement of a stable and constant earth. It is a state of *mutual and interdependent coöperation* between human animals, other animals, plants, and soils, which may be disrupted at any moment by the failure of any of them. . . . Land despoliation has evicted nations, and can on occasion do it again. . . .
>
> Unforeseen ecological reactions not only make or break history in a few exceptional enterprises—they condition, circumscribe, delimit, and warp all enterprises, both economic and cultural, that pertain to land.[3]

Out of ignorance, greed, callousness, shortsightedness, or desperation, man for thousands of years has violated the limits of nature's tolerance. Often he has wrecked his habitat through heedless, day-to-day responses to the need for food, shelter, and clothing. As Leopold sees it:

> All civilizations seem to have been conditioned upon whether the plant succession, under the impact of occupancy, gave a stable and habitable assortment of vegetative types, or an unstable and uninhabitable assortment. The swampy forests of Caesar's Gaul were utterly changed by human use—for the better. Moses' land of milk and honey was utterly changed—for the worse. Both changes are the unpremeditated resultant of the impact between ecological and economic forces.[4]

It has been held by some that religion, specifically the Judeo-Christian tradition of man's charge to subdue the earth, is the main historical root of our ecological crisis and that modern Western attitudes can be traced back to the Middle Ages.[5] But medieval attitudes had more ancient roots, for from earliest times man has usually lacked an ethic of respect for the natural world.[6] (That is to say, man as a species has been "an unecological animal."[7] A number of human subgroups—peoples with comparatively primitive technology and culture—have shown a far higher degree of ecological sensitivity than the human in modern industrial societies. Usually their religions, like those of the peoples of southern and eastern Asia, have not tended to separate man from nature, and this in itself has made them better stewards of the planet.)

Once man started his rape of the earth, a sense of inevitability seemed to follow. The attitude of the ancient Greeks might well have been summed by Pythagoras when he said, "Man is the measure of all things." This was corroborated by the Romans, as manifested by Aristotle's rationalism and the Roman orientation toward magnificent feats of engineering. The tempo quickened when the Genesis injunctions to "be fruitful, and multiply, and replenish the earth, and subdue it: and have dominion over every living thing that moveth upon the earth," and "create new heavens and a new earth: and the former shall not be remembered, nor come into mind," were taken literally and eventually coupled with post-Enlightenment ideas about growth and progress. The sentiments of those days were echoed recently by an American entrepreneur who, in damning the forests, proclaimed, "We have the directive from God: Have dominion over the earth, replenish it, and subdue it. God has not given us these resources so we can merely watch their ecological changes occur."[8]

Since ancient times the woodcutters of the world have, in fact, changed the face of the earth. Scientists note, for instance, that at one time the Sahara was probably blanketed with trees. When the forests were cut, exposing the lower vegetation and forest soil to the sun and wind, the desert took over.

While the forests vanished and soil fertility declined, the water and the air also suffered from man's arrogance or ignorance. Even before the first century B.C. the Romans realized that the sewage of the city's one million inhabitants was affecting the taste of the drinking water, so they built one of the first major municipal sewers in history. Yet, writing of Rome in the mid-1100s, Otto of Freising lamented, "The ponds, caverns, and ruinous places around the city were exhaling poisonous vapors and the air in the entire vicinity had become densely laden with pestilence and death." As for Venice, it remained, as Marshall Goldman puts it, "nothing but a sewer in search of a city."[9] (Until recently Venice was still flushing its sewage twice daily into the sea by using the natural flow of the tides.) In England during the Middle Ages, London had such stringent antipollution measures that at least one man was hanged for burning coal. Richard II in 1388 banned the dumping of filth into the Thames, but later generations got rid of their sewage whenever and wherever they pleased. In seventeenth-century London, rivers of filth coursed down the streets and a pall of smoke obscured buildings and fouled clothing.

In the burgeoning cities of America in the nineteenth century, conditions were not much better. A brilliantly unprophetic solution to some of the urban misery appeared in *Scientific American* magazine in July 1899:

> The improvement in city conditions by the general adoption of the motor car can hardly be overestimated. Streets clean, dustless, and odorless, with light rubber-tired vehicles moving swiftly and noiselessly

over their smooth expanse, would eliminate a greater part of the nervousness, distraction and strain of modern metropolitan life.

Earlier, in Lincoln's day, a map of the national capital noted that the canal behind the President's Park, along what is today Constitution Avenue, yielded "70 separate and distinct stinks" from sewage, dead animals, and trash.

Man's wanton use of the earth did not go unquestioned. As early as 1798, the English economist Thomas Malthus, with his book *An Essay on the Principle of Population,* initiated a controversy over the relation between population size and resources that still rages today. But the first book to describe in global terms what man was inflicting on the forests, water, soil, and wildlife was *Man and Nature,* published in 1864. Written by George Perkins Marsh, an American diplomat, lawyer, and philologist, it warned that wherever man "plants his foot the harmonies of nature are turned into discords. . . . Man has too long forgotten that the earth was given to him for usufruct [the right to use and enjoy all the advantages and profits of another's property without altering or damaging it in any way] alone, not for consumption, still less for profligate waste." This now-familiar theme, though it perhaps put a slight dent in public apathy, hardly created a groundswell of protest. It took a century for this message to excite general concern. Still, during that century, in America and in Europe, a few individuals ("eccentrics") struggled to protect the environment; occasionally they even chalked up a victory (see Selection 9).[10] John Muir, for one, singlehandedly saved hundreds of thousands of acres of virgin American land.

So, again, the question is, Why only now has the environment so dramatically become of worldwide concern? How, after thousands of years of man's mismanagement of the earth, could a global environmental crisis erupt practically overnight?

The answer is, it did not. For decades intermittent warnings of the ecological risks of disrupting natural systems had been greeted, for the most part, with a deafening silence. This was largely because most environmental problems were localized; and if an area became too polluted or otherwise unbearable, there were still fresh frontiers out there somewhere to move to. Moreover, the public-health measures introduced during this century cleaned up some of the more conspicuous forms of pollution.

Then two forces converged to confront mankind with a literally threatening dilemma: unrestricted modern technology accelerated the affronts to the natural world by enhancing man's ability to alter biochemical cycles and deplete natural resources; at the same time the global population began increasing at a phenomenal rate, and more was asked of the earth than it was able to give. Suddenly the earth was carrying unprecedented numbers of people who not only could desecrate it but had the means to do so wholesale. The technological changes that now made it possible to modify the biosphere usually exceeded the capacity of social, economic, and political institutions to adjust to the pace that was set by the proponents of growth and progress.

All this had been understood on an intellectual, rather than emotional, level for some time. Perhaps, as some believe, we were finally shamed into sharing a life-and-death concern for our earth by the first pictures from space, by the sight of the little blue planet floating out there, poignant in its aloneness.

Although many of the threats to the environment today are isolated and localized, they are all part of a pattern, a global mosaic of ecological crises that affect, directly or indirectly, physically and spiritually, everyone in the world. Poverty, war, famine, social injustice, urban blight, lackluster cultures are but the threads of the pattern. Ecological decay is all too often the *result* of economic, social, and political inadequacies; it is also increasingly the *cause* of poverty. The confluence of long-maturing events tied to man's unstable relationship with the earth has brought humanity to a crisis of crises.

Natural catastrophes, one of the major sources of crises in the past, would continue to assail man today no matter how much he corrected his ecological wrongs. But if he could unburden himself of these wrongs, he would be far better able to cope with destructive natural events. The earth has a remarkable capacity to heal itself of grievous wounds such as earthquakes, volcanic eruptions, and meteorological disturbances.[11] Many ask, however, How long can it continue to heal itself of both these wounds and those inflicted by man?

There is some doubt that natural phenomena are always the primary cause of so-called natural disasters. Although such disasters are thought by many to be inevitable, human activity all too often plays a role in their creation. It is tempting, of course, to place the full blame on nature when, say, a prolonged drought or excessive rainfall occurs, especially if you are of the opinion that "nature—in fact, the whole planet (and even, as far as we know, the entire universe)—is a floating crap game."[12] But the damage incurred in such cases is often drastic because inclement weather strikes a region made vulnerable by misdirected human activity. For instance, drought was the immediate cause of the famine that struck parts of Somalia, Ethiopia, and Haiti in mid-1975, but each of these regions was an ecological disaster zone well before drought beset it.

About 90 percent of all "natural" catastrophes are said to occur in the less developed countries. Uncoordinated aid and development programs sponsored by "*over*developed" nations sometimes unwittingly contribute to the Third World's predisposition to "natural" disasters by interacting in such a way as to generate disaster. For example, when projects designed to increase the health and life expectancy of a people are taken in conjunction with, but without reference to, agricultural programs that rely on ecologically unsound land use, they can—and in the case of the Sahel did—combine in precisely this unfortunate manner. Ill-conceived aid may not be all that industrialized nations are responsible for: it has been suggested that the by-products of their industrialization may cause climatic changes like that which led to the Sahelian drought (see Selection 2, p. 55).[13]

Now, having castigated man for undermining, in almost diabolical ways, the ultimate source of his own being, it is only fair to stress that human interventions into natural processes have often made for an interplay that has been both beneficial and creative. In the words of René Dubos, "All over the world, the changes brought about by man on the surface of the earth for the past few thousand years have brought out aspects of the wealth and beauty of nature that were hidden or had remained unexpressed in the state of wilderness."[14] In certain lands intensive agriculture, though practiced since time immemorial, has neither impaired their fertility nor sacrificed their scenery. Man can "create artificial environments from the wilderness and manage them in such a manner that they long remain ecologically stable, economically profitable, esthetically rewarding, and suited to his physical and mental health."[15]

Yet the human being has been called the world's worst predator, a "planetary disease,"[16] a parasite that the good earth could do very well without. His attempts to control nature far outdistance his attempts to control himself. All too often, he forgets or ignores or is unaware of the fact that he is an integral part of the ecological system—no more and no less. When he wars against nature, he wars against himself. As Alan Watts explains in one of his philosophical flights, "The world is your body."[17] By analogy, the world is your body in a physical sense, too. That is, we tend to take this incredible creation, the body, for granted: we overstuff it, drug it, abuse it in countless ways. And when it balks and gets out of kilter, we pay in one way or another—often by doling out inordinate sums to try to fix it up. Like the earth, too, the body has an amazing ability to heal itself—within limits.

Nature has its own immutable laws, and we attempt to violate and repeal them at our risk. In the seventeenth century, Francis Bacon inadvertently gave us an operational definition of ecology: "Nature to be commanded must be obeyed."

The earth, a comparatively minor planet in size, is from our perspective exceedingly large. Yet its shell of life—the biosphere—is so thin that, as John Livingston observes, it is virtually transparent.

> Slight and delicate though it is, the biosphere is sufficient for every life purpose. Miraculous? Perhaps not. Unique? Almost certainly not. In human terms, however, it is all there ever has been and all we can hope for. Earth—the product of sheer cosmic chance—may or may not be one in a million, but for our purposes it might just as well be.[18]

The regard for this earth as our only home is signaled by the root of the word "ecology": the Greek *oikos,* meaning home or house, which by extension came to mean the entire inhabited earth, or *oikoumene.*

The science of ecology, which has its origins in the practical environmental knowledge of our early ancestors; was first formally defined by Ernst Haeckel in 1869. (It may be of passing interest that the word "eco-catastrophe" was coined in 1969, exactly a century later.) Basically, ecology is the study of the interrelationships of organisms (including humans) with one another and with their nonliving environments, as well as the study of natural systems built upon these relationships.* Or, as one cynic has more loosely described it, "Ecology is the science which warns people who won't listen about ways they won't follow of saving an environment they don't appreciate."[19]

Lying at the core of all ecological thinking is the principle of holocoenosis—that is, the awareness of how natural factors exist as a vast complex and act as a whole rather than separately and independently (see illustration on title page of Part I). As Blaise Pascal expressed it more than 300 years ago, "Nature is an infinite sphere whose center is everywhere and whose circumference is nowhere."

---

*The study of how man interrelates with his home—the soils, the minerals, the water, the air, the animals and plants—is often called human ecology. The word "economics," incidentally, has the same root as "ecology" and originally meant management of the household. Although ecology (which deals essentially with minimizing long-term liabilities) has generally been considered a subsection of economics (which is concerned with maximizing short-term gains), history has proved again and again that it should be the other way around.

The Sioux Indians, whose land ethic was, "With all beings and all things we shall be as relatives," sensed this indivisible quality of all realms of life:

> Everything the Power of the World does is done in a circle. The sky is round, and I have heard that the earth is round like a ball, and so are all the stars. The wind, in its greatest power, whirls. Birds make their nests in circles, for theirs is the same religion as ours. The sun comes forth and goes down again in a circle. The moon does the same, and both are round. Even the seasons form a great circle in their changing, and always come back again to where they were. The life of a man is a circle from childhood to childhood, and so it is in everything where power moves.[20]

Barry Commoner has encompassed this in four simple ecological laws: (1) in a cycle everything is connected to everything else; (2) nothing ever goes away: it turns up somewhere—always; (3) "there is no such thing as a free lunch": in some way, at some turning, a cost must always be met; and (4) nature (not Du Pont) knows best.[21]

In a global context, what all this means is that, in the words of Garrett Hardin,

> The world can no longer afford to ignore what has been called the "ecological ethic." The ethical system under which we operated in the past was possibly adequate for an uncrowded world, though even this is debatable. But it is not adequate for a world . . . in which it is increasingly difficult for anyone to do anything at all without seriously affecting the well-being of countless other human beings. In castigating pesticide sprayers as monomaniacs, Rachel Carson [in *Silent Spring*]* was alerting the world to what has been called the fundamental principle of ecology, namely: *We can never do merely one thing.* . . .

---

*"The Other Road," the last chapter of *Silent Spring* (Boston: Houghton Mifflin Co., 1962), outlines a path toward a creative ecology. Only recently has the United States, which is on a "pesticide treadmill" and invests some $2.5 billion a year in insecticides, begun seriously to consider this path. A strategy called integrated pest control (IPC), already proved effective in other countries, is being initiated. ICP measures include biological control of pests through predators or parasites, the use of hormones and pheromones, sterilization of insects, and the development of insect-resistant crops. See G. R. Conway, "Better Methods of Pest Control," in William G. Murdoch, ed., *Environment: Resources, Pollution & Society* (Stamford, Conn.: Sinauer Associates, 1971), pp. 302–25, esp. pp. 304–5, 320–21; and "The Bugs Are Coming," *Time*, July 12, 1976, pp. 38–46.

> We can never do merely one thing, because the world is a system of fantastic complexity. Nothing stands alone. No intervention in nature can be focused exclusively on but one element of the system.[22]

The nations of the world have finally begun to recognize this, but they are still agonizingly slow in absorbing it. The sudden intense concern for the global environment has upset the traditional course of international relations. It has magnified the tensions between rich and poor nations, added new sources of potential conflict, made questions of trade and investment more complex, and snarled the unification processes within the European Community. But there is a bright side: it has opened many new channels of international cooperation.

The variety and complexity of environmental problems around the world stagger the mind (see Selection 33). Whereas in general those in the industrialized nations (not a few of whose inhabitants are poor and deprived) worry about air and water pollution, a possible energy shortage, urban blight and urban sprawl, the African or the Asian or the Latin American is likely to worry about hunger and malnutrition, disease, infertile soil, insufficient water, lack of education, and unemployment. But no matter where they live,

> the industrial worker, the farmer, the fisherman, the engineer and the scientist—every trade and profession virtually everywhere in the world—are already affected by new sets of restraints, priorities and fears for their livelihood. And we must all wrestle with such problems as: What degree of purity in air and water is enough and at what price? What qualities of life do we seek and how much are we willing to pay? How much growth is necessary and how much restraint is acceptable, whether it be in our procreation, our competition for resources or our devotion to the automobile?
>
> All these questions, and many more, must be answered by sovereign peoples with different cultures, values and goals. They must take action on the basis of insufficient knowledge and profound disagreements within and among themselves as to the extent and nature of our planet's peril.[23]

Man has two universal environments—the world of atmosphere and climates and the world of the oceans and seas. Their effective protection lies far beyond the realm of local decisions (see Selections 4, 17, and 18). For "even the sum of all local separate decisions, wisely made, may not be a sufficient safeguard and it would take a bold optimist to assume such general wisdom. Man's global interdependence begins to require, in these fields, a new capacity for global decision-making and global care."[24]

On a regional scale, particularly worrisome problems are the management of river basins and the control of the pollution that so blithely crosses frontiers. The water and the air, after all, have no politics. The once-exquisite Rhine has become "the sewer of Europe": at any time there are more than 1,000 pollutants present and these plus sewage make up 20 percent of the Rhine's contents when it eventually meets the sea. The disparity of cost/benefit between the polluting country and its downriver or downwind victim, along with the question of who is polluting whom, makes accommodation and resolution a slow process. However, the remarkably successful control of pollution in Oregon's Willamette River and in London's Thames (as well as the greatly reduced air pollution of London since the 1950s) points to what could be done on a regional scale.[25]

The recent landmark agreements to save the Baltic and Mediterranean seas (see Selections 17 and 18), as well as the Great Lakes Water Quality Agreement promulgated by the United States and Canada in 1972, are evidence that all is not lost—yet. The battle to clean up the air is being waged simultaneously, if desultorily, in many countries. (Not long ago north-flowing air from the Ruhr was so acrid and filthy that it caused black snow to fall on Scandinavia. More commonplace has been the acid rain bequeathed to Scandinavia by the coal-burning British and German industries.) "This campaign to clean up the air," Margaret Mead commented recently, "is the first time in human history where one group's loss is not another's gain."

In the past decade or so there have been more *major* international conferences on the environment than there are days in the year. But the most significant was the United Nations Conference on the Human Environment, held in Stockholm in 1972. It marked the first coordinated effort by nations all over the globe (110 of them) to alleviate their common ecological problems. It did not generate a policy consensus on the right way to deal with a vast variety of ecological problems, nor on an alternative to the techno-economic order that is in large part responsible for many of these problems. All was not calm, reasoned discussion, either, as Third World representatives spoke acrimoniously about the "advanced" nations' vast inroads on the earth's resources as well as their other crimes against the environment.

Still, the conference was a landmark, and gave birth to the United Nations Environment Program (UNEP). Headquartered in Nairobi, the UNEP stimulates and coordinates projects that reflect its philosophy of the interdependence of economic and social development and environmental protection policies. Its efforts are complemented by UNESCO's Program of Man and the Biosphere. Both programs stress cooperation, rather than self-protection or constraint, as the means of coping with environmental problems.

Although every nation has to some extent contributed to blemishing the earth, the United States is invariably and justifiably singled out as the greatest blemisher of all time. Granted, Americans created the first conservation movement as early as 1890; they set another precedent by establishing the Environmental Protection Agency and its Office of International Activities; they almost surely do more about pollution than the people of any other country; and they abandoned development of one of the foremost examples of technology gone insane, the SST. But the United States has had far more reason than any other nation to try to undo the environmental mess it has created.

Americans represent only about 6 percent of the world's population, yet they consume more than a third of its nonrenewable resources and a generous third of its energy, produce at least a third of the world's wastes and pollutants (Western Europe is a close second), and eat more food and use more water than their counterparts in any other country. Not only do many of them overstuff themselves (largely with foods that technology has tinkered with—adding chemicals, subtracting nutrients, and multiplying profits), but they waste it in incredible quantities. A survey in Tucson in 1974, for instance, revealed that in one year 9,500 tons of *edible* food, valued at about $10 million, ended up in the garbage can. Although the United States produces four times more food calories than it needs, it imports thousands of tons of protein-rich fishmeal from countries like Peru and Chile, whose neighbors suffer from protein deficiency, and feeds it to poultry, hogs, and cats at home. In 1970 each American generated an average seven pounds of solid waste daily (an amount that is said to increase 4 percent annually), for a combined total of 530,000 tons a day, over 500 billion pounds a year—or enough to fill the Panama Canal four times over. Some ingenious things are being done with some of this waste (in Chicago they made a ski hill out of a pile of refuse and named it Mount Trashmore) but at a snail's place.

It has been reported that we Americans use more electricity to run our air conditioners than mainland China uses for *all* its electrical needs. And to run an air conditioner,

> we will strip-mine a Kentucky hillside, push the dirt and slate down into a stream, and burn coal in a power generator, whose smokestack

contributes to a plume of smoke massive enough to cause cloud seeding and premature precipitation from Gulf winds which should be irrigating the wheat farms of Minnesota.[26]

In our short history, but especially in the past fifty years, we have managed to profane a continent that was some 6 billion years in the making. Which is why former Justice William O. Douglas once urged the Supreme Court to grant standing under the law to streams and mountains, arguing that, since they are at least as real as corporations and ships, they might as well hold rights already extended under the law to corporations and ships.

The nation's consciousness and its politics have indeed been changed by the disturbing awareness that our resources and our pollution solutions cannot keep up with our appetite, but our appetite remains unappeased. Environmental activists are winning some hard battles, but politics and business, while respecting the public-relations aspect of ecology, tend to encourage business-as-usual. Corporations plead in glossy ads that they are doing the best they can about pollution (they are not happy about a possible "effluent tax," and they're more concerned about penalties than incentives) and that Americans should give more consideration to energy requirements and jobs. But the point is that we do not really require all that energy, we just want it. And as for jobs, as Garrett Hardin commented during the SST controversy, "Don't tell us that it will employ people; the same can be said for peddling heroin."[27] It is possible to make a good living by healing the earth instead of wounding it.

Self-proclaimed "environmentalists" (as distinguished from environmental activists) abound in the United States. These are the people who rush to the seacoast in a gas-guzzling car to wring their hands over an oil spill. This is about as useful as an appendix transplant. Not only have big cars made a comeback, but during the first eight days of July 1975 (vacation travel days) Americans used up as much oil as the U.S. Armed Forces did in all of 1944, the most grueling year of World War II.

More perhaps than any other country, the United States is addicted to production and consumption. To measure national progress, it uses a frayed yardstick—the Gross National Product (GNP), which some have suggested is merely gross. Raising the GNP does not automatically guarantee well-being for one and all. For instance, if you are injured in a car wreck, the money that goes to the doctor, the hospital, the insurance adjuster, and the car-repair garage raises the GNP. And if you die, the funeral parlor and the coffin maker and the florist get into the act. You'll be raising the GNP even if it kills you.

All this in the name of progress. What the GNP measures is cost. What it does not measure is noise, crowding, murky air, poisoned streams.

After his first visit to Los Angeles, a congressman reported to his colleagues, "I've just seen the future, and it won't work." Los Angeles, with its notorious smog, unfortunately has plenty of sick-air rivals among other cities of the industrialized world. Yet, although the United States and other highly industrialized nations are the world's worst polluters, the air pollution in their cities is not as bad as that in some of the cities of the Third World, such as Seoul, Taipei, Mexico City, and Caracas. According to a NATO study, the most dangerously polluted air of any city on earth is Ankara, the capital of Turkey, not because it is the focus of heavy industry but because it lies in a natural bowl shaped by circling hills, and the smoke from soft-coal fires for home heating in winter cannot escape the low-lying clouds.

But the first town in history to be officially declared uninhabitable thanks to air pollution was the German village of Knapsack, the world's second largest phosphate producer. The fumes from its phosphate plant finally forced the residents of the 406-year-old village to move away. Their houses have been demolished; the town has disappeared.

Far worse was the evacuation in August 1976 of 100,000 residents from the Italian town of Séveso and its surrounding communities when an explosion at a nearby chemical plant released a highly toxic substance known as TCDD (a component of the defoliants used with such zest by the U.S. forces in Vietnam). The health of thousands of people was dangerously affected. Thousands of animals died, and the surviving ones were killed in an effort to halt the spread of contamination. Even with the adoption of a scorched-earth policy, no one was sure how to achieve a complete cleanup or what the long-term effects would be.

There are over half a million man-made chemicals in use today, and each year almost a thousand new ones are devised. Their effects, either singly or in combination, on human physiology and on supporting ecosystems are largely unknown. In the United States, a persistent and dangerous class of chemical compounds called polychlorinated biphenyls (PCBs) has been accumulating in the environment for over 45 years. By 1972 they had been found in every major U.S. river system. Some of the fish in the Hudson River recently contained so much of the substance that a person eating a 7-ounce portion would get 50 percent of his lifetime allowance of it in one serving. Observed one Hudson River fisherman, "Shopping in a fish market these days is like picking your way through a minefield."[28] Natural water in America is becoming as rare as natural air; and natural air no longer even exists in mainland United States.

It has been suggested that the trouble with pollution is that it has always been free. So why not put it on a pay-as-you-go basis? This has been tried in Germany, in the Emscher Valley of the Ruhr, where some 200 municipalities are involved, and it has worked.[29]

Every animal creates waste (the assimilation and recycling of waste is a basic ecological force), but only man makes products that nature cannot easily reclaim. It's true that the release of waste material into the environment does not necessarily mean that the environment is going to be polluted. A certain degree of organic enrichment, for example, can enhance the fertility of waterways. It is when wastes interfere with or defy the capacity of the environment to receive waste that pollution is created. Even where rivers are free of industrial chemicals and the air is pure, pollution may be rampant. In the developing countries, for instance, the major cause of water pollution is lack of sanitation.

Not only does increasing energy consumption mean more pollution, no matter what we do, but the very technology that is used to clean up the effects of technology can create more pollution. As a statement issued by the Field Museum of Natural History points out, "Even if we prevent or reverse pollution of the environment, we will not only retain *forever* the ability to pollute air, land, and water, but probably will develop new means of pollution which cannot be imagined today."

More than 70 percent of the planet's surface is covered by water, but *unsalted* water in an unfrozen state is rare—less than 1 percent of all the water on earth. And that small amount is unevenly distributed: more than a third of all fresh water is in Canada and the Soviet Union, nations with relatively sparse populations. Yet the variety of life that depends on fresh water far surpasses that found in all the oceans combined.

The massive use of water today makes for a vicious circle of destruction. It lowers water tables, thus depriving plant roots of water, destroying vegetation, and reducing transpiration. This in turn affects local climate, which further dries out the soil and causes additional dust to enter the atmosphere and affect worldwide climate. As loosened soil washes into streams, it smothers aquatic life and alters the quality of the water. The naked soil then accelerates the water runoff, which causes more erosion, reduces groundwater levels even further, kills off yet more vegetation, and quickens the cycle of devastation.

The world's affluent minority steadily increases its per capita use of water, although more than 3 billion people do not have water of acceptable quantity or quality. Food production is the primary user of water, and the rich world needs ten times more water to accommodate its dietary wants (not requirements) than the hungry world. The production of one pound of bread (counting the amount of water used to grow the wheat and to complete the milling process) consumes 2,500 pounds of water. The complete process of producing a pound of beef involves from 100 to over 500 tons of water. The mining of groundwater in many areas of the world has led to ground subsidence, and water tables have fallen down to 1,000 to 2,000 feet. Yet, according to the Food and Agriculture Organization, by the end of the century global demand for fresh water will increase 240 percent.

Recycled water is, of course, abundantly used; in some parts of North America, river water is used 50 times over before it reaches the sea. One water expert has stated that by 1980 Americans will be drinking sewer water. That is the good news. The bad news is that by 1985 there will not be enough to go around. As a matter of fact, in many cases

> we already hoist a glass of liquid that only recently went down somebody's toilet. . . . Windhoek, the capital of South West Africa [Namibia] recently had the distinction of becoming the world's first city to recycle its wastes directly into drinking water. . . . In the United States we are more squeamish than the people of Windhoek. We still prefer to drink other people's wastes.[30]

Mammoth engineering projects to rechannel rivers are altering entire landscapes in North America, the Soviet Union, and Western Europe, destroying ecosystems as old as the earth.[31] In the United States especially, sending out for more water ("the quick technological fix") is preferred to wise use of the available supply. While desalination has proved to be technologically feasible (Kuwait has a major desalination plant), the cost is said to be about 10 times too expensive at present for common use.

How will it be possible to replenish and improve the quality of natural waters faster than they deteriorate and are consumed? The special feature of water resources that lends encouragement is that they constantly renew themselves and that the intensity with which they do so is proportional to that with which they are used. But when water is withdrawn from the hydrologic cycle in one place, it returns to that place in only minute quantities. The renewal of resources occurs through an increase in underground runoff and through the greater precipitation caused by an increase in atmospheric humidity when water is extracted from the land.

A localized water-management scheme can touch off processes that affect the planet as a whole. Since all natural waters form a single entity, and are in constant interaction with the physical and geographical environment, their investigation and exploitation need to be integrated. That is, conservation and control must be organized not only at the level of river basins, but of continents and even groups of continents. This means, of course, that man will have to cooperate on a scale hitherto unknown, but that is the key to the alleviation of every problem in the man/environment sphere.[32]

It is commonly believed that there are not enough mineral resources to support all the world's people at a standard of living equal to that of Western Europe, at least not without significant new breakthroughs in technology. In Selections 2 and 34, this situation is examined from different viewpoints. What perhaps needs clarification is the meaning of "natural resources":

> Natural materials do not become resources until they are combined with man's ingenuity. Over time the record is impressive. Mineral resources have become more and more widely available despite (and partly because of) growing rates of consumption. This, in crude form, is the modern economic view of mineral resources, or the "cornucopian" view as it has been dubbed.
>
> Unfortunately, most descriptions of mineral resources include only the material that could be mined at today's prices and today's technology—what is properly termed "reserves"—and thus they understate the availability of these resources.[33]

Since the entire planet is composed of minerals, the literal notion of running out has been called ridiculous, for man could scarcely mine himself out. (Exceptions are a few substances that are discretely different from the rock masses containing them, such as natural gas and crude oil.) One source suggests that, even with no improvements in existing technology, there is not likely to be a shortage of raw materials until about A.D. 100,000,000.[34] The world, in this generous view, is not running out of mineral resources but out of ways to avoid the ill effects of high rates of exploitation. In the less generous view of many experts, however, the easily mined surface and subsurface deposits of a number of critical elements and minerals are already near exhaustion. Clearly, then, in choosing among alternative paths of growth and therefore among alternative rates of mineral-resource development, preferences and ethics will be important guidelines, as well as columns of numbers comparing reserves and consumption. Of course, as Barry Commoner points out, Americans in particular can slow the rate of exploitation by remedying their calamitously inefficient use of fuels—they waste fully 85 percent of the work potential in the oil, coal, and uranium they harness.[35]

The problems of a rapidly expanding world population, limited (or inequitably distributed) food supply, and staggering demands on arable land are noted again and again in the selections in this book (see particularly Selections 2, 3, 12, 13, 14, 22, 25, 27, 29, 35, and 36). As the population soars (see Selection 35), more and more people go hungry. The inevitability of this trend, however, has been seriously questioned (see Selection 36).

As a result of man's endless quest for food, over half of the earth's forest cover has disappeared, vast areas of grassland have been plowed, and major groundwater reserves have been dramatically lowered. In populous countries like India and China, forests and pastureland have been compressed to woefully reduced areas. Erosion, desiccation, waterlogging, and salination have ruined an inestimable amount of tilled land and still more is in jeopardy. Irrigation reservoirs are filling with silt at a rate 10 times faster than anticipated. "Such problems," says Georg Borgstrom, "are equally grave in numerous other countries that should be labeled biologically overdeveloped rather than designated underdeveloped."[36]

The earth has 8 billion acres susceptible to crop production, but 56 percent of the potentially arable land is not cultivated. There are serious limitations on a good deal of the land in South America, Australasia, and Africa because it is either desert or tropical rainforest. The limitations on food production are imposed, however, not only by temperature, available moisture, soil properties, and sunlight, but also by lack of money, the reluctance of farmers to use new agricultural practices, and the high price of fertilizers (which, anyway, have a point of diminishing return). Governments are often more interested in subsidizing an airline or building ostentatious capital cities than in aiding their people to improve their agricultural methods. The use of new genetic plant varieties, better cultivation and pesticide-management techniques, and more labor-intensive farming (especially on small holdings in the poorer nations) could raise yields considerably.

Unfortunately, even if an adequate supply of food is, or could be made, available, a major reason for the starvation and malnutrition in the world would not be solved—that is, maldistribution, especially of foods containing high-quality protein. These foods are inequitably distributed both between rich and poor countries and between the rich and the poor within countries. In fact, the net flow of protein is from the poor, protein-deficient nations to the rich, overfed ones. The problem stems from the inability of the poor to purchase food rather than from the inability of nations to produce it. So each day some 10,000 people worldwide die of malnutrition or starvation. According to Georg Borgstrom, "Agriculture has never been adjusted to nutrition considerations. It's based on what is profitable."[37]

A method does exist that not only can produce abundant food but can enhance the environment at the same time. Known as "biodynamic/French intensive gardening," it has proved beyond question that even an eroded, unproductive soil can, after a couple of years, be encouraged to produce two to ten times as much superior food as the most expert commercial farmers are getting today. This type of cultivation, which uses raised beds and deep spading, requires only hand tools and some organic fertilizing material. Although the crop must be well watered, it needs only half as much water as a commercial farmer uses. Moreover, the cultivator does not have to be an expert gardener or farmer. With this method, food for a completely balanced diet can be grown on a patch as small as 3,700 square feet per person, as compared with the some 21,500 square feet required in Japan under present farming methods. Apparently the main obstacle to the widespread use of this highly efficient technique, which is easy to learn, is that it calls for practically no investment of cash and is therefore totally unappealing to manufacturers of chemical fertilizers and farm equipment. Perhaps that is a clue to why no government, in either the rich countries or the poor ones, has as yet pushed this ingenious solution: it offers no industrial profits.

According to United Nations statistics, only 31 among the 120 developing nations of the world have policies favoring a lower rate of population growth. Worldwide expenditures for the direct control of fertility amount to around $3 billion a year (global military expenditures exceed $200 billion annually). The causes of and solution to the population problem, along with inherent ethical considerations, are frequently disputed, especially among three biologists who are eloquent on the subject: Barry Commoner, Garrett Hardin, and Paul Ehrlich.

Barry Commoner, for one, believes that poverty is the cause of overpopulation, and not the other way around.[38] Contending that socioeconomic disequilibrium is the core of the problem, he disagrees with the methods that have been proposed to achieve "zero population growth."

> Among the ones advanced in the past are: (a) providing people with effective contraception and access to abortion facilities and with education about the value of using them (i.e., family planning); (b) enforcing legal means to prevent couples from producing more than some standard number of children ("coercion"); (c) withholding of food from the people of starving developing countries which, having failed to limit their birthrate sufficiently, are deemed to be too far gone or too unworthy to be saved ([Hardin's] "lifeboat ethic").[39]

Commoner suggests that if the world's wealth were evenly distributed among the world's people, the entire population would have a low birthrate: "The poor countries have high birthrates because they are extremely poor, and they are extremely poor because other countries are extremely rich."[40]

As David Brower (see Selection 9, pp. 163 and 168) sees it, the flaw in this view is that boosting the rest of the world

> up to the average U.S. standard-of-using (averaging Park Avenue, Appalachia, Harlem, and Navajo country) would multiply the drain on world resources 2,000 per cent. The resources aren't there. . . .
>
> *Moral:* Less for those who are surfeited; more for those who have never known enough; a rational end to overdevelopment and a rational beginning of equity.[41]

Paul Ehrlich repeatedly warns that we are courting global disaster unless we put a halt to overpopulation, overproduction, and overconsumption. In his view,

> The birth of every American child is fifty times more of a disaster for the world than the birth of each Indian child. If you take consumption of steel as a measure of overall consumption, you find that the birth of each American child is 300 times more of a disaster for the world than the birth of each Indonesian child.[42]

The peoples of the developing countries, who comprise two-thirds of the world's population and contribute the bulk of the world's resources, have been largely denied the advantages of their own wealth; they have, in effect, been trying simply to live while the other one-third of the world has been living it up. They are justifiably leery of "advisers" from industrialized nations who, in the midst of all the plunder, tell them that they must now curb their populations and their wants.

On one side of the long and increasingly pertinent controversy over the race between population and resources are the Malthusians, who contend that in the not-too-distant future everything is going to be scarce—except, of course, people. On the other side are the Cornucopians, who believe that economic growth, wisely managed, can bridge and eventually close the gap between the wealthy nations and the deprived ones and that every land can be a land of plenty. As economist B. Bruce-Biggs puts it, "The neo-Malthusians would have us be more generous and unselfish and less greedy and materialistic. No decent man would disagree, but are they more persuasive than Confucius, Buddha, Isaiah, or Jesus Christ? Have computers greater authority than Scripture?"[43]

Gandhi once said that "there is enough for everyone's need, but not enough for everyone's greed." The environmental crisis is not so much a question of limited resources and of science and technology gone wild as it is a queston of cultural institutions that have distorted the values and priorities of peoples all over the world by promoting the idea of progress as simple accretion. As Gary Snyder notes, "True affluence is not needing anything."[44]

It has been said that the development of technology, especially in the United States, has led us "from know-how to nowhere."[45] In their despair, some misguided romantics insist that we should all go back to nature and, in essence, abandon the condominium for the cave. This is patently unrealistic; we cannot turn back the clock. Science and technology are here to stay. They need not be a continuing source of destruction and despoliation; they can just as well be a continuing source of beneficent change. Now that man has so efficiently deranged his habitat and his socioeconomic and political systems, he has an obligation to try to heal the disruptions just as efficiently. An ecological humanism would integrate science with ethics, economics with ecology, and philosophy with politics.[46] Granted, this is a tall order. But there is a slowly growing understanding that appropriate technologies might be found for each physical niche, appropriate not only for the demands of nature but also for each cultural and economic milieu. Integrity and diversity form a pattern of interdependence in nature. They can be applied to the works of man through a benign synthesis of complex and simple technologies and a re-ordering of values, thereby nurturing creative living on the part of people everywhere. "It is," Francis Bacon observed, "a secret both in nature and state that it is safer to change many things than one." It is not only safer; it is vital.

Anti-technologists who point to man's unecological past and see nothing but gloom and doom ahead have a legitimate concern. What they do not take into account is that people can learn from their errors, that the industrialized nations will not inevitably continue their wholesale misuse of technology and their disdain of its ecological hazards, and that the less-industrialized nations will not necessarily emulate their errant ways. If in the future man still has to dominate something, let it be technology, not nature. Technology was made to serve man. It should be appreciated for what it is—a servant, not a savior.

Among the assumptions about technology that are scientifically objectionable is the one that says new technologies and resources increase at linear rates (1, 2, 3, 4, 5, etc.) and that everything else increases exponentially (2, 4, 8, 16, 32, etc.). An often cited example of this fallacy goes like this. If in New York City in 1850 the level of horse manure in the middle of the streets averaged a half inch, by 1970 it would, if figured exponentially, have amounted to 2,048 inches—or 170 feet 8 inches, burying the population in a pile of manure about 14 stories high. The point is that, certainly, by the time the manure reached the second or third story, people would have begun to take notice and done something about it. Through technology they would have devised either a street-sweeping method to keep up with the level of manure or a transportation method that would obviate the need for horses.

As Daniel Callahan points out,

> . . . the problem of science and technology is one of degree, putting aside the views of those increasing number of jeremiads who see technology as the doom of us all. If they are right, it is too late anyway; if they are wrong, they are just misleading everyone. . . .
>
> . . . The only possibility of a check to an overinvestment of hope in technology is, paradoxically, the existence of checks against an overinvestment of hopes in nontechnology. Needed is a scaling down of technology and a rational control of it—not its deification and not its denial. . . . Technology has lacked sober support, where promises are modest but real, and sober criticism, where the attack is focused on real dangers and specific alternative forms of conduct are offered.[47]

If technology were rationally controlled, and if nature were allowed to provide the guidelines and limits of progress, tactics of desperation would not be needed. For example, there are probably only two readily available natural resources that we have barely used, though both are, as far as we know, unlimited: the sun and our (or all animal) intelligence. Yet, in the United States, instead of releasing substantial sums for research and development to utilize low-energy technologies and develop not only solar power but also wind power and other natural-energy sources, enormous sums are expended on developing nuclear energy, a source that may be neither desirable nor necessary. Of the $3.9 billion energy research budget in 1976, only about 4 per cent was funded for research and development of solar energy.

Solar energy could be a practical solution particularly for people in the developing countries that have a lot of sunlight but little coal or oil.

> Solar energy seems to promise unlimited benefits with no risks at all, no pollution at all. The fact that almost everyone in the scientific community continues to ignore it seems to indicate either that those who sell some kind of fuel—oil, gas, uranium—are in absolute control not just of our country but of the entire planet, or else that we are simply a bunch of lunatics bent on self-destruction.[48]

The longest war in history is the war of man against nature. It is a war we would do well to lose. For if we win it, we will be lost—yet another extinct species. The sweet air and sweet water of the planet's pristine past may be reborn—either with our help or without us. To ignore the ecological perils we have fomented by our cavalier treatment of the earth is to make doomsday prophecies self-fulfilling. In the words of Paul Shepard,

> Truly ecological thinking need not be incompatible with our place and time. It does have an element of humility which is foreign to our thought, which moves us to silent wonder and glad affirmation. But it offers an essential factor, like a necessary vitamin, to all our engineering and social planning, to our poetry and our understanding. There is only one ecology, not a human ecology on one hand and another for the subhuman. . . . For us it means seeing the world mosaic from the human vantage without being man-fanatic. We must use it to confront the great philosophical problems of man—transcience, meaning, and limitation—without fear. Affirmation of its own organic essence will be the ultimate test of the human mind.[49]

NOTES

1 Petr Beckmann, *Eco-Hysterics and the Technophobes* (Boulder, Colo.: Golem Press, 1973), p. 13.

2 Kenneth E. Boulding, "No Second Chance for Man," in Hal Borland and others/The Progressive, *The Crisis of Survival* (New York: William Morrow & Co., 1970), p. 168.

3 Aldo Leopold, "The Conservation Ethic," *Journal of Forestry,* October 1933, pp. 635, 637.

4 Ibid., p. 636. See also Eugene P. Odum, "The Strategy of Ecosystem Development," *Science,* 169 (April 18, 1969), pp. 262–70. Odum offers a theory of ecological succession in which he contrasts nature's strategy of protection, stability, and quality with man's strategy of production, growth, and quantity.

5 See, for instance, Lynn White, Jr., "The Historical Roots of Our Ecologic Crisis," *Science,* 155 (March 10, 1967), pp. 1203-7.

6 See Lewis W. Moncrief, "The Cultural Basis for Our Environmental Crisis," *Science,* 170 (October 30, 1970), pp. 508-12; John A. Livingston, *One Cosmic Instant: Man's Fleeting Supremacy* (Boston: Houghton Mifflin Co., 1973), pp. 147-71; J. Donald Hughes, *Ecology in Ancient Civilizations* (Albuquerque: University of New Mexico Press, 1975), esp. chap. 13, "The Ancient Roots of Our Ecological Crisis"; and Lynton K. Caldwell, *In Defense of Earth: International Protection of the Biosphere* (Bloomington: University of Indiana Press, 1972), pp. 18-20.

7 Caldwell, *In Defense of Earth,* chap. 1, offers an analysis of man's unecological behavior, plus abundant reference notes.

8 H. D. Bennett, executive vice-president of Appalachian Hardwood Manufacturers, quoted in Jack Shepherd, *The Forest Killers: The Destruction of the American Wilderness* (New York: Weybright & Talley, 1975), p. 5.

9 Marshall I. Goldman, "Pollution: The Mess around Us," in Marshall I. Goldman, ed., *Ecology and Economics: Controlling Pollution in the 70s* (Englewood Cliffs, N.J.: Prentice-Hall, 1972), p. 4.

10 See Robert McHenry with Charles Van Doren, eds., *A Documentary History of Conservation in America* (New York: McGraw-Hill Book Co., 1972).

11 For a survey and chronological chart of events—earth science, biological, and astrophysical—caused by the harsh turn of nature during the 1968-72 period alone, see Jean Shepard and Daniel Shepard, *Earth Watch: Notes on a Restless Planet* (Garden City, N.Y.: Doubleday & Co., 1973).

12 Shepherd Mead, *How to Get to the Future before It Gets to You* (New York: Hawthorn Books, 1974), p. 38.

13 See Nicole Ball, "The Myth of the Natural Disaster," *The Ecologist,* December 1975, pp. 368-71; Reid A. Bryson, "Drought in Sahelia: Who or What Is to Blame?" *The Ecologist,* October 1973, pp. 366-71; and Ibrahim Fall, "After the Drought," *Africa Report,* May-June 1975, pp. 37-40.

14 René Dubos, "The Despairing Optimist," *The American Scholar,* Spring 1974, p. 192.

15 René Dubos, "Humanizing the Earth," *Science,* 179 (February 23, 1973), p. 771.

16 Ian L. McHarg, "Man: Planetary Disease," in Robert T. Roelofs, Joseph N. Crowley, and Donald L. Hardesty, eds., *Environment and Society: A Book of Readings on Environmental Policy, Attitudes, and Values* (Englewood Cliffs, N.J.: Prentice-Hall, 1974), pp. 303-14.

17 Alan Watts, *The Book: On the Taboo against Knowing Who You Are* (New York: Pantheon Books, 1966), chap. 4.

18 Livingston, *One Cosmic Instant,* p. 33.

19 L. G. Heller, in a letter to the *New York Times,* April 2, 1970.

20 Black Elk, as told through John Neihardt, *Black Elk Speaks: Being the Life Story of a Holy Man of the Oglala Sioux* (Lincoln: University of Nebraska Press, 1961), pp. 198-99.

21 Barry Commoner, "The Social Use and Misuse of Technology," in Jonathan Benthall, ed., *Ecology in Theory and Practice* (New York: Viking Press, 1973), pp. 338-39. See also

Commoner's *The Closing Circle: Nature, Man and Technology* (New York: Alfred A. Knopf, 1971).

22 Garrett Hardin, "To Trouble a Star: The Cost of Intervention in Nature," *Bulletin of the Atomic Scientists,* January 1970, p. 17.

23 Philip W. Quigg, "Environment: The Global Issues," *Headline Series,* no. 217 (October 1973), p. 4.

24 Barbara Ward and René Dubos, *Only One Earth: The Care and Maintenance of a Small Planet* (New York: W. W. Norton & Co., 1972), p. 195.

25 Alex Morrison, "The Kiss of Life for the River Thames," and Kenneth H. Spies, "Restoration of Water Quality in the Willamette River," in *Environmental Accomplishments to Date: A Reason for Hope; International Symposium II . . . 1974,* edited by George M. Dalen and Clyde R. Tipton, Jr. (Columbus, Ohio: Battelle Memorial Institute, 1975), pp. 48–55, 59–63. This source offers many examples of environmental improvements around the globe.

26 Wayne H. Davis, "Overpopulated America," *The New Republic,* January 10, 1970, p. 13.

27 Hardin, "To Trouble a Star," p. 19.

28 "The Peril of PCBs," *Time,* May 10, 1976, p. 75.

29 See Edwin G. Dolan, *TANSTAAFL: The Economic Strategy for Environmental Crisis* (New York: Holt, Rinehart & Winston, 1971). TANSTAAFL is an acronym for "there ain't no such thing as a free lunch."

30 William Longgood, *The Darkening Land* (New York: Simon & Schuster, 1972), p. 69.

31 See, for instance, Philip P. Micklin, "Soviet Plans to Reverse the Flow of Rivers: The Kama-Vychegda-Pechora Project," *The Canadian Geographer,* 13(3), 1969, pp. 199–215.

32 See Tinco E. A. van Hylckama, "Water Resources," in Murdoch, *Environment,* pp. 135–55; and G. P. Kalinin and V. D. Bykov, "The World's Water Resources, Present and Future," *Impact of Science on Society,* 19(2), 1969, pp. 135–50.

33 David B. Brooks and P. W. Andrews, "Mineral Resources, Economic Growth, and World Population," *Science,* 185 (July 5, 1974), p. 13.

34 "A Little More Time," *The Economist,* June 29, 1974, p. 15.

35 Barry Commoner, *The Poverty of Power* (New York: Alfred A. Knopf, 1976).

36 Georg Borgstrom, "The Numbers Force Us into a World Like None in History," *Smithsonian,* July 1976, p. 73.

37 Georg Borgstrom, quoted in "How Far Can Man Push Nature in Search of Food?" *Conservation Foundation Letter,* November 1973, p. 8.

38 Barry Commoner, "How Poverty Breeds Overpopulation (and Not the Other Way Around)," *Ramparts,* August/September 1975, pp. 21–25, 58–59.

39 Ibid., p. 22. See also Garrett Hardin, "The Tragedy of the Commons," *Science,* 162 (December 13, 1968), pp. 1243–48; Garrett Hardin, *Exploring New Ethics for Survival* (New York: Viking Press, 1972); and Daniel Callahan, "Ethics and Population Limitation," *Science,* 175 (February 4, 1972), pp. 487–94.

40 Commoner, "How Poverty Breeds Overpopulation," p. 25.

41 David R. Brower, "What's Up (and Coming Up) in ECO," *ECO,* March 1974, p. 3.

42 Paul Ehrlich, in *The Listener,* August 30, 1970, p. 215. See also his *The Population Bomb* (New York: Ballantine Books, 1968); Paul Ehrlich and Anne Ehrlich, *Population, Resources, Environment* (San Francisco: W. H. Freeman and Co., 1970); and Dennis C. Pirages and Paul R. Ehrlich, *Ark II: Social Response to Environmental Imperatives* (New York: Viking Press, 1974).

43 Quoted in Stefan Kanfer, "Is There Any Future in Futurism?" *Time,* May 17, 1976, p. 51.

44 Gary Snyder, *Turtle Island* (New York: New Directions Publishing Corp., 1974), p. 97.

45 See Elting E. Morison, *From Know-How to Nowhere: The Development of American Technology* (New York: Basic Books, 1974).

46 Ecological humanism as a new world view is eloquently explained and advocated in Victor Ferkiss, *The Future of Technological Civilization* (New York: George Braziller, 1974).

47 Daniel Callahan, *The Tyranny of Survival and Other Pathologies of Civilized Life* (New York: Macmillan Publishing Co., 1973), pp. 62, 258.

48 Ruth Adams, *Say No! The New Pioneers' Guide to Action to Save Our Environment* (Emmaus, Pa.: Rodale Press, 1971), p. 122.

49 Paul Shepard, "Ecology and Man—A Viewpoint," in Paul Shepard and Daniel McKinley, eds., *The Subversive Science: Essays toward an Ecology of Man* (Boston: Houghton Mifflin Co., 1969), p. 10.

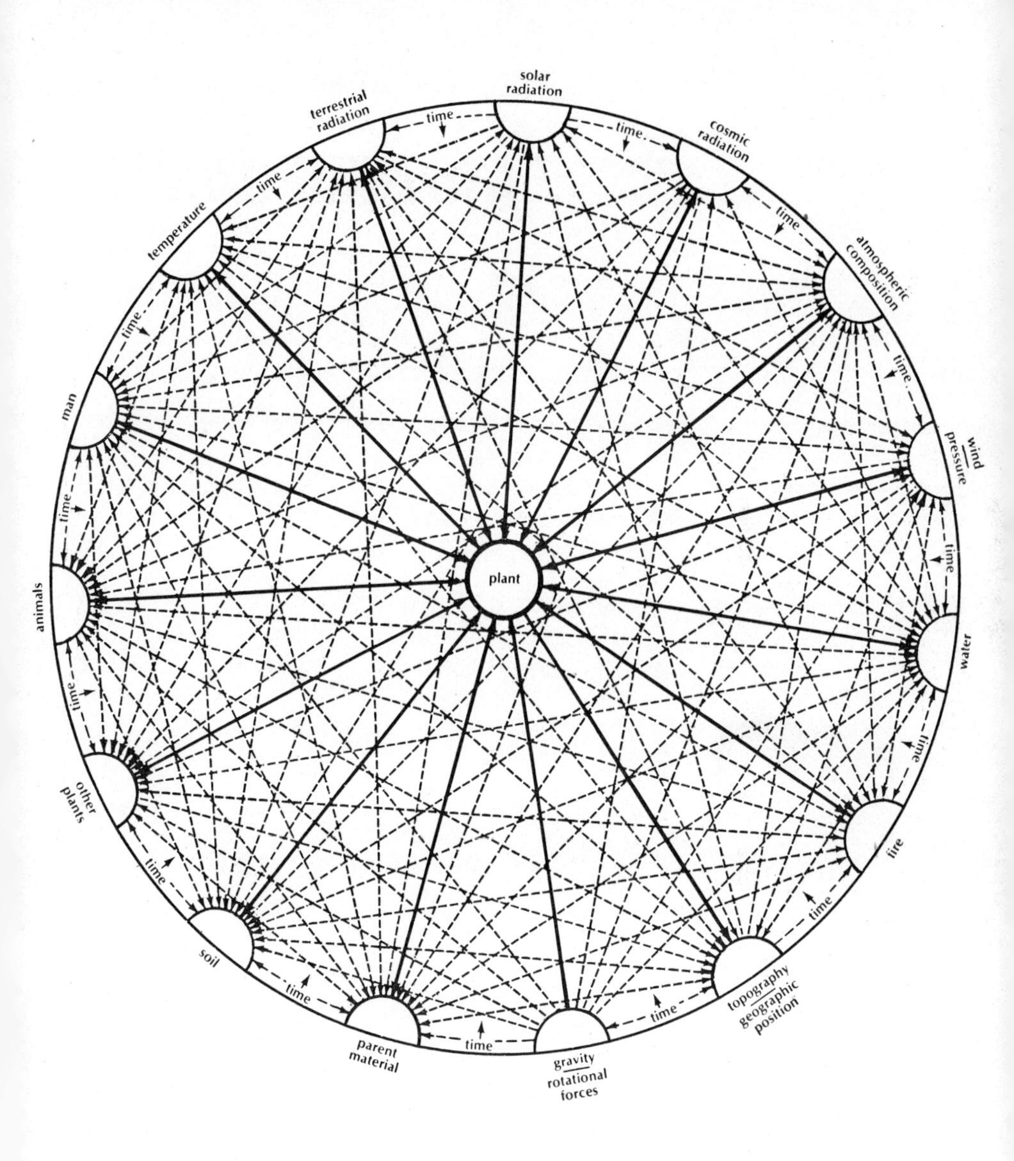

Part I Nature and Man: The Sources and Extent of a Global Disharmony

Introduction to

Part I

On reading some of the selections in this book, but perhaps especially those in this part, you may share the disconsolation of John Livingston as he reflects:

> Anyone who has spent the greater part of a lifetime enjoying and attempting to understand and preserve wild nature will have had the experience of witnessing his own species drift lower and lower on his personal scale of perfection. All the magnificence and nobility of our creativity cannot begin to compensate me for what my species has cost me. Shakespeare cannot compensate me for toxic pesticides. Bach cannot compensate me for Hiroshima, nor Michelangelo for the blue whale. Jesus Christ cannot compensate me for the brutal imposition of human power over nonhuman nature. Yet the total destruction of blue Earth may well precede any diminishment in human pride.[1]

The human community is so interdependent that the ecological mistakes of the past and of the present cannot be dismissed as local misfortunes. They are magnified on a global scale. Whether or not this is the eleventh hour, world ecology has become a common responsibility.

Robert Disch once noted that "the great irony of the environmental crisis is that the workings of natural phenomena—the ecological facts of life—in all their mysterious wonder, are utterly unconcerned with human illusions about man's place in the universe, his power over nature, his destiny, values, mystiques, and taboos."[2] Relentlessly man has ignored or skirted the vast implications of this ironic situation. Rather than treating the entire environment as a single but highly complex system, as ecologists do, those who promote "progress" through exploitation of the environment tend to separate it into two categories: the useful and the aesthetic, with emphasis on the former ("The Grand Canyon is pretty, all right, but what can you do with it?").

Although ecology is still an infant science, it has unequivocally shown that it is not possible to draw a line between what is utilitarian (or "economic") and what is merely aesthetic, because the biosphere displays an underlying unity. Despite enormous diversity and flexibility, life in every form is organized on the same fundamental principles. Yet so complex are ecosystems that we do not yet understand how they work; perhaps we never will. For, as Frank Egler has observed, "The web of the human ecosystem . . . is a highly complex phenomenon. It is not only more complex than we think. It is more complex than we *can* think."[3]

Selection 1 presents some of the basic principles of ecology, insofar as they are known. It shows how they apply to all biological systems, from the smallest ecological unit (the individual organism) to the collective ecosystems of the world (the biosphere) and explains how survival of any living organism depends upon its ability to maintain a steady state of equilibrium with its environment. The ecological facts of life discussed in this selection can and should be superimposed on every other selection in this book, and they can be underscored by three deceptively simple axioms:

> An organism without an environment is inconceivable.
>
> A successful organism without a suitable environment is an irrational thought.
>
> An organism unable to adjust to a *changing* environment becomes an extinct organism.[4]

In Selection 2 the significance, magnitude, and interactions of the earth's three great energy systems—biotic, industrial, and atmospheric*—are considered by Charles A. S. Hall. He gamely tackles a multitude of interconnecting subjects and suggests specific modifications of public policy on the scientific, the national, and the international levels.

The mounting social toll of ecological stress in the highlands of Asia, Latin America, and Africa is described, with disheartening case studies, in Selection

---

*As author Hall points out, world climate appears to be undergoing a cooling trend. In contrast to this chilling view, some climatologists foresee a worldwide warming trend that could lead to the total melting of polar ice caps. But whatever the long-term trends, scientists tend to agree that more widely varying conditions are in store and that their very unpredictability makes planning, especially for agriculture, difficult. For a challenging study of the impact of climate on the behavior of human societies, see Nels Winkless III and Iben Browning, *Climate and the Affairs of Men* (New York: Harper's Magazine Press, 1975).

3. As Erik Eckholm notes, the technical remedial measures that he outlines are thwarted by political and cultural factors.

"We treat the ocean as if it were not part of our planet—as if the blue water curved into space beyond the horizon where our pollutants would fall off the edge," writes Thor Heyerdahl* in Selection 4. And he shows how, by using the ocean as the ultimate dump, man has probably perpetrated the ultimate ecological folly. It is no comfort to note that the United Nations-sponsored Law of the Sea Conference, mentioned by Heyerdahl, continues its inconclusive meetings. Its negotiations have focused not on protection and wise management but on dividing the spoils (the marine life and the minerals) among nations.

In the humid tropics are forests that are the most complex in structure and diverse in species of any on earth. While millions of acres of these forests are felled for agriculture each year, foreign timber companies have started logging operations in tropical countries around the globe, particularly in Malaysia, Indonesia, the Philippines, and Brazil. These foreign concessionaires cut without restriction, and without any thought to forest management.[5] Selection 5 describes some of the incalculable values of the rainforests, their accelerating regression, and what might be done to halt it.

The extinction of species has been going on for eons. This extinction occurred wholly as a result of natural changes in the environment before the appearance of man, who then began, without nature's imprimatur, to exterminate animals at will and whim. It has been calculated that of all the animal species that ever lived on this planet, less than 1 percent exist today. Yet these animals, like the plants and biomes, can never be duplicated. In Selection 6 the authors trace the relationships between the extinction of animal species and the development of human culture, classify the endangered species, and point out the causes of extinction. And they explain why the preservation of endangered species is not just a matter of aesthetics or emotions but quite possibly a matter of our own survival.

---

*Alan Watts has pointed out that on Heyerdahl's *Kon-Tiki* expedition, Heyerdahl and his crew "drifted on a balsa-wood raft from Peru to distant islands in the South Seas just by going along with the natural processes of the ocean. Yet this intelligence is more than mere calculation and measurement. It includes that; but Heyerdahl's genius was that he had a basic trust in the unified system of his own organism and the ecosystem of the Pacific, and was therefore almost as intelligent as a dolphin." Alan Watts, with the collaboration of Al Chung-liang Huang, *Tao: The Watercourse Way* (New York: Pantheon Books, 1975), p. 121.

Man's effects on island ecosystems are pessimistically surveyed in Selection 7. F. R. Fosberg shows how especially vulnerable the world's islands are to the doings of man. There is a poignancy in the very uniqueness of islands—"when an island is degraded and its biota largely destroyed, the world has lost something irreplaceable." (See also Selections 15 and 31.)

The policies and technologies that have led the industrialized nations to their wealth and their ecological disgrace are now being foisted on, or demanded by, the developing countries. In Selection 8 Kenneth Dahlberg assesses the dire ecological effects of current development approaches and illustrates, by specific examples, not only the unexpected ecological costs but also the social and economic ones. He concludes with suggestions for aid, development, and regulatory policies quite different from those currently being followed. (See also Selection 28.)

A few years ago, in a provocative article on the comedy of survival, Joseph W. Meeker likened evolution to comedy: they are both a matter of muddling through. The assumption of superrationalists that

> nature is simple while civilization is complex is one of the sad legacies of romantic thought. . . . Ecology challenges us to vigorous complexity, not passive simplicity.
>
> If the lesson of ecology is balance and equilibrium, the lesson of comedy is humility and endurance. . . . Comedy illustrates that survival depends upon our ability to change ourselves rather than our environment, and upon our ability to accept our limitations with laughter rather than with shame or grim resignation. It is a strategy for living which agrees well with the demands of ecological wisdom. . . .[6]

NOTES

1 John A. Livingston, *One Cosmic Instant: Man's Fleeting Supremacy* (Boston: Houghton Mifflin Co., 1973), p. 88.

2 Robert Disch, in Robert Disch, ed., *The Ecological Conscience: Values for Survival* (Englewood Cliffs, N.J.: Prentice-Hall, 1970), p. 17.

3 Frank E. Egler, "Pesticides—in Our Ecosystem," *American Scientist*, 52 (March 1964), p. 120.

4 Ibid., p. 118. See also Barry Commoner, "The Ecological Facts of Life," in Huey D.

Johnson, ed., *No Deposit—No Return; Man and His Environment: A View toward Survival* (Reading, Mass.: Addison-Wesley Publishing Co., 1970), pp. 18-35.

5 See Jack Shepherd, *The Forest Killers: The Destruction of the American Wilderness* (New York: Weybright & Talley, 1975), pp. 376-86. In this book Shepherd angrily records how the U.S. timber industry is overcutting the public lands, leaving a shocking amount of waste in its wake, and yet demanding the right to increase this cutting.

6 Joseph W. Meeker, "The Comedy of Survival," *North American Review*, Summer 1972, p. 17.

THE BIOSPHERE AND ITS CONSTRAINTS

1 Ecology and the Individual

Rezneat M. Darnell

Rezneat M. Darnell is professor of oceanography and biology at Texas A & M University.

. . . In treating physiological phenomena, assimilation, respiration, growth, and the like, which have a varying magnitude under varying external conditions of temperature, light, supply of materials, &c., it is customary to speak of three cardinal points, the minimal condition below which the phenomenon ceases altogether, the optimal condition at which it is exhibited to its highest observed degree, and the maximal condition above which it ceases again. . . .

F. F. Blackman, 1905.

INTRODUCTION

Miscellaneous observations about the lives and habits of plants and animals have long been part of man's knowledge, and collectively they constitute the field of *natural history.* Observations concerning growing seasons, breeding habits, food habits, and behavior patterns fall into this category. The formal field of ecology has developed from the accumulated natural-history wisdom of the ages, but ecology is more than natural history.

Ecology seeks to describe nature in quantitative terms. Ecology is based upon numerical observations; it deals with ratios; and it is intimately concerned with rates at which natural processes occur. For many years ecology was known as quantitative natural history, but the field of ecology has additional parameters.

Quantitative information has led to the development of ideas about total functional systems of nature, and the natural history information has now been organized into a set of formal principles which govern the behavior of the natural systems. In some cases the principles can be expressed mathematically as differential equations. In other cases we still have to use words or, at best, mathematical approximations. Ecology is a field which is still growing and developing, but it represents man's best attempt to understand the living world. Its message cannot be ignored by any intelligent citizen, least of all by those citizens who through administrative, political, religious, or social position can influence the patterns by which groups of humans relate to the natural system.

BIOLOGICAL SYSTEMS

. . . Life is organized according to certain functional patterns which have been worked out by trial-and-error processes through the generations of time. Considering the fact that life has been on this earth at least 3 billion years (and that is a great many hours and minutes), and considering the fact that many billions of trial-and-error experiments are in progress throughout the world at any one particular moment, it is understandable that complex and workable life patterns have come into existence and that such patterns have themselves become organized into yet more complex entities. The functional systems of nature are thus arranged in a hierarchy of patterns from relatively simple to relatively complex. The functional systems of a given level of organization become the components of the next higher level. Certain laws or principles govern the operation of all biological systems, but as we shall see, at each level some generalizations and principles become evident which are unique to that particular level.

The levels of biological organization are illustrated in Figure 1. The individual organism is composed of organ systems which are, in turn, made

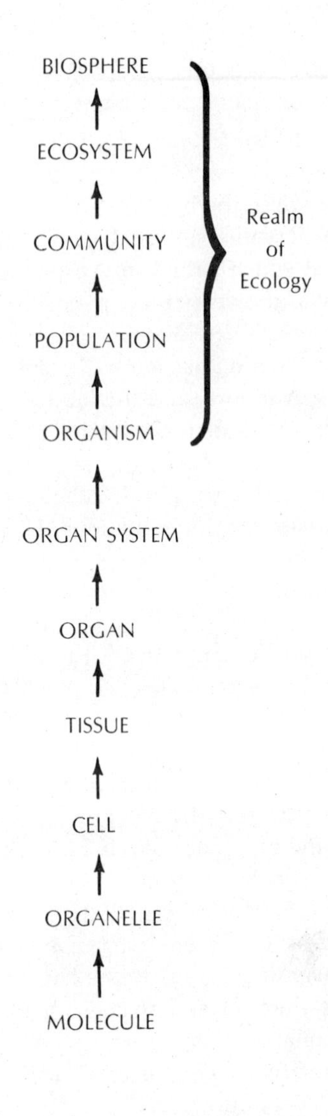

FIGURE 1 *Levels of biological organization. The field of ecology deals with biological systems at the* organism, population, community, ecosystem, *and* biosphere *levels.*

up of organs, these of tissues, and so on. Approached from the other direction, *individual organisms* (oak trees, zebras, sea gulls, etc.) make up single-species groups called *populations* (groves, herds, flocks, etc.). These live and interact with populations of other species and together constitute natural *communities* (forest, grassland, marine, etc.). A given community together with the environment with which it is normally associated makes up an *ecosystem* (forest, grassland, marine, etc.). All the ecosystems of the world considered as a collective unit comprise the *biosphere.* The field of ecology embraces those levels of organization from the *individual* through the *biosphere.* . . .

Nothing in the universe exists alone. Every drop of water, every human being, all creatures in the web of life and all ideas in the web of knowledge, are part of an immense, evolving, dynamic whole as old—and as young—as the universe itself. To learn this is to discover the meaning of joy.
DAVID CAVAGNARO

A biological system is a living unit of any size or degree of complexity which has a structural and functional integrity. It may be a cell or an ecosystem. Since it is a living unit, it displays certain properties which, taken together, distinguish it from all nonliving systems:

It embodies a recognizable structure.

It carries out orderly functional processes.

It undergoes a regular pattern of development and maturation.

It reproduces itself more or less exactly.

It is capable of long-range, orderly genetic change.

It survives only within a certain prescribed range of environmental conditions.

It maintains its integrity in the face of a hostile environment.

Its structure and function are finely adjusted to operate under the environmental conditions to which it is normally subjected.

It possesses internal regulatory mechanisms which permit it to offset or to recover from the effects of moderate deviations from the usual environmental conditions.

It cannot cope with severe deviations from the usual conditions. These result in permanent damage or death.

Although generally characteristic of all living systems, these principles vary in detail when applied to different systems of a given level and especially when applied to systems of different levels of biological organization.

THE INDIVIDUAL AS AN ECOLOGICAL SYSTEM

The smallest ecological unit is the *individual organism.* Except in the case of rare identical twins and individuals arising by certain asexual processes, all organisms are genetically unique. As a result, even closely related organisms differ in form, function, and ability to adjust or respond to conditions of the environment. Closely related organisms, though differing in some genes, share much of their genetic material in common. Hence, the small differences observed within a given group of organisms may be thought of as variation around a single theme. . . . This small variability, arising from gene mutation and recombination, is of great importance in permitting survival in the face of a variable environment.

During its life history, the individual passes through a regular series of events, beginning with fertilization, passing through a sequence of developmental stages to maturity, and followed by a period of senescence and death. At each stage the form and function are characteristic for that life period.

The Flow of Chemicals and Energy

Throughout its life each individual must obtain from the environment (or from stored food reserves) certain necessities. These include *inorganic substances* (water, nitrogen, sodium, potassium, calcium, magnesium, iron, etc.) as well as specific *organic chemicals* (carbohydrates, lipids, amino acids, and vitamins) for use as building blocks and chemical factories. In addition, every living organism must constantly derive from the environment a quantity of *energy* sufficient to support the various metabolic processes of the body. This energy is universally derived by chemical oxidation (i.e., burning) of organic compounds. All animals and some plants take in both fuel (organic compounds) and oxygen to provide the needed energy. Green plants, however, manufacture their own energy-rich fuel from carbon dioxide, water, and sunlight by a process known as *photosynthesis.* Since oxygen is a by-product of photosynthesis, the green plants even produce a portion of the oxygen which they need to burn the fuel. The end products of oxidation are carbon dioxide and water (and nitrates, phosphates, and sulfates if certain proteins and lipids are oxidized). A few species of bacteria and blue-green algae have developed the ability to oxidize organic compounds in the absence of oxygen. These species may employ iron, sulfur, and other elements as hydrogen acceptors in place of oxygen. . . .

As a result of its metabolic processes, every organism discards or *excretes* into the environment certain by-products or wastes. If oxidation has been complete, the organism will discharge the chemicals, carbon dioxide, water, nitrates, phosphates, and sulfates. Generally the oxidation is not complete, however, and many organisms also release urea, uric acids, other organic acids, ammonia, etc. Some bacteria and blue-green algae may release methane and hydrogen sulfide gases, as well as nitrites, sulfites, etc. Regardless of whether oxidation is complete or incomplete, all organisms release into the environment large quantities of "waste energy" in the form of heat.

Every living organism also releases into the environment certain chemical substances which are not "waste" in the usual low-energy sense, but they are high-energy organic compounds which seem to have important functional roles and which may be thought of as bodily *secretions.* These include

All things by immortal power
Near or far
Hiddenly
To each other linked are,
That thou canst not stir a flower
Without troubling of a star.
FRANCIS THOMPSON

such chemicals as antibiotics, mucus, and special odiferous substances, as well as certain organized structures such as skin cells, shed exoskeletons, feathers, hair, and gametes.

From the foregoing discussion we may derive the following important ecological principle. *Every living organism modifies its environment by the removal of certain substances and through the addition of others.* Different kinds of organisms do different things and at different rates, of course, but all modify their environments in one way or another. This means that all organisms

influence the environments of their neighbors. This is a cardinal point in ecology. It leads to success or failure when organisms attempt to live side by side.

MAINTENANCE OF LIFE IN THE HOSTILE ENVIRONMENT

Every living organism must maintain its chemical and physical integrity if it is to survive. To accomplish this the organism must selectively remove from the environment what it needs, selectively eliminate that which is no longer useful, and carry out these processes in spite of an ever-changing environment. The maintenance of a constant internal integrity entails a variety of adjustment and regulatory processes which collectively keep the system in a steady state equilibrium with its environment. This phenomenon, known as *homeostasis,* is one of the most remarkable features of living systems, and it is especially conspicuous at the level of the individual.

Consider the oyster, exposed to the air at low tide and submerged when the tide is in, subjected to fresh water after rainfall and salt water when there is no rain. It must endure summer heat and winter cold, the heavy silts of spring floods and the clear water of summer. Through it all the oyster must survive, grow, and reproduce its kind. This tenacity of life, the ability to keep going in spite of environmental change, is displayed to greater or lesser degree by all living organisms.

For each life history stage of an organism there is an optimum set of environmental conditions. So long as these conditions are met, the organism will grow, develop, and function in normal fashion. If one or more of the environmental factors deviates significantly from the normal range, then the organism is placed under stress. The stress condition may be sharp and sudden, or it may be low level and of long duration. It may be characterized by either an excess or a deficiency of a given environmental factor. In any case the organism must respond by some means which will aid in offsetting the effect of the stress agent.

Response to Environmental Stress

Individuals possess a remarkable portfolio of response mechanisms for dealing with environmental stress. Some simply move away to more favorable areas. Others become physiologically inactive and wait out the unfavorable period in a dormant or semidormant state. Others stay and tolerate certain internal body conditions which faithfully follow those of the external environment. All living organisms, however, show some ability to *regulate* the internal environment, and the ability to regulate is greatest among the higher forms of life.

Among those organisms which remain active, one of the first measurable signs of stress is the elevated requirement for oxygen. Respiration increases, but since the respiratory apparatus itself poses certain limitations, a smaller fraction of the organism's bodily resources remains available for normal metabolic activity when it is under stress. Thus, the organism is subjected to a *metabolic load* in proportion to the degree and duration of the stress factor.

To counteract the stress situation the organism must reorder the metabolic programs of the body. This often involves increased production of certain hormones and enzymes accompanied by the decreased production or inhibition of others. Certain metabolic pathways become favored over others. Body fluid concentrations are altered. Excretory products may be accumulated or voided in unmetabolized form. Long-continued stress may result in the harmful depletion of certain chemical and energy reserves, accumulation of salt deposits, hypertrophy of certain tissues and organs, atrophy of others, exhaustion of hormonal and enzyme capabilities, inhibition of reproduction, altered behavior patterns, and eventually death.

To protect the body from the potentially dangerous effects of stress most organisms have developed the ability to *acclimate.* This involves a series of metabolic readjustments which permit continued function with minimal deleterious effect in the face of long-enduring, low-level stress situations. A persistent high noise level can be "tuned out" and eventually ignored so long as the intensity remains reasonably constant. An environmental temperature increase can be tolerated much more readily if the temperature rise is sufficiently gradual to permit the high temperature adaptive enzymes to come into play. Organisms can adjust so long as the extremes are not too great and the rate of change is not too rapid. Acclimation thus permits the organism to carry on nearly normal bodily functions over a fairly wide range of environmental conditions, even into the stress zones.

It was pointed out above that stress may be occasioned by either the excess or deficiency of a given environmental factor. This concept may be generalized to the level of an ecological principle. *With respect to the factors of the environment, life is adjusted to some intermediate condition.* For each type of organism and for each stage of its development, there is a limited *range of tolerance* within which life is possible Somewhere within this total range of tolerance, there is an optimal zone or *range of normal activity.* Both above and below this optimal range and continuing to the limits of tolerance there are *zones of stress* wherein the organism either cannot carry on normal activity or in which body functions are impaired. . . .

. . . Requirements and tolerances vary from one species to another, but all have become adjusted to the general conditions of life in which the organism normally dwells. Those individuals which fail to adjust to their environments perish without leaving progeny. It is generally true that immature forms of a species (eggs, larvae, babies, etc.) are the stages least tolerant of environmental stress, and it is here that the highest rates of environmentally induced mortality occur.

All natural environments are capricious. They involve many factors acting together, and although the nature of the variation is to some extent predictable, unusual extremes do occur. Therefore, some degree of stress is a regular feature of existence of organisms in all environments. Environmental selection for the strong and the adaptable has been the overriding consideration throughout the history of life on this planet. It has been the primary sieve through which new genetic material has been strained, the anvil of adaptive evolution.

Earlier in this chapter it was pointed out that no two individuals are exactly alike, and the evolutionary significance of this fact now becomes clear. Since the environment varies from place to place and from time to time, it is of definite advantage to a species that its members possess a certain amount of variety in ranges of tolerance, zones of optimum, and so on. Thus, the group has greater potential than the individual, and group survival is insured in spite of individual hardships which are inevitable.

SUGGESTIONS FOR FURTHER READING

Blackman, F. F. "Optima and Limiting Factors." *Ann. of Bot.* 19, 1905.

Borradaile, L. A. *The Animal and its Environment.* London: Oxford University Press, 1923.

Buchsbaum, R., and Buchsbaum, M. *Basic Ecology.* Pittsburgh: Boxwood Press, 1958.

Burnett, A. L., and Eisner, T. *Animal Adaptation.* New York: Holt, Rinehart and Winston, 1964.

Darnell, R. M. *Organism and Environment: A Manual of Quantitative Ecology.* San Francisco: W. H. Freeman & Co., 1971.

Daubenmire, R. F. *Plants and Environment: A Textbook of Autecology.* New York: John C. Wiley & Sons, 1959.

Farb, P., and the editors of *Life. Ecology.* Life Nature Library. New York: Time Inc., 1963.

Odum, E. P., *Ecology.* New York: Holt, Rinehart and Winston, 1963.

2 The Biosphere, the Industriosphere, and Their Interactions

Charles A. S. Hall

Charles A. S. Hall, staff scientist at the Ecosystems Center, Marine Biology Laboratory, Woods Hole, Mass., is visiting assistant professor of biology at Cornell University. The author is grateful to D. Goodman, R. H. Whittaker, and G. M. Woodwell as well as many other friends and colleagues for their advice and criticism.

From the *Bulletin of the Atomic Scientists,* 31 (March 1975), pp. 11–21. Reprinted by permission of the author and the *Bulletin of the Atomic Scientists.*

Three great energy systems of the Earth—biotic, industrial, and atmospheric—provide man with the prerequisites for, and amenities of, life. In the Western world, particularly during these times of the "energy crisis," we tend to focus our attention on the industrial system. Less emphasis is placed on the larger, more important solar-powered systems of weather and life. This paper is an attempt to rectify this situation by considering the importance and magnitude of each of these systems, and the various ways in which they interact. . . .

The *biosphere* is the thin, discontinuous mantle of life that surrounds the Earth, and includes natural ecosystems, agricultural ecosystems and man. It is the product of roughly three and a half billion years of organic evolution that has made what is thermodynamically unlikely not only possible but both probable and highly organized. It is characterized by a virtually infinite, though diffuse, energy source—the Sun—and by negative feedbacks that tend to limit and stabilize its development. There is no reason to consider its future any less circumscribed in time than its past, but it has undergone dramatic changes and will continue to do so.

The biosphere contributes to human welfare by supplying man with food, fiber and wood as well as the fossil fuels, via ancient ecosystems, that power man's industrial systems. In addition, and in conjunction with the atmospheric systems, the biosphere provides man with a stable physical and chemical milieu. As man has learned to manage the biosphere through agricultural and related activities, his ability to create cities, arts, armies and the other trappings of civilization has increased, as have his own numbers.

The *industriosphere* is the series of fossil-fuel powered patches of human industrial activity located within, and supported by, the biosphere. It has a relatively short history, some 200 or 300 years at most, and is presently characterized by positive feedbacks that contribute to its growth. Its future is uncertain: if dependent upon the concentrated fossil fuels that have nurtured it so far, the industriosphere might continue for another 300 years, although other circumstances may increase or decrease this span.

The development of the industriosphere has contributed to the management of the biosphere and the increase in the material wealth of Western man. This has allowed the further development of cities, the arts and armies as well as an increase in population. However, the industriosphere has not only elevated but also degraded human existence through pollution and industrial squalor, and it has leaked its toxic products into the biosphere. The industriosphere has grown so large as to overtax the capacity of the biosphere to assimilate and disperse these by-products; the resulting and potential biotic degradations appear to be reaching critical dimensions on a global scale.

Among all the strange things that men have forgotten, the most universal lapse of memory is that by which they have forgotten they are living on a star.

G. K. CHESTERTON

Although we are used to thinking of the two systems—the biosphere and the industriosphere—in different conceptual frameworks, both systems use energy, principally derived from the Sun, to rearrange the chemicals of the Earth's crust to maintain their respective systems, and in so doing produce goods and services useful to man. The best common denominator is energy, as ecologists and industrial economists have both used units of energy to measure the major processes with which they are concerned, and energy use accompanies all physical and biological processes. Leith [1] has estimated the magnitude of the Earth's gross production—the total quantity of the Sun's energy fixed by green plants: gross $\cong 2 \times$ net—as being 14×10^{17} calories per year.

Global biological respiration is approximately the same. Total industrial consumption, on the other hand, is about 4.7×10^{16} calories per year or about 3.5 percent as large as the biospheric rate of energy production and use [2]. For the continental United States the ratio is about 13 percent and for New York City the industrial use of energy is more than 200 times the photosynthetic production rate.

In discussing how the biosphere and the industriosphere interact, I will develop the following concepts:

• Man is presently dependent upon the proper functioning of both of these systems.

• Considerations of the industrial energy crisis, world food problems or ecological purity cannot be made in isolation from one another if the desired end product is man's welfare, since his welfare is a function of the complicated interactions between the biosphere and the industriosphere.

• The future well-being of man requires intelligent and sometimes intensive management of the biosphere using limited inputs from the industriosphere instead of the present, ever more energy-intensive exploitation.

• Uncontrolled growth of the industriosphere presently threatens the productivity of the biosphere and, as a consequence, man's well-being. Given the seriousness of this existing and potential biotic destruction, given our very meager scientific knowledge of it, and given our limited industrial fuel resources, we would be wise to consider restraining the further industrialization of the presently overdeveloped nations. This is, admittedly, a very unlikely possibility.

As background for later considerations, let us briefly review man's historical relation to his biotic environment as expressed in his use of various forms of energy. Since man's principal interaction with his environment, other than the requirement for a reasonably stable physical and chemical milieu, is in the production and consumption of food, this account will emphasize the development of man's food-energy production in relation to other energies.

Prehistoric man was, of course, completely dependent upon the energy supplies of the natural ecosystems around him. Consider a hunting and gathering culture of the present-day tropical rain forest, a living representative of our preagricultural ancestors. The quantity of food energy that flows from the environment to man is small, some 0.4 kilocalorie per square meter per year [3]. However, man's need to use his own energy to control the means to his livelihood in this situation is likewise small, and the complex and diverse natural ecosystem provides this culture with all of life's requirements. Consider some of the natural processes that contribute to the survival of these people: reseeding through natural vegetative reproduction, soil structure maintenance by various arthropods and microorganisms, pest protection by the existence of vegetative diversity, and protein production via the natural animal food chains. These may seem like obvious and not particularly interesting aspects of the existence of these forest dwellers, but as we shall see the provision of these same services by modern society is extremely time and energy consuming, and has a number of other costs not presently assessed. The carrying capacity for man by an essentially unmanaged, reasonably fertile natural system appears to be on the order of one person per 3 square kilometers [3].

With the development of primitive agriculture the number of people that could be supported per unit of food-producing area was vastly increased. Human energies were used in farming efforts to redirect the natural biological productivity of the land to species that were more useful as food. In general, nonindustrial agriculture has been successful on a large, sustained scale only in temperate and/or floodplain regions where soil nutrients are plentiful and seasonal pulses enable man to insert his crops each year ahead of competing natural ecosystems. Where conditions are favorable the population density of man may reach 10,000 per square kilometer over large areas, particularly where conditions allow the development of labor-intensive irrigation systems [4, 5]. However, as Janzen, Vayda and others point out [4, 6], not all regions of the Earth (for example, some slash-and-burn agricultural regions of Borneo) are suitable for intensification of agriculture, and the attempt to increase yield may backfire.

Industrial man's agriculture is heavily dependent upon the use of fossil fuels. Relative to the conditions mentioned above, there is a considerably greater yield of edible calories, as much as 1,000 to 5,000 kilocalories per square meter per year, but there is also a proportionately much greater requirement for industrial energy. For example, genetic stocks of grains used in the "Green Revolution" are useless without such inputs from the industrial sector as fertilizers, irrigation and pesticides. For a number of situations in which this relationship has been studied, there appear to be from 0.3 to 5 (mean of about 1) industrial calories used on the farm to produce 1 edible plant calorie [3, 7-11].

Man is dependent upon the quality as well as the quantity of food, and the most critical component in most diets is protein. However, if man feeds 10 calories of a low protein cereal crop to domestic animals, he gets only 1 to 2 calories of high protein meat in return. The efficiency of energy transfer from one organism to another that feeds on it is about 10 to 20 percent and we do not eat all of a cow or a chicken. Thus our society uses about 5 to 10 industrial calories to produce a calorie of protein on the farm. About 15 percent of our diet is animal protein.

According to Steinhart and Steinhart [12] a conservative estimate of 9.5 industrial calories is required in the United States for each food calorie (including both plant and animal food) eaten by the human consumer, with about one-quarter of these industrial calories being used to grow the food. A large proportion of the remainder is used in food processing and distribution. The total energy used to grow food is about 3 percent of our total industrial energy consumption and at least 13 percent if we include energies used in processing and distribution. In the United States this large requirement for industrial energy in the food-production process has been in large part a result of the decrease in human (and draft animal) labor used on the farm. It also has been the result of creating suburbs out of the previously agricultural "green belts" that once supplied cities with fresh food daily without the requirement for a great expenditure of transportation or processing energies.

The increase in food production through agriculture and industry has allowed an increase in the individual standard of living in some countries; but this is not the case for the world as a whole, for population increases have generally kept up with or exceeded food production, especially protein production. This is particularly so in the less developed countries [13, 14].

In the United States and the world as a whole we are now in a period of diminishing agricultural returns per unit of industrial energy input. For example, although world food production increased 34 percent from 1951 to 1966, this increase was associated with an increase in the production of tractors by 63 percent, phosphates by 75 percent, nitrogen fertilizer (perhaps the major agricultural consumer of energy) by 146 percent, and pesticides by 300 percent [15]. Steinhart and Steinhart [12] have noted a logistic curve for total energy input versus food production in the United States from 1920 to the present. In the most recent years we have *increased* the industrial energy put into gaining about the *same* total food production.

UNMANAGED BIOLOGICAL SYSTEMS

In some regions man may obtain protein by allowing cattle or other beasts to graze natural, unmanaged grassland or by letting natural animal populations harvest natural food chains. Man then, in many cases, gets a high return of protein per industrial calorie input, since only harvesting energy must be invested. Fisheries, particularly anadromous and other coastal fisheries, may give 5 to 10 times more protein per industrial calorie input than domestic beef or chickens. Fish supply about 15 to 20 percent of man's animal protein [13, 14, 16].

Man's relation to the food biota of the Earth is relatively straightforward and reasonably quantifiable. However, our relation to other components of the biosphere is less apparent and the interactions are considerably more subtle.

Consider, for the moment, the "public service function" of natural systems. There are a number of ways in which processes essential for man's existence or well-being are provided through the normal functioning of natural ecosystems. These processes may in some cases be duplicated by man's industrial machinery although this is generally difficult, expensive and incomplete. A number of examples would include: storing of water, maintaining soil structure and productivity, maintaining the chemical equilibrium of air and water, maintaining an epidemic-free environment through natural predators and diverse plant communities, moderation of microclimates where natural vegetation is profuse, natural fertilization of agricultural land on undammed floodplains, and the cleansing of soil, air and water via natural biological metabolism.

In regions where human density is low, the natural systems are able to perform the essential "housekeeping" functions needed by man, and no municipal expenditures are needed to maintain a livable and productive habitat. However, as the human population density or per capita resource use increases, and critical natural areas are destroyed, it becomes necessary to carry on an increasing percentage of the work of natural systems with industrial machinery. Odum et al. [17], for example, have considered the useful work of natural watershed systems and have made approximate calculations of how much it would cost man, in dollars and calories, to do the same work with dams, reservoirs, sewage plants and piping. They concluded that the most desirable situation for man was an approximately equal combination of natural and developed systems, and not an excessive emphasis on one or the other.

It is important now that ecologists and others familiar with specific natural systems define and measure, perhaps in terms of "industrial equivalents," the public service function of these systems. As industrial energy becomes more expensive, and perhaps scarcer, it may be wise to reassess the potential value of natural ecosystems that provide services to man without requiring industrial energy.

INDUSTRIAL ENERGY USE AND BIOLOGICAL PRODUCTIVITY

Our concern thus far has been principally with the use of industrial energy to redirect biotic primary production to man. There are also ways in which global industrialization can change global rates of primary productivity. The include positive effects of carbon dioxide enhancement and local fertilization from smokestack nutrients as well as the negative effects of acid rain, potential climate changes, the changing of sunlight intensity by increased aerial particulates and aerosols, the destruction of oceanic fisheries by highly industrialized overfishing and coastal development, and the destruction or diminution of the "public service function" of natural ecosystems either by urbanization or by replacing natural biotic systems with intensively managed ones. Most of these interactions are global, fascinating, enormously important and poorly understood.

It has long been recognized that the combustion of fossil fuels is contributing to an increase in the concentration of carbon dioxide in the atmosphere [18-21], although Hutchinson [22] pointed out that the worldwide substitution of less productive—and hence less carbon dioxide absorbing—agricultural and urban systems of natural ecosystems also is important. Recent estimates of the increase put it at about 0.2 percent per year over the past decades, rising to 0.3 or 0.4 percent per year for the past few years. Much of these data are clear and free from noise, other than seasonal patterns of net photosynthesis, and stations as geographically remote as Hawaii and New York show similar patterns.

The increase in the concentration of carbon dioxide in the global atmosphere influences the biosphere in at least two ways. First, the carbon dioxide changes the balance of thermal energy entering and leaving the atmosphere due to the now well-known "greenhouse effect." Second, an increase in carbon dioxide concentration has been shown experimentally to increase the rate of photosynthesis. This relation appears to be approximately linear for values in the vicinity of present or projected atmospheric concentrations, at least for terrestrial ecosystems [23–26]. Thus we might consider increased carbon dioxide production by the burning of fossil fuels as a "good thing" for the biosphere at least within certain limits, for both the heating and the direct enhancement of available carbon dioxide would make the biosphere more productive [27].

However, many other global or continental effects of industrialization would act to decrease primary production. One such factor is the increasing acidity of rain due, apparently, to the increase in the concentration of oxides of nitrogen and sulfur in the atmosphere from combustion of fossil fuel. (Berner [28] has calculated that the organic enrichment of water may also be important in increasing the concentration of these compounds in the atmosphere due to increasing the area of reducing environment exposed to the atmosphere.)

The increased acidity of rain is a problem that is widespread and already of serious proportions: The pH of rain over large areas of northern Europe and the entire northeastern United States is now generally between 3.0 and 4.5 versus a normal 5.6 to 6.0.* The pH over large regions of Europe declined by about 1 unit from 1956 to 1965, although there are some indications that the pH in some areas has stabilized at from 3 to 4.5 [29, 30]. Localized acid rain has damaged buildings and crops, but the long-range effects have not been studied adequately. Although locally falling sulfate and nitrate particles may act as fertilizers, laboratory studies on the leaching rate of calcium indicate that below a pH of 4.0 rain is much more effective in leaching nutrients from soils [31]. This could be an important effect over long periods in determining regional rates of photosynthesis.

Both the heating and the photosynthetic enhancement of the biosphere that have resulted from atmospheric carbon dioxide enhancement have been compensated, at least partially, by another result of global industrialization—the loading of the upper atmosphere with dust from the burning of fossil fuels, and from mechanized agriculture. Overgrazing and slash-and-burn agriculture also increasingly contribute as human population grows. The data on dust loading are not as clear as those for carbon dioxide enhancement, and the atmospheric effects are very sensitive to particle size and height of injection. Changes that are potentially the result of man's activities are within the range of changes that have been observed before massive industrialization, changes principally due to volcanic eruptions.

---

*The symbol pH is used to denote acidity and alkalinity. The values on the pH scale run from 0 to 14; 7 indicating neutrality, numbers less than 7 increasing acidity and numbers greater than 7 increasing alkalinity. There is a tenfold change in relative acidity for each unit change: a pH of 5 is 10 times more acidic than a pH of 6, and 100 times more acidic than a pH of 7.

However, it does appear that upper atmospheric dust loading, either by volcanoes or man, decreases the temperature at the Earth's surface and increases the temperature in the upper atmosphere as a consequence of absorption and reflection.* While most of the evidence for this relation is circumstantial, we do have one "experiment" on at least the volcanic effects, performed gratis, as it were, by the 1963 eruption of Mount Agung in Bali, which threw large quantities of dust into the upper atmosphere. During the year or so that followed, upper atmospheric temperatures in the latitude of Agung increased by some 6 to 7 degrees Celsius due to the increased absorption of solar radiation by dust particles. There was a simultaneous decrease in sunlight reaching the Earth's surface [32].

Further validation of the relation between upper atmospheric dust loading and ground temperature is presented by Budyko [33]. There has been a strong correlation between decade-by-decade global temperature changes and the quantity of solar radiation reaching the Earth's surface. The quantity of insolation appears to be roughly correlated with past histories of volcanic eruptions: following major volcanic eruptions in 1882 and 1912, there was a decrease both in insolation and, with a slight time lag, temperature (for the northern hemisphere). Historically the years following major volcanic eruptions have been noted for spectacular sunsets, cold weather and high food prices.

The more recent decline in insolation and temperature is thought by some [33, 34] to be a reflection of man's atmospheric loading of dust—now greater than the long-term global volcanic average—as well as the effects of recent, relatively small-scale volcanic activity [33, 35–37]. However, it is important to emphasize again that man's dust effects may be different than the effects from volcanoes because of differences in particle size and height of injection. Nevertheless, there is some evidence that man's activities are decreasing insolation in the northern hemisphere by perhaps as much as 0.2 to 0.4 percent per year. In recent decades the rate may be as high as 0.3 to 0.4 percent per year. Anthropogenic particulates are also important in cloud formation [38]. Changes in cloud patterns could alter primary production by changing patterns of shading and rainfall distribution.

Although the increase in global temperature from the 1880s to about

---

*The loading of dust in the lower atmosphere (from volcanoes, industrial fuel burning or from the "natural" sources of windblown soil, tree derivatives, etc.) is probably much less important in determining climate than upper atmosphere dust loading.

1940 is attributable to carbon dioxide effects, it presently is thought that the effects of global dust loading have more than counteracted this effect since 1940, the net result being the temperature decline that we observe [33–39]. Most of the variability in temperature of the northern hemisphere can be attributed to carbon dioxide and dust effects alone [34]. Preliminary global models prepared by Watt [35] indicate potential decreases of 2° Celsius or more in the Earth's temperature by 2050 A.D. due to dust loading if present rates and patterns of industrialization continue. This would have marked effects on the length of growing seasons. Incidentally, the period 1915 to 1960 was an abnormally warm and agriculturally favorable period when compared with the past 1,000 years or so (with only one big volcano since 1900 versus a normal 5 or so per century). A return to "normal" times would probably result in some loss of biotic productivity [34, 40]. There is some recent indication that snow and ice are not melting in certain polar regions where they had melted in previous summers [41], and the Pacific Ocean has cooled recently.

Similarly, the effects of decreased insolation reaching the world's ecosystems would appear to decrease photosynthesis, again counteracting the increases due to carbon dioxide enhancement. Plant productivity is a function of sunlight intensity when whole plant communities are considered [42, 43]. However, this relation is confounded by two further factors: light inhibition at high intensities and the effect of light quality. Individual plants in the laboratory show a leveling-off and decline in photosynthesis at high light intensities, perhaps due to protein destruction at high intensities [42, 43]. But it is not clear that this phenomenon would occur for entire plant communities within the range of naturally occurring sunlight, for different species become important at different light levels and leaves lower in the canopy or water column receive higher levels of light as upper leaves become saturated.

In addition, the changes in quality of light resulting from dust loading—that is, the increased percentage of light that is scattered versus the percentage that strikes the plant directly—could favor photosynthesis (were intensity held constant). This perhaps is due to a decrease in plant respiration since the temperature of the leaf would be less, or due to increased canopy penetration [44].

The interaction of all these factors is not understood, particularly the effects on whole ecosystems with light- and dark-adapted leaves, different foliage levels, different physiological adaptations, and so forth. With the enormous amount that we have learned in the laboratory about plant physiology it is remarkable how little information we can bring directly to bear on answering these important questions about the biosphere. We obviously need carefully controlled, large-scale studies on whole ecosystems before we can have much faith in what these sunlight changes mean.

Anthropogenic dust and carbon dioxide loading may also affect regional primary production by changing rainfall patterns in monsoon climates. Bryson [45] has suggested that the dust loading and carbon dioxide increases caused by industrialization have contributed to the drought conditions in the Sahelian region of Africa where millions of people have died, or may die, due to a failure of monsoon rains for an unprecedented six years. The rains fall on the oceans instead. This change in monsoon patterns may be occurring in a number of monsoon regions of the world. About one-half of the people in the world live in monsoon climates.

The relations between industrial and ecological systems mentioned above are a direct result of the burning of fossil fuel and are emphasized in this paper. However, there are in addition a series of indirect interactions in the biosphere that result from a leakage of various secondary materials and concepts from industrial operations to the biosphere. These include DDT and other pesticides, PCBs (polychlorobiphenyls) and similar persistent, toxic industrial chemicals.

We have considered a series of individual interactions by which global primary production could be increased or decreased by the activities of the industriosphere. To try to determine the aggregate result of all interacting at once is mind-boggling. Our attempt to put these relations into a Fortran model gave us widely divergent results depending upon our assumptions. Something is happening; it is important, but predicting the magnitude or even the direction of the outcome would appear to be a nearly impossible task. However, there exist several empirical ways, presently tentative, by which the aggregate effect of these interactions can be checked in some approximate way. Although it may not be possible yet to separate industrial effects from natural or stochastic events, there are some interesting recent patterns. First, trees leave a record of their growth in annual rings. Tree-growth records over the past 1,000 years are roughly correlated with climatic patterns [40]. The only data that I have available which are applicable to the recent past are those of Whittaker et al. [46], Bolin [47], and some preliminary analyses from Cogbill [48] for northern New Hampshire, Sweden, and northern New York, respectively. Although analyses of these data are complicated by the drought pattern of the north-eastern United States during the mid-1960s, all three studies show a reduction in annual tree-ring increments in the most recent decade or so when compared with earlier decades. This reduction is apparently not completely attributable to the drought, and whether or not the reduction indicates a decline in forest productivity is not yet certain. However, the pattern could, and should, be checked by a global series of tree-ring surveys to see if the pattern is general.

Another empirical approach to the assessment of potential changes in biospheric primary production is based on an analysis by Hall, Ekdahl, and Wartenberg [49] of the intensive 15-year record of atmospheric carbon dioxide at the Mauna Loa high altitude weather station in Hawaii. After a careful analysis of this data they concluded that there has not been a consistent change in either photosynthesis or respiration over this 15-year period. One explanation might be that the biosphere is just too big to be influenced yet by global industrialization. However, we think a better explanation is that the lack of changes in hemispheric metabolism is due to there being an approximate balance between the photosynthetically-enhancing effects of industrialization and those effects that would tend to decrease photosynthesis. It may be that, as in the case of global temperatures, man has "lucked out" in that a tenuous and precarious balance of two industrial effects is maintaining the global equilibrium he is dependent upon.

Less dramatic but perhaps more immediate changes in world protein production arise from over-fishing and from alterations of the coastal environment. These alterations are in large part due to the effects of large amounts of industrial energy at man's disposal. Estuarine regions currently are directly or indirectly the major producers of our more important commercial fishes during part or all of their life cycles. For example, there are currently about 6.6 billion pounds of fish harvested annually from the coastal waters of the eastern United States, enough fish to meet the minimum protein requirements of all Americans—except that the fish are being harvested mostly by Russians and eastern Europeans. Of this 6.6 billion pounds, about two-thirds are from species that are believed to be "estuarine dependent" or at least associated with coastal waters during some parts of their life histories [50]. Stroud [50] divided the quantity of estuarine-dependent fish harvested each year from the Atlantic by the area of shallow-water estuaries and found that more than 500 kilograms (wet weight) of fish are contributed each year to the fisheries from each hectare of estuary.

Estuaries are being rapidly destroyed by commercial and industrial development. Long Island, for example, has lost approximately 50 percent of its tidal marshlands within the past 20 years; and, partly as a result, the local commercial fisheries and shellfisheries are producing but a fraction of former yields.

However, let us consider a more direct relation between industrial energy and marine protein production. Power plants, particularly the extremely large units currently being built, are being placed on coastal and estuarine regions with increasing frequency, for these regions are often the largest available sources of cooling water near population centers. The quantities of water required for these plants are immense, millions of cubic meters per day for even a modest-sized plant. To give an idea of what this means, the 10 or so plants that have been built or proposed on the mid-Hudson River would require the equivalent of more than a 3-kilometer segment of the Hudson River each day for cooling. This water, when passed through the

The world is too much with us; late and soon,
Getting and spending, we lay waste our powers:
Little we see in Nature that is ours;
We have given our hearts away, a sordid boon!
The Sea that bares her bosom to the moon;
The winds that will be howling at all hours,
And are up-gathered now like sleeping flowers;
For this, for everything, we are out of tune . . .
WILLIAM WORDSWORTH

condenser tubes of the generating systems, kills larval fish by thermal and physical shock. In addition, during the winter months when fish cannot swim rapidly, hundreds of thousands of larger fish may be killed in one day on the trash screens. Neither process can be avoided with present or likely technology.

The striped bass (*Morone saxatalis*) is an important sport and a commercial fish of the northeast that (except for several minor populations) spawns only in the Hudson River and the Chesapeake estuary system. Computer models have been constructed of the interrelation of the striped bass' life history and the operation of the various power plants on the Hudson [51, 52]. The models indicate that the operation of all plants would kill 60 to 80 percent of the young stripers each year. This in turn would snowball over the years as fewer spawning fish produced fewer larvae that would still be subject to the high mortality rate. The results of this work predict a large reduction in fisheries dependent upon the Hudson, although the assumptions of the model have been challenged by consultants to the power industries [53]. Other species would also be affected, although not as severely.

. . . The problem of fish kills also exists at many other sites and for many other species, and this will become more serious as power plants are increasingly built on coastal regions. . . .

"Fish kills" is an emotion-laden term, and it is necessary to assess the potential fisheries loss within a broader context. One suggestion might be to assess the marginal cost, in calories, of locating a power plant *not* on the coast against 5 times the calories of fish expected to be lost from the fishery, since that is the minimum energy cost of terrestrial protein production in our society.

Another important way that the use of industrial energy interacts with fish is to make overfishing easy with petroleum-intensive fisheries [16]. World fish catches that used to increase every year now show signs of leveling off, despite enormous increases in fishing efforts. Although this situation is potentially subject to rational analysis and management, the results of international oceanic fisheries management to date have not been encouraging.*

The future world food production will be limited by two things: the amount of available productive land and water, and the degree to which industrial energies can be applied to food production systems without concomitantly degrading them. Kenneth Watt [54] has made an interesting study of the future agricultural potential in the United States. The amount of food that can be grown is at least as dependent on our housing (and population) policies as on our agricultural technology, since we use about 2.4 acres (about 1 hectare) of land for each new person born into the United States for housing, schools, highways and so forth. Although we do have an immense quantity of land not presently under agriculture, little food can be expected from much of it due to its poor quality.

It is clear that with a shrinking amount of land available for agriculture,

---

**Editor's note* Some biologists believe that the oceans could eventually produce a harvest of more than 220 million tons of fish a year. New techniques are being devised to increase the fish catch without harming their habitat, to hasten the growth of some species, and to exploit certain fish that many consumers now consider unappealing or even inedible. More promising is a source of extra fish protein—fish farming, or aquaculture, which up to now has contributed only a small fraction of the world's protein.

with a population that is expected to increase by at least 50 percent before leveling off, and with energy availability for the next century uncertain, we have no good reason for complacency. In many other countries, of course, these limits to food production have already been met or will be soon. Unfortunately, it seems unlikely that voluntary birth control will have any particular effect worldwide in this century. In any country with limited land resources and a growing population either starvation or the industrialization of agriculture will occur.

In addition, turning stable natural ecosystems into intensively managed agriculture or forestry is normally done with some or large loss of the "public service function" of the natural systems, resulting in either a requirement for increased industrial energy use by municipalities or a degradation of the conditions of human existence. Presently about 25 percent of the world's land is managed for agriculture or domestic grazing, and about 0.02 percent is urbanized. These are lands that no longer contribute directly to the "public service function" that man receives from natural systems, and in the case of urbanized regions food production as well is lost.

Known reserves of energy available to man are quite limited, particularly if exponential growths in energy use continue. In many considerations of industrial energy available to man, there has been no separation between "gross" and "net" energy available. In other words, corrections must be made for energies required to find, extract, and process the energy used by society as well as the energy required to construct capital equipment and to subsidize the development phase of any new energy source, such as breeder reactors.

Today the quantities of energy required for all these processes is perhaps 40 to 50 percent of the total energy delivered [55]. But as reserves of readily accessible oil and gas diminish, and if it becomes necessary to develop offshore and Alaskan oils, shale oils, and tar sands, for example, the efficiency of yield will decrease considerably and a pollution ratio will increase —(the pollution from energy burning): (useful energy delivered to society). For example, it has been estimated that approximately 50 percent of the energy extracted from oil shales will be needed just to separate the oil from the rock [56]. In addition, approximately 10 times the land area must be worked per calorie derived, compared to strip mining coal. If the region has any agricultural or other value, further calculations must be made on the total cost/benefit to society for all forms of energy involved, including biotic.

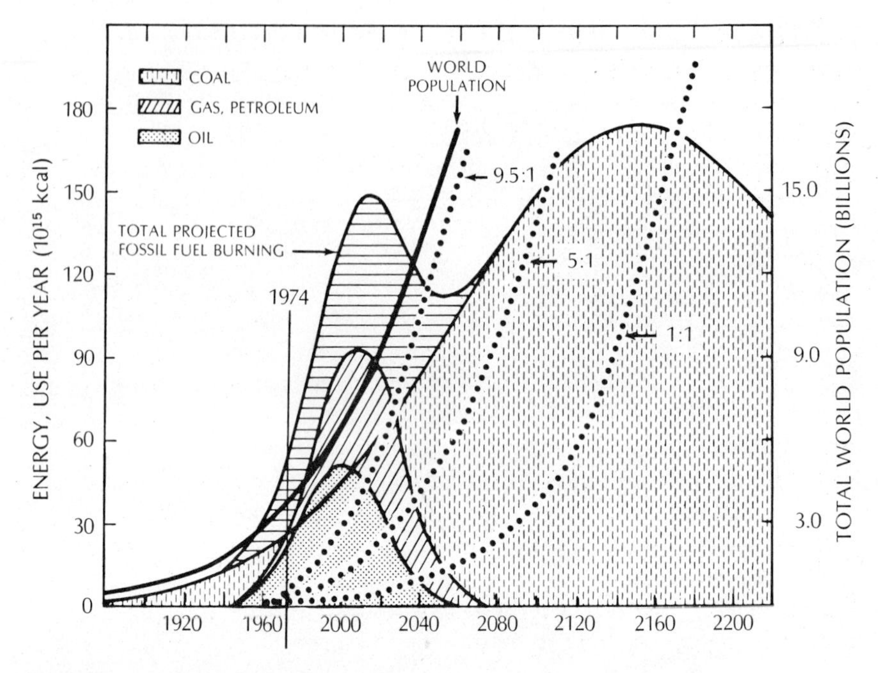

Limits to growth? *If present world population growth continues (—line), how much industrial energy will be needed for food by man?*

I have calculated this energy requirement (. . . lines) under two assumptions: 3 billion people is about the maximum number that can be supported with non-industrial agriculture, and all people above 3 billion will require industrial energy to raise the productivity of existing farmland or to grow food on soils not presently cultivated.

Three estimates are used of the energy required for feeding these people: 9.5 industrial calories per food calorie (U.S. average), 1 industrial calorie per calorie of food (minimum estimate for a food producing, processing and distribution system with no animal protein), and an intermediate, perhaps most likely, value of 5 to 1. These are plotted along with very rough estimates of potential world use of fossil fuels. (Hubbert [16]).

The intersection of the energy required for food production and the available industrial energy might be considered as some sort of indication as to the maximum ability of the Earth to support man.

There are various alternative sources available for the industrial energy used by the United States and other highly industrialized nations; the important ones currently are oil, natural gas and coal. Presently nuclear and hydroelectric sources together are only about 5 percent of U.S. total energy use, and hydroelectric power is unlikely ever to be very important. There are arguments for and against using each of these fuel sources to meet the anticipated increase in demand.

Oil and natural gas are relatively clean and flexible, but domestic supplies are limited (a 20- to 30-year supply at present rates of use [2]) and imports are increasingly costly. Coal is abundant in the United States but is less flexible to use, and mining and burning of coal create greater human and environmental problems than oil or gas. Nuclear power produces less air and water pollution than fossil fuel; but the development and construction of atomic power plants is, at least presently, very heavily subsidized by fossil fuels. In addition, sources of concentrated uranium are not abundant; various problems exist with radioactive wastes and potential accidents; and a large-scale breeder program, which would help to compensate for uranium shortages, would commit us to a national energy economy based in large part on plutonium—a chemical noted for its remarkable toxicity, radioactivity, and long half-life. Large-scale fusion energy appears remote at this time.

Regardless of which source of industrial energy one may consider the *most* desirable, the use of *each of them* (that is, coal, oil, gas, and nuclear), including whichever sources are considered *least* desirable, has been increasing in this country and the world with a doubling time of about 15 years. (Domestic oil production, however, has declined since 1970 and domestic coal production has recently increased after a long, stable period.)

As we have seen recently, gains in air quality are among the first casualties when energy supply falls behind demand, a condition expected to be chronic in the United States for at least several years. This condition negates some plans for more power with less pollution, as technical feasibility may have little to do with institutional implementation. As we use up the bulk of our remaining domestic natural gas, oil and cheap uranium within the next two or three decades, it is likely that coal, and possibly breeder systems, will take on an increasing share of our industrial energy production. Each of these fuel systems is "dirtier" than the systems it replaces, and it would appear that negative interactions with the biosphere will increase.

There are many technical solutions to pollution. However, most of them cost energy, and hence require more total fuel to be mined and burned for the same net energy delivered to society. And cleaning up one pollutant often means making another worse; for example, cleaning up particulates in automobiles and smokestacks is often accompanied by an increase in the output of nitrates, a principal component of acid rain [29, 30]. However, it seems unlikely that technical pollution abatement can be implemented on a national scale while *all* energy sytems are expanding to capacity. Gas from coal, an example of a technical cleaning of fuel, could produce only a small proportion of our *present* natural gas consumption within the next two or three decades; but by that time we could easily be using twice as much total gas if energy growth continues. In addition, more land would have to be strip mined or deep mined per useful calorie delivered to society, since energy is required to form the gas from coal.

If the industriosphere continues to grow, the per unit impact upon the biosphere is likely to increase. This will be particularly true with developing nations that cannot afford pollution-cleaning devices.

IN SUMMARY

Modern man has become dependent upon large industrial systems superimposed upon the biological systems that have nurtured him since his inception. The biotic systems provide man with food, fodder, fiber, and a stable physical and chemical milieu, all absolute prerequisites for life. The industrial systems have provided some of these and in addition many of the various amenities and disamenities that we call civilization.

But the most important effect, both historically and presently, that industrial systems have had on man has been to redirect, through interactions with the biosphere, the net production of ecosystems from natural consumers to man. This has increased the wealth of some groups of men dramatically, although for the world as a whole the major effect has been to increase the number of people. More food generally has meant more mouths, not fuller bellies.

The present large global human population is dependent upon massive inputs of industrial energy applied to biotic systems in order to provide enough food, since natural systems do not provide sufficient edible yield. However, recently there has been an increase in the industrial energy used per unit of biospheric management. In addition, the redirection of ecosystem productivity into managed-for-man systems has been accompanied by a loss of the "public service function" of many natural systems. This often has required the additional use of industrial energy to provide the necessary services to man once provided by the unmanaged natural systems.

A second major interaction between industrial and biotic systems appears to be developing. Massive industrialization of increasingly large areas of the world has the potential to change the productivity of large areas of the biosphere, perhaps of the biosphere as a whole. Since the principal real wealth of the world is biological, any major changes in global photosynthesis would have large effects upon the wealth of man, particularly those living in nations that are primarily dependent upon the biota as the basis of their economies. At present global biological science is too poorly developed to predict the magnitude, or even the direction, of these changes. Some empirical evidence suggests that world photosynthesis has been decreasing recently; other evidence indicates that there has been no change.

The arguments for limiting industrial expansion, at least in the developed nations, are simple but powerful. The first is that the most important sources of the energy used for this expansion will be essentially exhausted within a very few decades, and fuels used for expansion today are fuels that will not be available in the future for more important functions such as agriculture or defense. The second argument is that through increasing industrial wealth, we run the risk of reducing the biotic wealth of some nations and perhaps of the world as a whole. Our knowledge of these effects lags far behind the production of the effects themselves.

Within this broad framework I wish to suggest a number of specific modifications of public policy.

Scientific Policies

The science of biology, remarkably successful in recent decades, appears unable to answer the important questions as to whether the rate of global photosynthesis will be altered by the continued increase in global industrialization, or if temporal and spatial patterns are likely to be affected. The reductionist philosophy of science, championed by John Platt among others, rejects the very principles of holism and multiple interaction that must be understood in order to comprehend these processes, if indeed they are to be understood at all. Large-scale research, including large-scale experimental manipulations of natural and agricultural ecosystems, is needed to understand how limiting factors (including industrially derived limiting factors) operate at ecosystem and biome levels.

- Our global monitoring systems should be expanded to include a close watch on primary production and on the factors that are believed to be influencing rates and patterns of productivity. One critical need, for example, is for expanded, continuous atmospheric carbon dioxide monitoring. Another is for an expanded program of tree-ring analysis.

• Agricultural, technical and industrial energy development schemes must be subjected to new cost/benefit procedures based on quantities of fossil fuel energy required for development, capital equipment and operating costs. For example, how much industrial energy is required for Alaskan oil operations? For offshore oil production? Green Revolutions? Nuclear power cycles? These costs need to be computed, evaluated and made public.

National Public Policy

It should be made clear to each American that there is only about 20 to 40 thousand gallons of readily extractable domestic petroleum remaining per capita, compared to a consumption of the energy equivalent of over 2,200 gallons per capita per year. How should this be used? Automobiles? Food? Jobs? Growing grain to trade for oil? Investments in new energy resources? Military reserves? At present there is no policy for the wise use of this finite resource besides governmental efforts to maintain a full and expanding economy, and there are no well developed analyses which determine costs and benefits resulting from alternative use schemes.

• The present administrative policy of advocating national "energy independence" *and* continued economic expansion should be reconsidered. A likely result of this policy is virtual exhaustion of our liquid and gaseous petroleum supplies within the next two or three decades. This could place us at that time among the most foreign-energy dependent nations of the world.

There exists a potential for the development of a new conservation ethic for times of limited industrial energy supplies. Land in its natural state does not require any petroleum to fuel the services it provides society. Remaining fossil fuels are needed to maintain the present physical structure of society, and should not be squandered on massive attempts to "reform" nature, such as by large-scale river channelization.

• Our economic and political systems are not prepared for a voluntary or forced (by fuel shortage) changeover to a stable or reduced energy regime. In a nation conditioned by the media and the government to desire and expect an ever-expanding level of material affluence, national psychosis and political chaos could result. New government and economic policies should be developed to prepare the public for these likely conditions.

International Policies

Any increases in world agricultural production will probably be principally a function of the level of industrialization that can be applied to crop management, rather than "miracle science" alone or increases in the amount of land cultivated. Because of population pressure, most Third World nations do not have the option of reducing industrialization even if it were demonstrated to be in their best self-interest.

• However, any development policies for or by overcrowded, underdeveloped nations must be based on the recognition that cheap fossil fuel and unusually mild weather patterns over the past 50 years have allowed man's population to grow to the present level, a level that is almost undoubtedly above the long-term carrying capacity of the Earth. Both the cost of fuels and the climate are changing, both partly as a result of continued growth in the use of fossil fuels. In addition, there is some basis in speculating that another very large volcanic eruption, by now overdue, would decrease the photosynthesis that man is dependent upon. Pro-natalist policies obviously would make the effects upon man of any such changes even worse.

• As the most heavily industrialized nations increasingly deplete their own petroleum and other resources, the resource-exporting nations increasingly will be able to control the level of global industrialization. If increasingly heavy industrialization is found to decrease global photosynthesis and depress the economies of Third World nations dependent upon biotic production, the resource-exporting nations may find it in their own best interest to develop policies that would limit industrial fuel burning in the most heavily industrialized nations, or that would, by agreement, substitute relatively less polluting oil for more polluting coal.

NOTES

1 H. Leith, *Journal of Geophysical Research,* 68:13 (1967), 3887.

2 E. Cook, *Technology Review,* 75 (1972), 16.

3 H. T. Odum, in *The World Food Problem: A Report of the President's Science Advisory Committee* (Washington, D.C.: Government Printing Office, 1967), vol. III.

4 A. P. Vayda, *Environment and Cultural Behavior* (Garden City, N.Y.: Natural History Press, 1969).

5 S. McNaughton and L. Wolf, *General Ecology* (New York: Holt, Rinehart and Winston, Inc., 1973), ch. 20.

6 D. H. Janzen, "Tropical Agroecosystems," *Science,* 182:4118 (Dec. 21, 1973), 1212–1219.

7 H. T. Odum, *Environment, Power and Society* (New York: Wiley-Interscience, 1971).

8 C. C. Delewiche, *Scientific American,* 223 (1970), 137.

9 J. M. Perelman, *Environment,* 14, (1972), 8.

10 T. C. Foin, Jr., in *Systems Analysis and Simulation in Ecology,* ed. B. C. Patten (New York: Academic Press, 1972), vol. II.

11 D. Pimentel, et al., "Food Production and the Energy Crisis," *Science,* 182:4111 (Nov. 2, 1973), 443–449.

12 J. S. Steinhart and C. E. Steinhart, "Energy Use in the U.S. Food System," *Science,* 184:4134 (April 19, 1974), 307–316.

13 P. R. Ehrlich and A. H. Ehrlich, *Population, Resources, Environment* (San Francisco: W. H. Freeman and Co., 1970).

14 G. Borgstrom, *Too Many: A Study of the Earth's Biological Limitations* (London: Macmillan Co., Collier-Macmillan Ltd., 1969).

15 Study of Critical Environmental Problems, *Man's Impact on the Global Environment: A Report of the SCEP* (Cambridge, Mass.: MIT Press, 1970).

16 W. E. Ricker, in National Academy of Science, National Research Council, *Resources and Man* (San Francisco: W. H. Freeman and Co., 1969).

17 H. T. Odum, C. Littlejohn and W. Huber, *An Environmental Evaluation of the Gordon River Area of Naples, Florida and the Impact of Development Plans* (Gainesville, Fla.: Environmental Engineering Dept. University of Florida, 1972).

18 B. Bolin and W. Bischof, *Tellus,* 22 (1970), 431.

19 L. Machta, *Brookhaven Symp. Biolo.* 24 (1973), 21.

20 C. A. Ekdahl and C. D. Keeling, *Brookhaven Symp. Biolo.,* 24 (1973), 51.

21 G. M. Woodwell, R. A. Houghton, and N. R. Tempel, *Journal of Geophysical Research,* 78:6, p. 932.

22 G. E. Hutchinson, in *The Earth as a Planet,* ed. G. P. Kuiper (Chicago, Ill.: University of Chicago Press, 1954).

23 J. D. Hesketh, *Crop Science,* 3 (1963), 493.

24 H. Leith, *Journal of Human Ecology* (forthcoming).

25 I. Zelitch, in *Harvesting the Sun: Photosynthesis in Plant Life,* ed. A. San Pietro, F. A. Green, and T. J. Army (New York: Academic Press, 1967).

26 D. Wilson and J. P. Cooper, *New Phytology,* 68 (1969), 627.

27 D. Botkin, J. F. Janak, and J. R. Wallis, *Brookhaven Sym. Biol.,* 24 (1973), 328.

28 R. A. Berner, *Journal of Geophysical Research,* 76 (1971), 6597.

29 G. E. Likens and F. H. Bormann, *Environment,* 14:2 (1972), 33.

30 G. E. Likens and F. H. Bormann, "Acid Rain: A Serious Regional Environmental Problem," *Science,* 184:4142 (June 14, 1974), 1176–1179.

31 L. N. Overrein, *Ambio,* 1:4 (1972), 145.

32 H. T. Ellis and R. F. Pueschel, *Science,* 172 (1972), 33.

33 M. I. Budyko, *Tellus,* 21:5 (1969), 612.

34 R. A. Bryson, *Climatic Modification by Air Pollution,* Report no. 1 (Madison, Wis.: Institute for Environmental Studies, University of Wisconsin, 1972).

35 K. E. F. Watt, *Land Use, Energy Flow and Decision-Making in Human Society* (Davis, Calif.: University of California, 1973).

36 H. H. Lamb, *Proceedings of the Royal Society,* Series A, 266 (1970), 425.

37 J. M. Mitchell, Jr., in *Global Effects of Environmental Pollution,* ed. S. F. Singer (Boston, Mass.: Reidel Publishing Co., 1970).

38 P. V. Hobbs, H. Harrison, and E. Robinson, "Atmospheric Effects of Pollutants," *Science,* 183:4128 (March 8, 1974), 909–915.

39 S. I. Rasool and S. H. Schneider, *Science,* 173 (1971), 138.

40 V. C. LaMarche, Jr., "Paleoclimatic Inferences from Long Tree-Ring Records," *Science,* 183:4129 (March 15, 1974), 1043–1048.

41 G. J. Kukla and H. J. Kukla, "Increased Surface Albedo in the Northern Hemisphere," *Science,* 183:4126 (Feb. 22, 1974), 709–714.

42 L. H. Allen, Jr., S. E. Jenson, and E. R. Lemon, *Science,* 173 (1971), 256.

43 Y. Aruga, *Botanical Magazine,* 78 (1965), 280.

44 R. Lister, Ph.D. dissertation, Cornell University, 1974.

45 R. A. Bryson, *Climatic Modification by Air Pollution,* Report no. 9 (Madison, Wis.: Institute for Environmental Studies, 1973).

46 R. H. Whittaker et al., *Ecology Monographs,* 44 (1974), 233.

47 B. Bolin, ed., *Air Pollution Across National Boundaries: The Impact on the Environment of Sulfur in Air and Precipitation,* Report of the Swedish Preparatory Committee for the U.N. Conference on Human Environment (Norstedt, Sweden: U.N. Conference on Human Environment, 1971).

48 C. Cogbill to author.

49 C. A. S. Hall, C. A. Ekdahl, and D. E. Wartenberg, unpublished.

50 R. H. Stroud, in *A Symposium on the Biological Significance of Estuaries* (Houston, Texas: 1971).

51 Atomic Energy Commission. *Final Environmental Statement Related to Operation of Indian Point Nuclear Generating Plant Unit No. 2* (Washington, D.C.: Government Printing Office, 1972), vol. I.

52 C. A. S. Hall, in *Ecosystem Analysis and Prediction,* ed. S. A. Levin (Society for Industrial and Applied Mathematics, forthcoming).

53 AEC, *Final Environmental Statement Related to Operation of Indian Point Plant No. 2,* vol. II.

54 K. E. F. Watt to author.

55 F. Salzano, *Brookhaven Lab. Internal Report,* 1973.

56 K. O. Emery to author.

THE FAR REACHES OF ECOLOGICAL PERIL

3

The Deterioration of Mountain Environments

Erik P. Eckholm

Erik P. Eckholm is a member of the research staff of Worldwatch Institute, Washington, D.C. He is author of the *Losing Ground: Environmental Stress and World Food Prospects* (1976), from which this article is adapted.

An unusual meeting was convened in Munich, Germany, in December 1974. Any organizing principle, any common thread among the participants, would have eluded an outsider. The group included biologists, anthropologists, foresters, ecologists, economists, geographers, businessmen, and civil servants . . . from Europe, North and South America, Africa, and Asia. What drew this disparate group together was a shared concern for a problem that has scarcely been recognized as one deserving attention in its own right: the deterioration of mountain environments in the poor countries (*1*). . . .

Highlands occupy about one-fourth of the earth's land surface, but provide a home for only a tenth of the world's people (*2*). Still, it is curious that mountain ecosystems have been ignored so long in comparison to other natural areas, for history has repeatedly shown that when ecological changes take place in the highlands, changes soon follow in the valleys and the plains. And while only 10 percent of the human population lives in the highlands, another 40 percent lives in the adjacent lowland areas, and their future is intimately bound to developments on the slopes and plateaus above.

For all the diversity that characterizes the land and peoples of the three major mountain ranges of the developing regions—the Himalayas, the Andes, and the East African highlands—they present a rather uniform set of environmental and economic challenges. What strikes the casual observer of ranges like these is their stark immutability, their massive grandeur; but in fact they are among the most fragile ecosystems on earth. Steep mountain slopes can seldom sustain the degree of cropping, woodcutting, and grazing that is customary in flatter areas. Yet increases in all three practices are forced by rising populations throughout the mountains. And the inherent difficulty of adapting cultural practices to rapidly changing environmental circumstances is exacerbated by the tendency of mountain populations to be those with the least income, the least education, and the least political power in any country.

When the environment starts to deteriorate on steep mountain slopes, it deteriorates quickly—far more so than on gentler slopes and on plains. And the damage is far more likely to be irreversible. The mountain regions are not only poor in economic terms; many areas are rapidly losing any chance of ever prospering as their thin natural resource base is washed away. Degenerating economic and ecological conditions in the mountains, in turn, often push waves of migrants into the lowlands, leaving behind an aged, dispirited population incapable of reversing the negative spiral.

The net result of current trends, as evaluated by the international committee of experts recently established by Unesco to study mountain environments, is "accelerating damage to the basic life support systems" today in practically every mountainous region of Asia, Africa, and Latin America. "Within the last decade there has been a marked increase in the destructive clearance of forests, in flood damage and silting, in soil erosion and the explosive spread of pests. . . . In sum, human pressures on tropical high mountain ecosystems are increasing nearly everywhere. . ." (3).

Mountain regions often have great economic potential as sources of hydroelectric power, of valuable timber and minerals, and of scenic natural refuges which are in growing demand and diminishing supply. But one must take care to distinguish between the potential toward which societies need to work, and the present-day realities. For without a massive effort to preserve and restore the ecological integrity of the mountains, within a few decades they will not be idyllic vacation spots but, rather, barren eyesores that perennially present the lowlands with devastating torrents and suffocating loads of silt.

TRAGEDY IN SHANGRI-LA

There is no better place to begin an examination of deteriorating mountain environments than Nepal. In probably no other mountain country are the forces of ecological degradation building so rapidly and visibly. This kingdom of 12 million people is minuscule by Asian standards, but it forms the nucleus of one of the more strategic ecological nerve centers in the world. The Himalayan arc, stretching from Afghanistan through Pakistan, India, Nepal, and Bhutan to Burma, forms an ecological Gibraltar whose fate will affect the well-being of hundreds of millions. From the Himalayas flow the

Mountains grow and decay, they breathe and pulsate with life. They attract and collect invisible energies from their surroundings: the forces of the air, of the water, of electricity and magnetism; they create winds, clouds, thunderstorms, rains, waterfalls, and rivers. They fill their surroundings with active life and give shelter and food to innumerable beings. Such is the greatness of mighty mountains.

LAMA ANAGARIKA GOVINDA

major rivers of the Indian subcontinent—the Indus, the Ganges, and the Brahmaputra—which annually bring life, and sometimes death, to Pakistan, India, and Bangladesh.

Nepal has an exotic facade of romance and beauty, but behind it are the makings of a great human tragedy. Population growth in the context of a traditional agrarian technology is forcing farmers onto ever steeper slopes, slopes unfit for sustained farming even with the astonishingly elaborate terracing practiced there. Meanwhile, villagers must roam farther and farther from their homes to gather fodder and firewood, thus surrounding most villages with a widening circle of denuded hillsides. Ground-holding trees are disappearing fast among the geologically young, jagged foothills of the Himalayas, which are among the most easily erodable anywhere. Landslides that destroy lives, homes, and crops occur more and more frequently throughout the Nepalese hills (*4, 5*).

Topsoil washing down into India and Bangladesh is now Nepal's most precious export, but one for which it receives no compensation. As fertile soil slips away, the productive capacity of the hills declines, even while the demand for food grows inexorably. Some terraces are expertly managed and ecologically stable; others continue to be farmed despite their waning productivity, and others reach the point of no return and are abandoned. In the country's most densely populated region, the Eastern hills, as much as 38 percent of the total land area consists of abandoned fields (*6*).

While the acceleration of the heavy natural erosion is the chief threat, the declining fertility of the hills stems in part from another problem. Nepalese farmers have always assiduously applied the available animal manures to their fields as fertilizer, but in some regions the fields now receive less manure than in the past—well below the full amount necessary to preserve high fertility (*7*). This is partly because herd sizes have not grown as rapidly as the cultivated area; the hills are already overgrazed, and fodder of any kind, whether tree leaves or forage crops, is scarce. Even more ominously, farmers facing an unduly long trek to gather firewood for cooking and warmth have seen no choice but to adopt the self-defeating practice of burning dung for fuel.

The average hectare of arable land in Nepal's hills must now support at least nine people. This is a man-to-land ratio comparable to that in Bangladesh or Java, but those countries are blessed with far more fertile soils and a climate which permits several crops a year on the same land. And in the hills, the Nepalese government realizes that "there is absolutely no scope whatsoever for bringing new land under agriculture" (*5*).

If Nepal's borders ended at the base of the Himalayan foothills, the country would by now be in the throes of a total economic and ecological collapse. Luckily the borders extend farther south to include a strip of relatively unexploited plains known as the Terai, which is an extension of the productive Indo-Gangetic Plain of northern India. The Terai suffers seasonal floods and is heavily vegetated, and historically the high incidence of malaria precluded heavy settlement. Once an effective malaria eradication program got under way in the 1950's, however, a rush to colonize the region was inevitable.

The Nepalese government knows that the controlled settlement of the Terai is an essential step to help take some of the pressure off the hills, but it lacks the capacity to keep the land rush under any meaningful control. In the decade from 1964 to 1974, 77,700 hectares of Terai forest land were officially distributed by the government to settlers. But more than three times that amount was cleared illegally by migrants from the hills or, perhaps even more significantly, from India (*5*).

The presence of undeveloped arable land in the Terai does provide Nepal with some breathing space in which to reverse the downward spiral of population growth, land destruction, and declining productivity in which it is now caught. Yet the length of this reprieve is frequently exaggerated, both by outsiders and by some Nepalese. Analysis of satellite imagery indicates that less than half the remaining three-quarters of a million hectares of Terai forestlands will be suitable for cultivation (*8*). If migration down into the Terai continues at the pace of the last 10 years, all the good farmland will be occupied in little more than a decade.

Agricultural modernization—better seeds and innovative techniques, land reform, extension services, and so on—will quickly have to replace the extensive spread of farming to new lands if Nepal is to avoid an acute food crisis. The potential for raising yields may be greatest in the Terai, but the hills cannot be neglected any longer. Migration is a temporary palliative, not a long-term solution; a large share of the country's citizens will always live in the hills regardless of the quality of life they provide, and soon there will be no place for the people to go. Furthermore, the devastation in the hills is exacting a heavy, if as yet unmeasured, price on the potential productivity of the Terai. The incidence of flooding by swollen rivers coming down from the mountains is increasing, according to those who have lived near their banks for decades. Government soil conservation officials observe that the bed level of many Terai rivers is rising from 6 inches to 1 foot every year (*5, 9*). This not only guarantees wider floods from even normal volumes of water in the monsoon season; it is also causing the river courses to meander about, often destroying prime farmland as they go.

The cultures of Nepal have not historically faced a serious problem of land scarcity, and while terracing has always been necessary to farm the hills, the development of a national conservation ethic was never before essential to their survival. By now, however, the pace of destruction is reaching unignorable proportions, and this, in combination with the continuing efforts of dedicated individuals both in and out of the Nepalese government, is stirring new interest in the integrity of the environment. The country's influential National Planning Commission recently expressed its concern with an urgency unsurpassed by any party. Soil erosion, the commission fears, is "almost to the point of no return. . . . It is apparent that the continuation of present trends may lead to the development of a semi-desert type of ecology in the hilly regions" (5).

Translating official awareness into meaningful programs on the ground is no mean task, particularly in a country with such limited resources and unique problems of transportation and communication. In August 1974, a Department of Soil and Water Conservation was finally formed within the Ministry of Forests. This department hopes to establish demonstration projects in villages scattered about the country, but in early 1975 the department included only 67 employees, of which fewer than a third had college-level professional training. According to national development policy, this department is slated to expand to 167 employees over the next 5 years—still far from adequate.

INDIA'S WORRIES

Faced with the inevitability of absorbing many of the consequences of ecological degradation in Nepal, Indian officials have encouraged the Nepalese to attack the problems and have even provided limited assistance in land management research (6). The Indians are worried about environmental trends in Nepal and with good reason, but the fact is that virtually identical problems plague even larger hilly expanses within India itself in such states as Himachal Pradesh, Uttar Pradesh, Assam, and Jammu and Kashmir. In large mountain regions in these states, the fertile valley floors have long been overcrowded, and cultivation is constantly pushed onto steeper slopes by population growth in the absence of nonagricultural employment opportunities. On millions of hectares there is no longer any topsoil at all, just a rocky substratum lacking organic matter or fertility. Forests are receding under the combined pressures of shifting cultivators; uncontrolled herds of goats, sheep, and cattle; and wood-gathering for home consumption or sale

(*10*). In many areas, firewood is beyond the reach of village populations, so dung is burned instead.

Westward in the Himalayas, first in northern Pakistan and then Afghanistan, the outlook is hardly more encouraging. Both countries are mainly arid and desertlike, which means that the limited available forests in the mountains must bear an especially heavy burden. Pakistan today, for example, only classifies 3.4 percent of its land as forest, nearly all of which is concentrated in its hilly northern provinces. Large stretches in these hills have been visibly deforested within the last century (*11*). On top of the growing pressures from agriculture and overgrazing, the remaining forests must satisfy the burgeoning national need for wood for construction, industry, furniture, and fuel. While farmers lop the branches off trees to provide their animals with fodder and their homes with fuel, timber concessionaires respond to some of the world's highest lumber prices in the cities below by clearing large stands of timber, often without concern for the environmental consequences.

Through these northern hills, after passing through India, flows Pakistan's jugular, the Indus River system. The water of the Indus and its six major tributaries is about all that stands between a bustling, densely populated civilization and a deserted, sandy wasteland. With erosion rampant in the uplands, the exceptionally heavy silt load carried by these rivers is rendering the country's expensive new reservoirs useless with startling rapidity and has become a favorite subject of editorials and political speeches over the last decade. Pakistani foresters are pressing ahead with an excellent series of forest resource and land management surveys in the mountain regions, but obtaining the funds and political commitment needed to act on their research findings is proving more difficult.

The barren, infertile landscape presented by most of Afghanistan, and by the Zagros and Elburz mountains farther west in Iran, is fair warning of what lies ahead for parts of Pakistan, India, and Nepal, if prevailing trends are not reversed. A United Nations team visiting Afghanistan in the mid-1960's found the country's river basins "remarkable for their sparseness of vegetation and the paucity of animal life. The upper catchments are often bare rocky mountains with almost no soil cover and very little vegetation." A German forester who worked 8 years in Paktia province, where most of Afghanistan's few remaining trees are standing, laments their imminent demise by writing, "In Afghanistan the last forest is dying—and with it the basis of life for an entire region" (*12*). Afghanistan was never blessed with ample rain and fertile soils; even many centuries back, the forests covered only a small area. By now, however, they have been reduced to less than 0.1 percent of the country.

STRESS IN THE LAND OF THE INCAS

On the far side of the world from the Himalayas, dominating the west coast of South America, stand the Andes mountains. The longest range in the world, these mountains form what looks like the misplaced spine of an entire continent, stretching from Venezuela through Colombia, Ecuador, Peru, Bolivia, and Chile almost to Cape Horn. This massive ridge rises with remarkable rapidity from a deep trough in the Pacific, then just as quickly falls almost to sea level and the tropical forests in which the Amazon's tributaries begin their 5000-kilometer course to the Atlantic.

The major temperate zone mountain ranges, such as the Alps and the Rockies, have always been thinly settled in comparison with the surrounding lowlands. Just the opposite is true of the Andes and the Ethiopian Highlands, which have been densely populated for many centuries, and are bordered by sparsely populated tropical rainforests or deserts. Though rugged, the Andean uplands present a natural environment far more hospitable to permanent agriculture than the surrounding areas.

From Ecuador southward much of the narrow coast separating the Andes from the Pacific Ocean is a leafless desert, striped with the irrigated oases which surround each of the 60 or so rivers flowing down the westward side of the mountains. Some of these flow only seasonally, but most flow year-round, and for centuries have supported intensive agriculture and compact human settlements. Across the mountains to the east, in the jungle of the upper Amazon Basin which constitutes more than half of Peru and Bolivia and parts of Ecuador and Colombia, tropical diseases and the limited carrying capacity of tropical agricultural systems have kept the population low. Over the last few decades, however, this humid tropical zone has been on the receiving end of one of the great human migrations of the century: the movement of highland people throughout Latin America down into the humid lowlands. The spread of modern medicine to the jungles has been the prerequisite for this procession; its roots, however, lie in the slopes above. Most of its participants are best described as refugees from the deteriorating agricultural systems and the exploitative social systems of the mountains.

The pressure of human numbers on the environment is not a new phenomenon in the Andes, nor is knowledge of the farming techniques necessary to save the soil from accelerated erosion. Some Andean valleys doubtlessly provided sustenance for more people 500 years ago than they do today, and they did so at less cost to their longterm fertility than is exacted by present-day farmers. The unusual history of ecological deterioration in the Andes begins long ago, in the millennia preceding the European discovery of the New World.

By about A.D. 500 one of the major cultures of pre-Columbian America was emerging in the coastal valleys and upland plateaus and gorges of Peru. In relatively independent small settlements ruled strongly by priestly and warrior castes, the fruits of many centuries of technological and political development began to crystallize in societies able, like the earlier Egyptians and Sumerians in the Old World, to master the hydraulic cycle and thus create a large, complex civilization. These ancient Peruvians created an intensive agricultural system in the arid valleys of the coast, and on the slopes and plains of the mountains, by performing what McNeill has termed "extraordinary feats of water engineering," and building "terraces as elaborate as any the world has ever seen" (*13*). The terraces held down erosion, while fallowing was practiced to preserve the fertility of the soil. . . .

The Inca Empire fell to Pizarro's invasion in 1532. It had successfully created, in a fragile environment, a sustainable agricultural system that minimized the damage to the productivity of the land. Yet by the latter years of the empire, there is some evidence that the pressure of population on the limited arable area was beginning to show. . . .

. . . Even before this time, the central Andes were largely deforested and most mountain residents were dependent on the dung of llamas for cooking fuel (*14, 15*). . . .

. . . Historians disagree about the population of the Inca Empire at its height; estimates range from several million to more than 16 million. In any case, the combined influence of wars, forced labor in unhealthy silver and mercury mines, and European diseases sharply reduced the population of the Andes—probably by as much as three-fourths in the first century of colonial rule (*14–16*).

While this decimation of the population obviously reduced any immediate pressures on the land it contributed, ironically, to the emergence of quite unfavorable conditions for land conservation in the following centuries. A low population and wide open spaces facilitated the establishment of huge

European-owned estates, including cattle ranches on the highest plateaus and crop-growing haciendas in the middle and lower valleys. . . . By the mid-20th century many of the Andean Indians remained in a state of near serfdom to great landed proprietors. And outside the large estates on which many have lived and worked are overcrowded, fragmented lands unable to provide adequate sustenance for their populations.

Unfortunately, the decimation of the Inca population and social order was accompanied by a disastrous loss of the conservation ethics and know-how of the former empire. With scarce labor concentrated in the mines and on the hacienda fields of fertile valley floors, most of the terraces and irrigation facilities constructed over previous centuries fell into ruin. Although in a few areas the very terraces constructed by the Incas are still in use today, this basic soil conservation technique has almost completely disappeared in the Andes (*15–17*).

If the population of the Andes had remained at the reduced level of the 17th century, the absence of conservation practices would not be so threatening. With land abundant, farmers could afford to exploit the slope for a year or two, watch its topsoil and fertility wash away, and move on to another hill. But over the last century, the population in the Andes has overtaken that of the Inca Empire, with devastating consequences for the land and those whose livelihood depends on it.

Peru's population was about 4 million at the turn of the century, 9 million by 1950, and 15 million by 1975. Colombia's population grew from about 3 million in 1900 to 11 million in 1950, and 26 million in 1975. The ecological and social consequences of this mounting pressure on the Andes become more apparent with each passing decade. Farmers are driven onto slopes so steep that erosion is a serious problem from the moment cultivation begins. Lands which need to rest for 8 years, 12 years, or longer to regain their fertility can now be left fallow for only a few years. Agricultural output is generally stagnant in the Andes, and in some areas is declining (*17*).

Land reforms in Bolivia since 1953, and in Peru since 1964, are benefiting many mountain residents by improving the ownership patterns in these two countries, but cannot, of course, create new land. Nor does land reform guarantee land-management practices that preserve the productivity of the soil, although it may be an essential first step. In the near feudal economic system that has characterized the Andes, land redistribution is certainly a prerequisite of improved farm productivity and modernized farming practices. But a look at Bolivia's mounting soil erosion problem in the decade and a half after the 1953 reforms suggests the necessity of safeguarding their benefits for future generations by also introducing new farming systems and curbing the pressures of population.

Overgrazing and overcropping have been depleting the fertility and vegetation of the Bolivian Altiplano, a treeless tundra, and surrounding valleys, at least since the beginning of Spanish rule; but fresh gulleys, increasingly frequent abandonment of once fertile fields, and a fall in crop yields on the steeper fields in recent years all indicate that the scale of damage has accelerated since the early 1950's. Preston argues that the increased access—for grazing, cropping, and firewood gathering—to areas once controlled by large landowners is an important cause of this acceleration in Bolivia, since this new access was not accompanied by a comprehension of soil conservation needs and practices (*18*).

Families in the more densely settled areas of Peru and Bolivia, such as the Lake Titicaca Basin, have as little as ½ to 2 hectares of land available for their use. This area does not provide enough food to meet even the modest needs of the family. Continual food shortages impel overexploitation of the soil and hence extensive soil erosion, and finally result in the abandonment of farms and seasonal or permanent migration (*17*).

In Colombia, erosion, landslides, and sedimentation—all natural problems now accelerated by the pressures of humans and livestock on the land—are major obstacles to national development. Cornell University engineers working in a representative area, the Cauca region of the southern mountains, have described the readily visible economic damage and heavy loss of life resulting each year from the deteriorating environment. "Landslides are so common that socially important slides occur every few months," they write. Sediment deposits regularly block the Cali and Canaveralejo rivers and cause major flooding problems in the city of Cali. The lower Anchicaya reservoir has filled with silt in only 7 years, and the multimillion dollar hydroelectric plant it was built to support now runs on the river flow alone—at one-third of its planned capacity (*19*).

MIGRATION TO CITY SLUMS AND TROPICAL LOWLANDS

With life untenable for so many residents of the Andes, large-scale migration to other areas is inevitable (*20*). As usually occurs when human populations are forced to flee an impossible situation, the movement out of the mountains has given rise to many new social problems. Some migration has been seasonal and keyed to available jobs, with able-bodied family members finding work in plantations, mines, or construction jobs along the coast or in Argentina. But the mushrooming shantytowns of the major cities of the Andean region attest to the fact that many uprooted families face conditions in their new homes little better than those they left. Unemployment, poverty, and despair are widespread.

The eastward movement of settlers to the jungles of the upper Amazon Basin can appropriately be labeled anarchic. Every government has colonization programs which have met with varying degrees of success or failure, but for the most part, the evacuees of the highlands are ill-prepared to face the unfamiliar conditions of the humid tropics. The frequent result is double disaster: a costly depletion of lumber resources as trees are cleared without regard to the soil's suitability for agriculture, followed shortly thereafter by the abandonment of the unproductive farms (*17, 21*).

There is, of course, no doubt that further resettlement of mountain residents in the lowlands will be necessary over the coming decades, as will the expansion of nonagricultural employment throughout the Andean region. The questions to be confronted are: How many people can the jungles and cities safely absorb? And will the spread of roads, infrastructure, and technical know-how in the eastern lowlands catch up with a continuing flow of settlers? Already, suggests Peru's Office for the Evaluation of Natural Resources, the most fertile agricultural soils of the country's vast jungle region are being cultivated. New colonists will have to settle on increasingly inferior soils (*17, 21*).

With the exception of Chile, which is more advanced economically than the others, the Andean countries have population growth rates among the highest in the world. At the current pace the populations of Peru, Bolivia, Colombia, Ecuador, and Venezuela all will double in size within three decades. Oil, copper, and tin will certainly continue to boost the economic development and diversification of the Andean countries. Yet, at this point, it is difficult to imagine where these unborn citizens will live, where they will work, and where their food will grow.

DETERIORATION IN EAST AFRICA

Africa's principal mountainous zone, the highlands of the eastern side, present a topography quite different from that of the Andes. These highlands, which stretch southward from Ethiopia through Kenya, Uganda, Tanzania, and beyond, are not a sharply rising mountain range in the customary sense, but rather a wide bulky massif. The huge Amhara Plateau, which constitutes most of Ethiopia and is the most extensive mountain region of any African country, rises abruptly from the surrounding arid plains, and its general level is about 2000 meters above them. Yet the term plateau is also misleading in this case, for on these highlands are superimposed other mountains which rise to more than 4500 meters above sea level. Jagged, intricate, and fantastic, the Ethiopian Highlands share with the Himalayas and Andes the unfortunate distinction of being among the most erosion-prone areas on earth.

With a population estimated at 28 million in 1975, Ethiopia is the third most populous country in Africa, surpassed in numbers only by Nigeria and Egypt. The country's population growth rate of 2.4 percent annually is not so high as that of Tanzania, at 3.0 percent, and Kenya, at 3.3 percent. This is not due to a lower birth rate, but to one of the world's highest death rates—a rate surpassed only in Bangladesh, Mali, and Sikkim. Low agricultural productivity, and the need to give a high share of the land's produce to the landholding elite, chronically leaves much of the peasantry at a bare subsistence level (*22*). In the bad years, such as those that have visited the northern provinces of Wollo and Tigre in the early and mid- 1970's, the line between bare subsistence and famine for tens or even hundreds of thousands of people is quickly crossed.

The Ethiopian plateau receives a good rainfall, and as much as three-fourths of the country was once forested. Clearing for cultivation, burning to create pasture-lands, and tree felling to meet fuel and timber needs have reduced the forest area to a mere shadow of its former domain; significant stands of timber now cover less than 4 percent of the country. The tempo of destruction has quickened since mid-century, and by the early 1960's, reported an American research team, natural woodlands were disappearing at a rate of 1000 square kilometers per year. The country's major watersheds and steep mountain slopes are not being spared. "Even to the casual visitor to Ethiopia," a top-level soil conservation adviser recently wrote, "the extent of soil erosion seen in many parts of the country . . . will leave a lasting impression of desolation and impending disaster" (*23*).

The extent of deforestation and erosion varies by region according to population density and historical length of settlement. Leslie Brown, a prominent ecologist with decades of experience in East Africa, points out that one can observe the historical progress and results of land degradation in Ethiopia by journeying from north to south (*24*). In the oldest inhabited areas to the north, such as the provinces of Tigre and Eritrea, some of the steepest slopes no longer even carry grass or shrubs. People extract what produce they can from eroded, infertile patches. Many streams have dried up except during the rainy season, when they are prone to violent torrents.

Following the depletion of the soils and forests to the north, the center of Ethiopian civilization moved southward around the 10th century (*25*). The central highlands, including the region surrounding Addis Ababa, the nation's capital, have thus been increasingly exploited over the last millennium. Through much of this region the high forest cover has been replaced by bush, grass, and scrub; springs have often dried up and silty rivers flow erratically. Traveling farther to the south, where cultivators have more recently penetrated the forests, the streams tend to carry clear water and run constantly even in dry seasons.

A dramatic alteration in environmental quality in the hills surrounding Addis Ababa has been visible within the span of a single lifetime. When the capital was founded in 1883 by the Emperor Menelik II, it was still surrounded by remnants of rich cedar forests and reasonably clear streams. Deforestation and erosion were immediately spurred by the influx of humans. In the ensuing nine decades, virtually all the available land in the region has been cultivated, while charcoal producers cut trees within a hundred-mile radius for sale in the city. Only the widespread planting of eucalyptus trees to provide fuel has allowed Addis Ababa to survive. Now the waters of the nearby Awash River and its tributaries are thick with mud, and waterways are shifting their courses more markedly and frequently than in the past. A United Nations research team has expressed fears that the upper Awash Basin may become a "rocky desert" (*24, 25*).

Farther south in the Gamu Highlands, Oxford University geographers recently witnessed the incipient breakdown of a sustainable agricultural system. When they visited this area in 1968, erosion was not yet a serious problem. The steeper slopes had been saved for grazing animals rather than plows, and the people showed an awareness of the erosion hazard by terracing the hillsides and constructing drainage channels on slopes to carry off excess rain. Animal manure was carefully collected and applied to the fields, while crop rotations and fallowing also helped to preserve the soil's fertility.

Under the pressure of a mounting population, however, farmers have started plowing up lands formerly reserved for grazing. This has accentuated overgrazing and consequent erosion on the remaining pastures, and has also resulted in a lower population of cattle. Fewer cattle mean less manure, which, in turn, means lower yields and greater requirements for arable land, which will then necessitate further inroads into the pastureland, thus completing the cycle of degeneration. The villages in this area are violating their own land management rules and they know it, but they see no alternative (*26*).

Until the announcement of sweeping land reforms in early 1975 by the country's new military government, which had just unseated Emperor Haile Selassie, most of Ethiopia's best farmlands were owned by the Church, the Royal Family, or the powerful landed aristocracy. The land was worked by peasants in a state of bondage, whose daily lives were circumscribed by an elaborate castelike system of social and economic stratification (*22, 27, 28*). There is no question that the economic and political conservatism inherent in this land tenure system slowed the modernization of agriculture in Ethiopia, and reduced the incentive for the peasantry to manage soils properly. Whether the new regime will successfully carry out its announced reforms—and whether they will be accompanied by a new concern for soil conservation—remain to be seen.

The highlands of Kenya, Tanzania, and Uganda have fertile soils and a pleasant climate that have proved attractive to both Africans and Europeans. It is no accident that Nairobi, Kenya's capital in the heart of the highlands, is the nexus of industry and economic development in East Africa. Above undulating plains, deep valleys, and numerous smaller mountain ranges tower Mt. Kenya and Mt. Kilimanjaro, Africa's highest peak.

These highlands include some of the most productive and densely settled farmlands of Africa. They are bounded by zones infested with the tsetse fly, which prohibits keeping the cattle so highly prized by most of East Africa's peoples, and to the north by semiarid lands best suited to grazing. The concentration of people on the more fertile lands, such as around Lake Victoria, on the slopes of Uganda's Mt. Elgon, and in the vicinity of Nairobi, is more than 200 per square kilometer, and in places it is double that. In the colonial era the pressure of the African population on available lands in Kenya was intensified by the reservation of a fourth of the arable area, including many of the most fertile portions, for use by whites only. Since 1960 these lands have been gradually resettled by Africans.

Between the efforts of land-hungry cultivators and charcoal makers, the East African highlands have been largely deforested except for the most inaccessible mountain areas and occasional government-protected reserves. Particularly where cultivators have moved up steeper mountain slopes, and where the combination of population density and traditional techniques has run down fertility, erosion is on the rise (*28, 29*). Not surprisingly, migrants from the fragmented, overcrowded farming areas are pouring into the cities of East Africa, where they frequently wind up subsisting on whatever occasional work can be found. The future of East Africa's magnificent game reserves is also jeopardized as the public pressure for new farmlands grows.

TECHNICAL ANSWERS, POLITICAL NEEDS

All in all, little hard data are yet available on the true scale and the nature of environmental deterioration in the mountains. However, there is a rather broad consensus among the various scientists and governmental agencies cited above, as well as many others, about the general direction of prevailing mountain trends. On the basis of already available knowledge, it is no exaggeration to suggest that many mountain regions could pass a point of no return within the next two or three decades. They could become locked in a downward spiral from which there is no escape, a chain of ecological reactions that will permanently reduce their capacity to support human life.

This possibility is very real, but it is not inevitable. There is no major mountain problem for which technological solutions are not already known. If the existing negative trends are not abruptly reversed within the coming decades, it will be because human institutions have failed to adapt themselves to environmental necessity.

While every mountain zone has its peculiar problems and solutions, some general considerations with wide application can be noted. Undoubtedly, the most important need in virtually every case is an intensification of food production on the best farmland—in the lowlands, the valleys, and the gentler mountain slopes. Only when this occurs can the self-defeating pressure to move onto even steeper hillsides be countered. Reforms in land tenure and in the distribution of extension and credit services—political, not technical tasks—are in many cases the prerequisites of agricultural progress.

Where hillsides must be farmed, the adoption of soil conservation techniques is essential. Often erosion can be curbed markedly through the simplest of measures. In parts of Nepal, for example, many terraces are poorly constructed with an outward rather than inward slope and an inadequate buttress of stone to help them survive the annual monsoon deluge. In the Andes, a restoration of the lost art of terracing would bring immediate benefits to farmers, the land, and cities downstream. Terracing is not always the answer; soil experts in the Uluguru Mountains of Tanzania, for example, found that terracing encourages landslides there, and it exposes too much infertile subsoil. They did find that strategic tree planting, farming on the contour, and other simple changes in cultivation practices could greatly reduce erosion (*30*). Measures like these have not been widely adopted because farmers are either not aware that they are possible, or they are not convinced of the production benefits they bring. Many mountain farmers have never seen an agricultural extension agent.

Where population pressures do not permit a return of mountain slopes to forest, which might be the ecological ideal, the introduction of permanent tree crops like apples, apricots, nuts, or timber plantations may be a good compromise. Tree crops combine many of the ecological advantages of the forest with employment and incomes for the former farmers. An apparently successful United Nations watershed improvement project in Pakistan has utilized foreign-donated food aid as wages in road construction and planting activities to help tide farmers over until their new orchards start producing income (*23*).

More extensive reforestation programs are needed throughout the mountains of Asia, Latin America, and Africa. Trees are required not only to protect vulnerable slopes and soils, but also to provide firewood, and thus halt the spreading use of manures for cooking fuel. Putting manure back onto the fields will in turn help boost their productivity, reducing the pressure to spread cultivation onto unsuitable slopes. Virtually every government in the mountain regions has demarcated forest reserves in especially strategic locations such as above important rivers. But it is only when adequate food and fuel are available from other sources that these "protective forests" can be protected.

Greater opportunities to earn a living outside of agriculture can also reduce pressures on the land. Mining and related industries are already a source of jobs and money in the Andes. However, the environmental consequences of these operations must be carefully monitored and controlled, lest their impact be self-defeating. It has become something of an axiom in many quarters that the poor countries cannot afford the luxury of pollution controls on their industries; but the gap between environmental protection needs in the rich and poor countries may be narrower than many think. In Peru, a government agency points out that air pollution is killing vegetation on thousands of mountainous hectares surrounding mines and refineries, resulting in "truly spectacular" soil erosion (*21*).

Tourism, too, at once poses a great potential and a threat for the mountains. With their fascinating scenery and cultures, countries like Nepal, Peru, and Ethiopia clearly can expand their tourist trade severalfold. Yet planning will be essential to prevent the degradation of their natural resources by visiting sightseers. The soaring number of trekkers in the high Himalayas of Nepal over the last decade has created a booming firewood business for some mountain people, but it has grown at the expense of the forests and particularly fragile ecosystems of the upper slopes.

The central threat to the future of the mountains is the burden of the burgeoning human numbers they must bear. Planned migration to the less crowded lowlands will be important, but can only buy a brief reprieve. There is no escaping the need to rapidly bring population growth to a halt in the mountains; their limited carrying capacity will assert itself in no uncertain terms over the next few decades.

It is generally easy to recommend technological answers to ecological problems. Political and cultural factors are invariably the real bottlenecks holding up progress. Changing the relationship of man to land in the mountains, as anywhere else, invariably involves sensitive changes in the relationship of man to man. Developmental funds and talents spent in the mountains are resources denied the cities and the plains. In the end, this may be the greatest challenge of all: how to convince the people of the plains that the future of the mountains cannot be isolated from their own.

REFERENCES AND NOTES

1 International Workshop on the Mountain Environment, Munich, December 1974, sponsored by the German Foundation for International Development. For further discussion of deteriorating mountain environments see E. P. Eckholm, *Losing Ground: Environmental Stress and World Food Prospects* (Norton, New York, [1976]).

2 "Highlands" is defined as all areas more than 1000 meters in altitude.

3 Unesco, Programme of Man and Biosphere, *Working Group on Project 6: Impact of Human Activities on Mountain and Tundra Ecosystems, Final Report, Lillehammer, 20 to 23 November 1973* (MAB Report Series No. 14, Paris, 1974), pp. 20–21.

4 R. G. M. Willan, Chief Conservator of Forests, *Forestry in Nepal* (UNDP, Kathmandu, 1967).

5 Government of Nepal, National Planning Commission Secretariat, *Draft Proposals of Task Force on Land Use and Erosion Control* (Kathmandu, August 1974).

6 Government of India, Department of Agriculture, *Soil and Water Conservation in Nepal: Report of the Joint Indo-Nepal Team* (New Delhi, November 1967), p. 5.

7 J. C. Cool, *The Far Western Hills: Some Longer Term Considerations* (mimeograph) (U.S. Agency for International Development, February 1967), p. 6.

8 International Bank for Reconstruction and Development and International Development Association, *Agricultural Sector Survey, Nepal, Volume II* (Washington, D.C., 3 September 1974), p. 7.

9 Government of Nepal, Ministry of Forests, *Introduction of Department of Soil and Water Conservation* mimeograph (Kathmandu, no date).

10 D. C. Kaith, "Forest practices in control of avalanches, floods, and soil erosion in the Himalayan front," *Fifth World Forestry Congress Proceedings* (Seattle, 1960), vol. 3.

11 World Food Programme, *Interim Evaluation of Project Pakistan 385. "Watershed Management in the Kaghan and Daur Valleys,"* draft, 29 October 1974 (FAO, Rome, October 1974).

12 FAO/U.N. Special Fund, *Survey of Land and Water Resources. Afghanistan,* vol. 1, *General Report* (FAO, Rome 1965), p. 22; K. J. Lampe, *Forest in Paktia, Afghanistan* (Federal Agency for Development Assistance, Frankfurt am Main, 1972).

13 W. H. McNeill, *The Rise of the West* (Mentor, New York, 1963), p. 455.

14 F. Monheim, "The population and economy of tropical mountain regions, illustrated by the examples of the Bolivian and Peruvian Andes," presented to Munich Workshop on the Development of Mountain Environments, 8 to 14 December 1974 (German Foundation for International Development, Feldafing, December 1974), p. 6.

15 J. A. Mason, *The Ancient Civilizations of Peru* (Penguin Books, Harmondsworth, England, 1957), p. 139.

16 A. Metraux, *The History of the Incas,* G. Ordish, Translator (Random House, New York, 1969), p. 166.

17 R. F. Watters, *Shifting Cultivation in Latin America* (FAO, Rome, 1971).

18 D. A. Preston, *Geograph. J.* 135 (1), 1 (March 1969).

19 "Erosion, Landslides, and Sedimentation in Colombia," proposal to the U.S. Department of State, Agency for International Development—Bogota by Cornell University, Corporacion Antonoma del Valle de Cauca, and Universidad del Valle, J. A. Liggett, Principal Investigator (December 1974).

20 F. Monheim (*14,* p. 10) provides migration data for Peru.

21 Republica del Peru, Officina Nacional de Evaluacion de Recursos Naturales, and Organizacion de los Estado Americanos, *Lineamientos de Politica de Conservacion de los Recursos Naturales Renovables del Peru* (Lima, May 1974), pp. 8–14.

22 G. Nicolas, "Peasant rebellions in the socio-political context of today's Ethiopia," *Fifteenth Annual Meeting of the African Studies Association,* Philadelphia, 8 to 11 November 1972, p. 11.

23 World Food Programme, "Project summary: Reforestation and soil conservation in the province of Wollo," *Ethiopia Project No. 2097, WFP/IGC: 26/9 Add. 11* (FAO, Rome, July 1974);

"Reforestation and soil conservation in the province of Tigre," *Ethiopia Project No. 769, WFP/IGC: 23/10 Add. 2* (FAO, Rome, March 1973); U.S. Department of the Interior, Bureau of Reclamation, *Land and Water Resources of the Blue Nile Basin, Ethiopia, Appendix VI, Agriculture and Economics* (prepared for U.S. Agency for International Development, Washington, D.C., 1964), p. 140; H. F. Mooney, "The problem of shifting cultivation with special reference to Eastern India, the Middle East, and Ethiopia" *Fifth World Forestry Congress Proceedings* (Seattle, 1960), vol. 3, p. 2023; quote from W. D. Ware-Austin, "Soil erosion in Ethiopia: Its extent, main causes and recommended remedial measures," mimeograph (Institute of Agricultural Research, Addis Ababa, May 1970), p. 1.

24 L. Brown, *East African Mountains and Lakes* (East African Publishing House, Nairobi, 1971), pp. 70–71.

25 FAO/U.N. Special Fund, *Report on Survey of the Awash River Basin,* vol. 2, *Soils and Agronomy* (FAO, Rome, 1965), p. 114.

26 R. T. Jackson, P. M. Mulvaney, T. P. J. Russell, J. A. Forster, *Report of the Oxford University Expedition to the Gamu Highlands of Southern Ethiopia, 1968* (School of Geography, Oxford University, Oxford, 1968), pp. 34–37.

27 J. M. Cohen, "Land reform in Ethiopia: The effects of an uncommitted center on the rural periphery," presented at the Sixteenth Annual Meeting of the African Studies Association, Syracuse, 31 October to 3 November 1973.

28 W. A. Hance, *The Geography of Modern Africa* (Columbia Univ. Press, New York, 1964).

29 A. Warren, "East Africa," in *Africa in Transition,* B. W. Hodder and D. R. Harris, Eds. (Methuen, London, 1967), p. 117; M. S. Parry, "Progress in the protection of stream-source areas in Tanganyika," *East African Agricultural and Forestry Journal,* vol. XXVII, special issue, March 1962, p. 104; W. T. Lusigi, "Some environmental factors in food production in Kenya," prepared for the United Nations World Food Conference, 5 to 16 November 1974 (National Environment Secretariat, Office of the President, Nairobi, 1974).

30 P. H. Temple, in *Studies of Soil Erosion and Sedimentation in Tanzania* A. Rapp, L. Berry, P. H. Temple, Eds. (University of Dar es Salaam, Bureau of Resource Assessment and Land Use Planning, and Department of Physical Geography, University of Uppsala, 1972), pp. 110–155; A. Rapp, V. Axelson, L. Barry, D. H. Murry-Rust, in *ibid.,* pp. 255–318.

4 How to Kill an Ocean

Thor Heyerdahl

Thor Heyerdahl, a Norwegian archeologist, has led numerous ocean expeditions in his studies of the migrations of cultures. He is the author of *Kon-Tiki, Aku-Aku, The Ra Expeditions,* and, most recently, *Fatu-Hiva.*

Reprinted from the *Saturday Review,* November 29, 1975, pp. 12–18, by permission of Saturday Review/World, Inc.

Since the ancient Greeks maintained that the earth was round and great navigators like Columbus and Magellan demonstrated that this assertion was true, no geographical discovery has been more important than what we all are beginning to understand today: that our planet has exceedingly restricted dimensions. There is a limit to all resources. Even the height of the atmosphere and the depth of soil and water represent layers so thin that they would disappear entirely if reduced to scale on the surface of a common-sized globe.

The correct concept of our very remarkable planet, rotating as a small and fertile oasis, two-thirds covered by lifegiving water, and teeming with life in a solar system otherwise unfit for man, becomes clearer for us with the progress of moon travel and modern astronomy. Our concern about the limits to human expansion increases as science produces ever more exact data on the measurable resources that mankind has in stock for all the years to come.

Because of the population explosion, land of any nature has long been in such demand that nations have intruded upon each other's territory with armed forces in order to conquer more space for overcrowded communities. During the last few years, the United Nations has convened special meetings in Stockholm, Caracas, and Geneva in a dramatic attempt to create a "Law of the Sea" designed to divide vast sections of the global ocean space into national waters. The fact that no agreement has been reached illustrates that in our ever-shriveling world there is not even ocean space enough to satisfy everybody. And only one generation ago, the ocean was considered so vast that no one nation would bother to lay claim to more of it than the three-mile limit which represented the length of a gun shot from the shore.

It will probably take still another generation before mankind as a whole begins to realize fully that the ocean is but another big lake, landlocked on all sides. Indeed, it is essential to understand this concept for the survival of coming generations. For we of the 20th century still treat the ocean as the endless, bottomless pit it was considered to be in medieval times. Expressions like "the bottomless sea" and "the boundless ocean" are still in common use, and although we all know better, they reflect the mental image we still have of this, the largest body of water on earth. Perhaps one of the reasons why we subconsciously consider the ocean a sort of bottomless abyss is the fact that all the rain and all the rivers of the world keep pouring constantly into it and yet its water level always remains unchanged. Nothing affects the ocean, not even the Amazon, the Nile, or the Ganges. We know, of course, that this imperviousness is no indicator of size, because the sum total of all the rivers is nothing but the return to its own source of the water evaporated from the sea and carried ashore by drifting clouds.

The sea surges and recoils, thunders in boundless confidence and then lies low, conquers and retreats, heaves and broods and pounds furiously against its limits—the faithful image of our spirit as it struggles onward between hope and doubt.

The sea belongs to the dreamer. It is beyond the reach of the skeptic. It speaks to us with a voice that no weight of scientific evidence can reduce to a meaningless mechanical tumult; it keeps under our eyes a promise of something that can never die. That is perhaps why it has been put there.

ROMAIN GARY

What is it really then that distinguishes the ocean from the other more restricted bodies of water? Surely it is not its salt content. The Old and the New World have lakes with a higher salt percentage than the ocean has. The Aral Sea, the Dead Sea, and the Great Salt Lake in Utah are good examples. Nor is it the fact that the ocean lacks any outlet. Other great bodies of water have abundant input and yet no outlet. The Caspian Sea and Lake Chad in Central Africa are valid examples. Big rivers, among them the

Volga, enter the Caspian Sea, but evaporation compensates for its lack of outlet, precisely as is the case with the ocean. Nor is it correct to claim that the ocean is open while inland seas and lakes are landlocked. The ocean is just as landlocked as any lake. It is flanked by land on all sides and in every direction. The fact that the earth is round makes the ocean curve around it just as does solid land, but a shoreline encloses the ocean on all sides and in every direction. The ocean is not even the lowest body of water on our planet. The surface of the Caspian Sea, for instance, is 85 feet below sea level, and the surface of the Dead Sea is more than 1,200 feet below sea level.

Only when we fully perceive that there is no fundamental difference between the various bodies of water on our planet, beyond the fact that the ocean is the largest of all lakes, can we begin to realize that the ocean has something else in common with all other bodies of water: it is vulnerable. In the long run the ocean can be affected by the continued discharge of all modern man's toxic waste. One generation ago no one would have thought that the giant lakes of America could be polluted. Today they are, like the largest lakes of Europe. A few years ago the public was amazed to learn that industrial and urban refuse had killed the fish in Lake Erie. The enormous lake . . . was polluted from shore to shore in spite of the fact that it has a constant outlet through Niagara Falls, which carries pollutants away into the ocean in a never-ending flow. The ocean receiving all this pollution has no outlet but represents a dead end, because only pure water evaporates to return into the clouds. The ocean is big; yet if 10 Lake Eries were taken and placed end to end, they would span the entire Atlantic from Africa to South America. And the St. Lawrence River is by no means the only conveyor of pollutants into the ocean. Today hardly a creek or a river in the world reaches the ocean without carrying a constant flow of non-degradable chemicals from industrial, urban, or agricultural areas. Directly by sewers or indirectly by way of streams and other waterways, almost every big city in the world, whether coastal or inland, makes use of the ocean as mankind's common sink. We treat the ocean as if we believed that it is not part of our own planet—as if the blue waters curved into space somewhere beyond the horizon where our pollutants would fall off the edge, as ships were believed to do before the days of Christopher Columbus. We build sewers so far into the sea that we pipe the harmful refuse away from public beaches. Beyond that is no man's concern. What we consider too dangerous to be stored under technical control ashore we dump forever out of sight at sea, whether toxic chemicals or nuclear waste. Our only excuse is the still-surviving image of the ocean as a bottomless pit.

It is time to ask: is the ocean vulnerable? And if so, can man survive on a planet with a dead ocean? Both questions can be answered, and they are worthy of our attention.

First, the degree of vulnerability of any body of water would of course depend on two factors: the volume of the water and the nature of the pollutants. We know the volume of the ocean, its surface measure, and its average depth. We know that it covers 71 percent of the surface of our planet, and we are impressed, with good reason, when all these measurements are given in almost astronomical figures. If we resort to a more visual image, however, the dimensions lose their magic. The average depth of all oceans is only 1,700 meters. The Empire State Building is 448 meters high. If stretched out horizontally instead of vertically, the average ocean depth would only slightly exceed the 1,500 meters that an Olympic runner can cover by foot in 3 minutes and 35 seconds. The average depth of the North Sea, however, is not 1,700 meters, but only 80 meters, and many of the buildings in downtown New York would emerge high above water level if they were built on the bottom of this sea. During the Stone Age most of the North Sea was dry land where roaming archers hunted deer and other game. In this shallow water, until only recently, all the industrial nations of Western Europe have conducted year-round routine dumping of hundreds of thousands of tons of their most toxic industrial refuse. All the world's sewers and most of its waste are dumped into waters as shallow as, or shallower than, the North Sea. An attempt was made at a recent ocean exhibition to illustrate graphically and in correct proportion the depths of the Atlantic, the Pacific, and the Indian oceans in relation to a cross section of the planet earth. The project had to be abandoned, for although the earth was painted with a diameter twice the height of a man, the depths of the world oceans painted in proportion became so insignificant that they could not be seen except as a very thin pencil line.

The ocean is in fact remarkably shallow for its size. Russia's Lake Baikal, for instance, less than 31 kilometers wide, is 1,500 meters deep, which compares well with the average depth of all oceans. It is the vast *extent* of ocean surface that has made man of all generations imagine a correspondingly unfathomable depth.

When viewed in full, from great heights, the ocean's surface is seen to have definite, confining limits. . . . The astronauts have come back from space literally disturbed upon seeing a full view of our planet. They have seen at first hand how cramped together the nations are in a limited space and how the "endless" oceans are tightly enclosed within cramped quarters by surrounding land masses. But one need not be an astronaut to lose the sensation of a boundless ocean. It is enough to embark on some floating logs tied together, as we did with the *Kon-Tiki* in the Pacific, or on some bundles of papyrus reeds, as we did with the *Ra* in the Atlantic. With no effort and no motor we were pushed by the winds and currents from one continent to another in a few weeks.

After we abandon the outworn image of infinite space in the ocean, we are still left with many wrong or useless notions about biological life and vulnerability. Marine life is concentrated in about 4 percent of the ocean's total body of water, whereas roughly 96 percent is just about as poor in life as is a desert ashore. We all know, and should bear in mind, that sunlight is needed to permit photosynthesis for the marine plankton on which all fishes and whales directly or indirectly base their subsistence. In the sunny tropics the upper layer of light used in photosynthesis extends down to a maximum depth of 80 to 100 meters. In the northern latitudes, even on a bright summer's day, this zone reaches no more than 15 to 20 meters below the surface. Because much of the most toxic pollutants are buoyant and stay on the surface (notably all the pesticides and other poisons based on chlorinated hydrocarbons), this concentration of both life and venom in the same restricted body of water is most unfortunate.

What is worse is the fact that life is not evenly distributed throughout this thin surface layer. Ninety percent of all marine species are concentrated above the continental shelves next to land. The water above these littoral shelves represents an area of only 8 percent of the total ocean surface, which itself represents only 4 percent of the total body of water, and means that much less than half a percent of the ocean space represents the home of 90 percent of all marine life. This concentration of marine life in shallow waters next to the coasts happens to coincide with the area of concentrated dumping and the outlet of all sewers and polluted river mouths, not to mention silt from chemically treated farmland. The bulk of some 20,000 known species of fish, some 30,000 species of mollusks, and nearly all the main crustaceans lives in the most exposed waters around the littoral areas. As we know, the reason is that this is the most fertile breeding ground for marine plankton. The marine plant life, the phytoplankton, find here their mineral nutriments, which are brought down by rivers and silt and up from the ocean bottom through coastal upwellings that bring back to the surface the remains of decomposed organisms which have sunk to the bottom through the ages.

When we speak of farmable land in any country, we do not include deserts or sterile rock in our calculations. Why then shall we deceive ourselves by the total size of the ocean when we know that not even 1 percent of its water volume is fertile for the fisherman?

Much has been written for or against the activities of some nations that have dumped vast quantities of nuclear waste and obsolete war gases in the sea and excused their actions on the grounds that it was all sealed in special containers. In such shallow waters as the Irish Sea, the English Channel, and the North Sea there are already enough examples of similar "foolproof" containers moving about with bottom currents until they are totally displaced and even crack open with the result that millions of fish are killed or mutilated. In the Baltic Sea, which is shallower than many lakes and which—except for the thin surface layer—has already been killed by pollution, 7,000 tons of arsenic were dumped in cement containers some 40 years ago. These containers have now started to leak. Their combined contents are three times more than is needed to kill the entire population of the earth today.

Fortunately, in certain regions modern laws have impeded the danger of dumpings; yet a major threat to marine life remains—the less spectacular but more effective ocean pollution through continuous discharge from sewers and seepage. Except in the Arctic, there is today hardly a creek or a river in the world from which it is safe to drink at the outlet. The more technically advanced the country, the more devastating the threat to the ocean. A few examples picked at random will illustrate the pollution input from the civilized world:

French rivers carry 18 billion cubic meters of liquid pollution annually into the sea. The city of Paris alone discharges almost 1.2 million cubic meters of untreated effluent into the Seine every day.

The volume of liquid waste from the Federal Republic of Germany is estimated at over 9 billion cubic meters per year, or 25.4 million cubic meters

Some sublethal effects of petroleum products on marine life

Type of organism and species	*Type of petroleum product*	*Concentration*	*Sublethal response*
MARINE FLORA Marsh plants	crudes and refinery effluents	Single or successive coatings	Inhibition of germination and growth. Repeated coatings cause disappearance of some plants (increasing order of tolerance: shallow rooted plants, shrubby perennials, filamentous green algae, perennials, perennials with large food reserves)
phytoplankton	crude Naphthalene	1 ppm 3 ppm	Suppress growth Reduction of bicarbonate uptake
phytoplankton	oil	10^{-1} 10^{-4} ppm	Inhibition or delay in cellular division
phytoplankton	kerosene	3 ppm; 38 ppm	Depression of growth rate
phytoplankton	Kuwait crude	1 ppm	Depression of growth rate
phytoplankton	Kuwait crude; dispersant emulsions	20–100 ppm	Inhibition of growth; reduction of bicarbonate uptake at 50 ppm
Kelp	Toluene	10 ppm	75 percent reduction in photosynthesis within 96 hours
LARVAE AND EGGS Pink salmon fry	Prudhoe Bay crude	1.6 ppm	Avoidance effects; could have effect on migration behavior
Black sea turbot	oil	0.01 ppm	Irregularity and delay in hatching—resulting larvae deformed and inactive
Plaice larvae	BP 102	0–10 ppm	Disruption of phototactic and feeding behavior
Cod fish larvae	Iranian crude	Aqueous extracts	Adverse effect on behavior, leading to death

Lobster larvae	Venezuelan crude	6 ppm	Delay molt to fourth stage
Sea urchin larvae	extracts of Bunker C	0.1–1ppm	Interference with fertilized egg development
Barnacle larvae	oil	10–100 μ1/1 (ppm)	Abnormal development
Crab larvae	oil	10–100 μ1/1 (ppm)	Initial increase in respiration
FISH Chinook salmon Striped bass	Benzene	5, 10 ppm	Initial increase in respiration
CRUSTACEANS Lobster	crude, kerosene	10 ppm	Effects of chemoreception, feeding times, stress behavior, aggression, grooming
Lobster	La Rosa crude	Extracts	Delay in feeding
MOLLUSKS Mussel	crude	1 ppm	Reduction in carbon budget (increase in respiration; decrease in feeding)
Snail	kerosene	0.001–0.004 ppm	Reduction in chemotactic perception of food
Clam	No. 2 fuel oil	Collected from field	Gonadal tumors
Oyster	Bleedwater		Reduced growth and glycogen content
Snail	BP 1002	30 ppm	Significant inhibition to growth
Oyster	oil	0.01 ppm	Marked tainting
Mussel	No. 2 fuel oil	Collected from field after spill	Inhibition in development of gonads

Courtesy of the U.S. Environmental Protection Agency, Environmental Research Laboratory at Narragansett, R.I.

per day, not counting cooling water, which daily amounts to 33.6 million cubic meters. Into the Rhine alone 50,000 tons of waste are discharged daily, including 30,000 tons of sodium chloride from industrial plants.

A report from the U.N. Economic and Social Council, issued prior to the Stockholm Conference on the Law of the Sea four years ago, states that the world had then dumped an estimated billion pounds of DDT into our environment and was adding an estimated 100 million more pounds per year. The total world production of pesticides was estimated at more than 1.3 billion pounds annually, and the United States alone exports more than 400 million pounds per year. Most of this ultimately finds its way into the ocean with winds, rain, or silt from land. A certain type of DDT sprayed on crops in East Africa a few years ago was found and identified a few months later in the Bay of Bengal, a good 4,000 miles away.

The misconception of a boundless ocean makes the man in the street more concerned about city smog than about the risk of killing the ocean. Yet the tallest chimney in the world does not suffice to send the noxious smoke away into space; it gradually sinks down, and nearly all descends, mixed with rain, snow, and silt, into the ocean. Industrial and urban areas are expanding with the population explosion all over the world, and in the United States alone, waste products in the form of smoke and noxious fumes amount to a total of 390,000 tons of pollutants every day, or 142 million tons every year.

With this immense concentration of toxic matter, life on the continental shelves would in all likelihood have been exterminated or at least severely decimated long since if the ocean had been immobile. The cause for the delayed action, which may benefit man for a few decades but will aggravate the situation for coming generations, is the well-known fact that the ocean rotates like boiling water in a kettle. It churns from east to west, from north to sourth, from the bottom to the surface, and down again, in perpetual motion. At a U.N. meeting one of the developing countries proposed that if ocean dumping were prohibited by global or regional law, they would offer friendly nations the opportunity of dumping in their own national waters—for a fee, of course!

It cannot be stressed too often, however, that it is nothing but a complete illusion when we speak of national waters. We can map and lay claim to the ocean bottom, but not to the mobile sea above it. The water itself is in constant transit. What is considered to be the national waters of Morocco one day turns up as the national waters of Mexico soon after. Meanwhile Mexican national water is soon on its way across the North Atlantic to Norway. Ocean pollution abides by no law.

My own transoceanic drifts with the *Kon-Tiki* raft and the reed vessels *Ra I* and *II* were eye-openers to me and my companions as to the rapidity with which so-called national waters displace themselves. The distance from Peru to the Tuamotu Islands in Polynesia is 4,000 miles when it is measured on a map. Yet the *Kon-Tiki* raft had only crossed about 1,000 miles of ocean surface when we arrived. The other 3,000 miles had been granted us by the rapid flow of the current during the 101 days our crossing lasted. But the same raft voyages taught us another and less pleasant lesson: it is possible to pollute the oceans, and it is already being done. In 1947, when the balsa raft *Kon-Tiki* crossed the Pacific, we towed a plankton net behind. Yet we did not collect specimens or even see any sign of human activity in the crystal-clear water until we spotted the wreck of an old sailing ship on the reef where we landed. In 1969 it was therefore a blow to us on board the papyrus raft-ship *Ra* to observe, shortly after our departure from Morocco, that we had sailed into an area filled with ugly clumps of hard asphalt-like material, brownish to pitch black in color, which were floating at close intervals on or just below the water's surface. Later on, we sailed into other areas so heavily polluted with similar clumps that we were reluctant to dip up water with our buckets when we needed a good scrub-down at the end of the day. In between these areas the ocean was clean except for occasional floating oil lumps and other widely scattered refuse such as plastic containers, empty bottles, and cans. Because the ropes holding the papyrus reeds of *Ra I* together burst, the battered wreck was abandoned in polluted waters short of the island of Barbados, and a second crossing was effectuated all the way from Safi in Morocco to Barbados in the West Indies in 1970. This time a systematic day-by-day survey of ocean pollution was carried out, and samples of oil lumps collected were sent to the United Nations together with a detailed report on the observations. This was published by Secretary-General U Thant as an annex to his report to the Stockholm Conference on the Law of the Sea. It is enough here to repeat that sporadic oil clots drifted by within reach of our dip net during 43 out of 57 days our transatlantic crossing lasted. The laboratory analysis of the various samples of oil clots collected showed a wide range in the level of nickel and vanadium content, revealing that they originated from different geographical localities. This again proves that they represent not the homogeneous spill from a leaking oil drill or from a wrecked super-tanker, but the steadily accumulating waste from the daily routine washing of sludge from the combined world fleet of tankers.

The world was upset when the *Torrey Canyon* unintentionally spilled 100,000 tons of oil into the English Channel some years ago; yet this is only a small fraction of the intentional discharge of crude oil sludge through less spectacular, routine tank cleaning. Every year more than *Torrey Canyon's* spill of a 100,000 tons of oil is intentionally pumped into the Mediterranean alone, and a survey of the sea south of Italy yielded 500 liters of solidified oil for every square kilometer of surface. Both the Americans and the Russians were alarmed by our observations of Atlantic pollution in 1970 and sent out specially equipped oceanographic research vessels to the area. American scientists from Harvard University working with the Bermuda Biological Station for Research found more solidified oil than seaweed per surface unit in the Sargasso Sea and had to give up their plankton catch because their nets were completely plugged up by oil sludge. They estimated, however, a floating stock of 86,000 metric tons of tar in the Northwest Atlantic alone. The Russians, in a report read by the representative of the Soviet Academy of Sciences at a recent pollution conference in Prague, found that pollution in the coastal areas of the Atlantic had already surpassed their tentative limit for what had been considered tolerable, and that a new scale of tolerability would have to be postulated.

The problem of oil pollution is in itself a complex one. Various types of crude oil are toxic in different degrees. But they all have one property in common: they attract other chemicals and absorb them like blotting paper, notably the various kinds of pesticides. DDT and other chlorinated hydrocarbons do not dissolve in water, nor do they sink: just as they are absorbed by plankton and other surface organisms, so are they drawn into oil slicks and oil clots, where in some cases they have been rediscovered in stronger concentrations than when originally mixed with dissolvents in the spraying bottles. Oil clots, used as floating support for barnacles, marine worms, and pelagic crabs, were often seen by us from the *Ra,* and these riders are attractive bait for filter-feeding fish and whales, which cannot avoid getting gills and baleens cluttered up by the tarlike oil. Even sharks with their rows of teeth plastered with black oil clots are now reported from the Caribbean Sea. Yet the oil spills and dumping of waste from ships represent a very modest contribution compared with the urban and industrial refuse released from land.

That the ocean, given time, will cope with it all, is a common expression of wishful thinking. The ocean has always been a self-purifying filter that has taken care of all global pollution for millions of years. Man is not the first polluter. Since the morning of time nature itself has been a giant workshop, experimenting, inventing, decomposing, and throwing away waste: the incalculable billions of tons of rotting forest products, decomposing flesh, mud, silt, and excrement. If this waste had not been recycled, the ocean would long since have become a compact soup after millions of years of death and decay, volcanic eruptions, and global erosion. Man is not the first large-scale producer, so why should he become the first disastrous polluter?

Man has imitated nature by manipulating atoms, taking them apart and grouping them together in different compositions. Nature turned fish into birds and beasts into man. It found a way to make fruits out of soil and sunshine. It invented radar for bats and whales, and shortwave transceivers for beetles and butterflies. Jet propulsion was installed on squids, and unsurpassed computers were made as brains for mankind. Marine bacteria and plankton transformed the dead generations into new life. The life cycle of spaceship earth is the closest one can ever get to the greatest of all inventions, *perpetuum mobile*—the perpetual-motion machine. And the secret is that nothing was composed by nature that could not be recomposed, recycled, and brought back into service again in another form as another useful wheel in the smoothly running global machinery.

This is where man has sidetracked nature. We put atoms together into molecules of types nature had carefully avoided. We invent to our delight immediately useful materials like plastics, pesticides, detergents, and other chemical products hitherto unavailable on planet earth. We rejoice because we can get our laundry whiter than the snow we pollute and because we can exterminate every trace of insect life. We spray bugs and bees, worms and butterflies. We wash and flush the detergents down the drain out to the oysters and fish. Most of our new chemical products are not only toxic: they are in fact created to sterilize and kill. And they keep on displaying these same inherent abilities wherever they end up. Through sewers and seepage they all head for the ocean, where they remain to accumulate as undesired nuts and bolts in between the cogwheels of a so far smoothly running machine. If it had not been for the present generation, man could have gone on polluting the ocean forever with the degradable waste he produced. But with ever-increasing speed and intensity we now produce and discharge into the sea hundreds of thousands of chemicals and other products. They do not evaporate nor do they recycle, but they grow in numbers and quantity and threaten all marine life.

We have long known that our modern pesticides have begun to enter the flesh of penguins in the Antarctic and the brains of polar bears and the blubber of whales in the Arctic, all subsisting on plankton and plankton-eating crustaceans and fish in areas far from cities and farmland. We all know that marine pollution has reached global extent in a few decades. We also know that very little or nothing is being done to stop it. Yet there are persons who tell us that there is no reason to worry, that the ocean is so big and surely science must have everything under control. City smog is being fought through intelligent legislation. Certain lakes and rivers have been improved by leading the sewers down to the sea. But where, may we ask, is the global problem of ocean pollution under control?

There is endless merit in a Man's knowing when to have done.
THOMAS CARLYLE

No breathing species could live on this planet until the surface layer of the ocean was filled with phytoplankton, as our planet in the beginning was only surrounded by sterile gases. These minute plant species manufactured so much oxygen that it rose above the surface to help form the atmosphere we have today. All life on earth depended upon this marine plankton for its evolution and continued subsistence. Today, more than ever before, mankind depends on the welfare of this marine plankton for his future survival as a species. With the population explosion we need to harvest even more protein from the sea. Without plankton there will be no fish. With our rapid expansion of urban and industrial areas and the continuous disappearance of jungle and forest, we shall be ever more dependent on the plankton for the very air we breathe. Neither man nor any other terrestrial beast could have bred had plankton not preceded them. Take away this indispensable life in the shallow surface areas of the sea, and life ashore will be unfit for coming generations. A dead ocean means a dead planet.

5 Whither the Tropical Rainforest?

Lawrence S. Hamilton

Lawrence S. Hamilton, of the Department of Natural Resources, Cornell University, conducted the research for this report.

Reprinted from the *Sierra Club Bulletin,* April 1976, pp. 9-11, by permission of the author and the Sierra Club.

While the picture is not absolutely clear, it appears that north of the Orinoco River, Venezuela has lost roughly one-third of its tropical rainforest in the past twenty-five years. Moreover, the rate of disappearance is accelerating. This information comes from a study recently carried out by the Sierra Club through its International Environment Office under a contract with the United Nations Environment Programme and with the cooperation of the Venezuelan government.

South of the Orinoco, vast areas of primary, untouched rainforest remain protected in part by their inaccessibility. But roads are being pushed into this region and frontier settlements are being established "to fill in the blank spaces on the map" and "to open up the jungle." Similar developments are occurring throughout the tropical world. The activities bring with them the slash-and-burn, unstable cropping system which can, on many sites, initiate a chain reaction of degradation following which tropical rainforest may never return to the area (or only after the lapse of centuries). On other sites selected by land-capability surveys, clearing of rainforest is necessary for grazing and cropping, as part of economic development, and some forest must be lost. But there are far too many instances of clearing the wrong sites, not only in Venezuela but throughout the tropical world. Clearing for agriculture is not the only reason this exuberant and magnificent forest biome is disappearing. Wood-hungry developed countries are exploiting it as a vast, scarcely tapped storehouse of timber and wood fibre. Global monitoring carried out by the Food and Agriculture Organization of the UN will soon give us a picture of how fast the tropical rainforest is regressing, but meanwhile FAO estimates that in South America, only about thirty-six percent of what was originally moist forest exists today.

The Sierra Club project has provided specific information on the regression of rainforest for one large region (the Western High Llanos) of one country. Moreover, it has identified the causes of this unfortunate loss and suggested ways of developing a sound conservation program in Venezuela, which might in turn provide leadership not only for Latin America but for the tropical world in general.

The tropical rainforests have many values which are realized locally, nationally and internationally. Their conservation should be of concern to all people. . . . A few of these values may be briefly described.

The tropical rainforests of the world are the richest and most exuberant expression of life on land. They have developed over millennia in a relatively nonstressful environment in which temperature and moisture fluctuations are the smallest of anywhere on earth where plants grow—except in the oceans and in some caves. As a consequence, large numbers of very unusual species of plants and animals have evolved in these ecosystems. They are in

Man in the rain forest is just another rather simple animal. . . . He gains a new perspective in this complex world which he has not yet been able even to catalogue, let alone control. It is not a hostile world, but it is very indifferent to human needs and human purposes.
MARSTON BATES

a very real sense our richest gene pool, literally the genetic cradle of evolution. For instance, of trees alone there may be fifty to 200 species per hectare in a primary tropical rainforest, compared to twenty per hectare in a rich temperate forest. This diversity in itself can be a liability, since it means that on any given area (say ten hectares), there are few individuals of any one species. Thus, heavy exploitation of any one species (or a few) can lead to local extinction because many species have heavy seeds that do not colonize areas far removed from the parent tree. The seed in most cases also has short-term viability. Primary rainforest species are not aggressive in general, and are therefore quite vulnerable to disturbance. Pockets of endemism (a plant or animal species limited in distribution to a small region) are common. Thus, destruction of rainforest can cause rarity or even extinction of species because of their diversity and limited distribution characteristics. It is important to note that not all tropical rainforests show this amazing species diversity, for there are special soil or flood conditions where species are reduced to a relatively few. This is more the exception than the rule and, in general, the world's tropical rainforests are characterized by floristic and faunal abundance of forms and species as well as interactions.

Scientists are continually searching for and finding new drugs, resins, fibers, foods, and other plant products in this rich reservoir of life. Not only is the tropical rainforest an important and scarcely explored source of new products, but it is also locale for the discovery of new knowledge about unusual plant and animal interrelationships. It is the world's major gene pool and the evolutionary cradle of terrestrial life. It is imperative to slow the rate of disappearance of primary tropical rainforest before some of these scientific values are lost forever.

Some aspects of climate are influenced by the presence or the removal of rainforest. Albedo (reflected heat radiation) is increased when this forest is replaced by other kinds of land uses, thus affecting the global heat budget. While more research is needed, there is scientific concern that large-scale removal of tropical rainforest could result in temperature shifts in the tropics and resultant shifts in world wind patterns and rainfall. There is already much empirical evidence that clearing of rainforest has led to local changes in rainfall, with attendant disruption to established land uses. Further investigation and quantification of this rainfall-forest relationship are needed, but prudence would indicate a "go slow" policy. While there has been concern expressed about the effect of forest removal in reducing the oxygen supply of the atmosphere, there is as yet no concrete scientific evidence that this is the case.

Rainforests have other local hydrological benefits. Cloud forests serve to capture water from the atmosphere and make it available as underground and surface-flow streams (and thus of service to man as water supplies). Moreover, these forests and those on steep upper watershed serve to regulate the timing of stream flow and maintain the quality of water. As cover for erosive soils, they keep soil where it belongs and it does not end up in streams, reservoirs, lakes, etc., to reduce the usefulness of these bodies of water to agriculture, power production, navigation, municipal water supply and fishing.

Because of the luxuriant vegetation, the soils beneath tropical rainforests would be expected to be high in fertility, but the reverse is true in most cases. Fertility exists primarily in the vegetation, not in the soils (some alluvial soils being exceptions). Nutrients are cycled rapidly in this moist, warm, nonstressful environment, and circulate in a virtually leak-proof system. When the forest is removed, as when it is cut and burned for agriculture, this efficient recycling system is shattered. The nutrient additions flushed to the soil from the ash and organic material are usually quickly leached or oxidized after one or two years of cropping. Unless fallowed, or unless large amounts of organic material or fertilizer are added, the soil becomes impoverished. In some lateritic soils, impervious crusts may develop. Where slopes are steep, serious erosion and increased water discharge occur under the high rainfall. A chain of degradation follows, aggravated by weed and pest invasion, and true tropical rainforest may never return to these sites. There are, it is true, some types of soils under tropical rainforest where these distressing changes do not occur when the land is cleared, and such areas can be identified in advance by soil and land-capability surveys. It is important to note that because of these soil limitations, plus problems of pests and weeds, tropical rainforest soils do *not* have the vast potential to meet world food production needs some people are wont to believe they have. These forests are not the world's last large breadbasket.

There are many useful species of woods in the tropical rainforests, though the properties of many others are scarcely known. Growth rates for many species are excellent. Furthermore, the forests are a renewable resource and if harvested wisely can provide a sustained flow of products. They are thus an important part of economic development to meet the country's timber needs and even as material for foreign commerce. They can provide employment and rural stability. On nutrient-poor soils, these forests are the only productive crop that can utilize the energy of the site effectively.

Rainforests are the habitat for some of the world's most unusual and interesting wildlife. As with the flora, there exist the same characteristics of high diversity of species but relatively low numbers of any one species. To prevent wildlife species extinction, maintenance of the forest habitat is necessary.

People native to rainforests have evolved a way of life that depends on a stable relationship with their forest environment, and thus are truly ecosystem people. Industrialized man can learn from their cultures and adopt many attitudes and practices that would perhaps aid him in coming to terms with his own environmental constraints. At the least, any development of rainforest must take into account the presence and rights of these peoples.

6 Breaking the Web

George Uetz and Donald Lee Johnson

George Uetz is assistant professor in the Department of Biological Sciences, University of Cincinnati. Donald Lee Johnson is associate professor in the Department of Geography, University of Illinois, Champaign-Urbana.

Reprinted and abridged from *Environment*, 16(10) (December 1974), pp. 31–39, by permission of the Scientists' Institute for Public Information.

In the course of geologic time, the world's animal populations are constantly changing. We know from cave drawings, and other ancient finds, that many animals seen by humans in prehistoric times do not exist today. We know from Indian writings and ancient scrolls that the abundance and distribution of much of the world's fauna (animals) has changed considerably. We know from recorded history that numerous animals have disappeared from the face of the earth in the past few centuries.

The disappearance of some animal species, and the appearance of new ones, is part of the process of evolution—a process in which plant and animal species constantly adapt themselves to changing environmental stresses. Most of the organisms that have ever lived on this planet are now extinct, and it has been learned from the fossil record that extinction is the eventual fate of all species. Changes in the physical environment (climate, soil, and so on) or the biological environment (food sources, competitors, predators) usually initiate the process in nature. Extinction occurs when a species is not able to adapt to such changes and maintain its numbers. As these species disappear, they are gradually replaced by others.

Before the appearance of humans in the biosphere, extinction occurred entirely as a result of natural changes in the environment. With the advent of humans, however, came new stresses on the physical environment and its inhabitants. Some data suggest that the rate of extinction has increased as a result of human activities. Human stress on the environment takes a variety of forms—agricultural and forestry practices, for instance, change, and sometimes destroy, entire ecosystems; hunting, and predator and pest control, directly eliminate some animal species; and pollution by chemical contamination can cause a decline in others. Often, it is the interaction of a number of stresses which finally determines whether or not a species will become extinct.

In a time when problems of all sorts besiege our collective consciousness, however, it is difficult to mobilize concern for the passing of seemingly remote species of animals. Most people have never seen, and may never see, a whooping crane, a blue whale, a California condor, or any of the species on the endangered list. Many people question the value of these species to mankind and wonder whether we would be any worse off if any or all of them should become extinct.

Much of the concern for endangered species is based on aesthetic reasons or is part of an emotional commitment to the integrity of our natural heritage. Although these reasons are legitimate in themselves, it is necessary to understand the importance of endangered species and their fate in light of the ecological consequences—and in terms of the interrelatedness of organisms in the biosphere. If animal species become extinct, it is possible that the ecological communities in which they live may be functionally damaged, and may even collapse at some future time. Our biosphere is incredibly complicated and has taken more than four *billion* years to develop as it is today. The extinction of species we are presently witnessing is the erosion of this system in a very short period of time—much too short for any evolutionary adjustments to occur.

ENDANGERED SPECIES

Since 1600, when accurate record-keeping began, approximately 225 to 250 species of animals have ceased to exist.[1] The causes of these extinctions are numerous, yet nearly all are at least partially attributable to people and their increasing impact on the biosphere. The number of animal extinctions increased rapidly from 1600 to 1900, and there is the potential for an even greater increase in the future. By the 1920s, the number of animal extinctions in this century was close to that of the previous 100 years. Conservation efforts begun in the 1920s and 1930s may have halted the increase in extinct species for the present, but numerous species today are quite literally on the verge of extinction. As man's impact on the globe continues to grow, it is probable that many more extinctions will occur.

To make our projections, we have used information from the *Red Data Books* of the International Union for the Conservation of Nature and Natural

Resources (IUCN). The IUCN maintains an international list of endangered, rare, and threatened species, which at present includes 903 species or subspecies of vertebrate animals (see Table 1). The IUCN pools information from areas all over the world, monitoring the status of wildlife populations. Animals on the lists are placed in categories relative to their extinction status. These classifications include:[2]

Endangered. Species in immediate danger of extinction, for example, the blue whale, peregrine falcon, black-footed ferret.

Critically Endangered. Survival of the species unlikely without immediate intervention by man, for example, the California condor, whooping crane, and oryx.

Vulnerable or Threatened. Species still abundant in some parts of their range, but under threat due to depletion of numbers, for example, the mountain lion, sandhill crane, and green turtle.

Rare. Species with a population that is neither endangered nor vulnerable, but is subject to risk because of its small size, for example, the Galápagos tortoise and Komondo dragon.

In making our projections for the future, we consulted only the endangered, critically endangered, and rare lists. We found that in cases where counts were available, animal numbers were extremely low. Species listed

TABLE 1 *Number of species and subspecies of vertebrate animals (worldwide) already extinct or endangered as of 1974*

Animal group	*Already extinct**	*Critically endangered***	*Endangered and/or rare*
Reptiles	28	34	152
Amphibians	—	6	34
Birds	130	66	346
Mammals	68	73	292
Fishes	—	6	79
TOTALS	226	185	903

*Data provided by J. A. Davis, New York Zoological Society, the Zoological Park, Bronx Park, New York, Jan. 1972.

**Data from the International Union for the Conservation of Nature and Natural Resources, *Red Data Books*, Lausanne, Switzerland.

as critically endangered had world populations from as few as 3 to 5 individuals, to no more than a few hundred. Others on the endangered species list ranged from about 300 to 5,000 in worldwide numbers. Understandably, the number necessary to maintain a viable population varies between species. However, the great reduction in numbers, rapid shrinkage of habitat, and severity of threats documented in the *Red Data Books* show these animals to be likely candidates for extinction unless there is human intervention.

The facts speak for themselves. We stand an excellent chance of losing a substantial portion of the world's animal species, and with them innumerable strands of the interconnected web of life that supports our existence on this planet. . . .

HUMAN IMPACT

Studying the relationship between the development of human culture and the extinction of animal species is a sobering experience. The spread of civilization has been accompanied by a drastic reduction of animal numbers and by reduced diversity of animal species. Fairly accurate records are available since the sixteenth century, when the western world became conscious of its impact on animals. During this period, the number of worldwide animal extinctions increased considerably. This pattern is likely to continue unless an effort is made to halt the process.

Using data from several sources, we have examined extinction rates and the reasons for extinction during the past four centuries. The spread of cities, towns, and farms accompanied the increase in human population during this time. Human beings spread over the globe, exploring and colonizing many previously uninhabited areas, resulting in the demise of many animal species. Just as man's early migrations may have precipitated a wave of extinction on each new continent that was settled, the arrival of explorers and colonists from western societies decimated animal populations in Africa, the Americas, Australia, and on islands everywhere. Island faunas are particularly sensitive to human disturbance because they are often adapted to live only in an ecologically unique environment. With resources limited by the area of the island or archipelago they occupy, the animals are likely candidates for extinction. Few areas of the world, then, have escaped the direct and indirect effects of modern man.

The geography of animal extinction documents the expansion of modern civilization and provides mute testimony to the development of technology. The sharp rise in bird extinction in the late 1700s and early 1800s, for instance, can in part be correlated with the development of the musket. The causes of extinction, however, are more complex than this. Research in this area has been interesting, revealing distinct patterns and trends in the ways people affect animals.

Figure 1 illustrates trends in the primary civilization-related causes of extinction in the past. The most obvious trends are the decrease in the role of hunting for food and an increase of hunting for commercial products and sport. These are likely due to more sophisticated pursuit and hunting operations, as well as an increased demand for luxuries. Ecosystem alteration has increased, along with a general increase in other indirect means of extinction, such as the effect of introduction of feral (domestic animals that have reverted to a wild state) animals. Overall, there has been an increase in the

Our whole moral and legal tradition is founded on the assumption that there is an unbridgeable gap between man and animals, giving us the right to own and exploit them without reference to their best interests. If our scientists now tell us that animals are different from us only in degree, then our exploitation carries far more serious implications than we have been willing to admit.

STEPHEN I. BURR

diversity of ways species are threatened, creating a complex web of extermination closely related to the technological development of civilized man. In the seventeenth century, 86 percent of extinctions were caused directly by human activities (hunting), while 14 percent were indirect (habitat destruction). By the nineteenth century, the pattern was almost reversed—24 percent direct and 76 percent indirect. Introduced animals accounted for 28 percent of the extinctions during the last century, but during the first 70 years of this century they accounted for nearly 50 percent.

CAUSES OF EXTINCTION

Prominent causes of extinction include ecosystem alteration, hunting, introduction of species, predator and pest control, capture of wild animals for legitimate and illegal purposes, and pollution. Ecosystem alteration is one of the more significant causes of extinction. Alone, or in combination

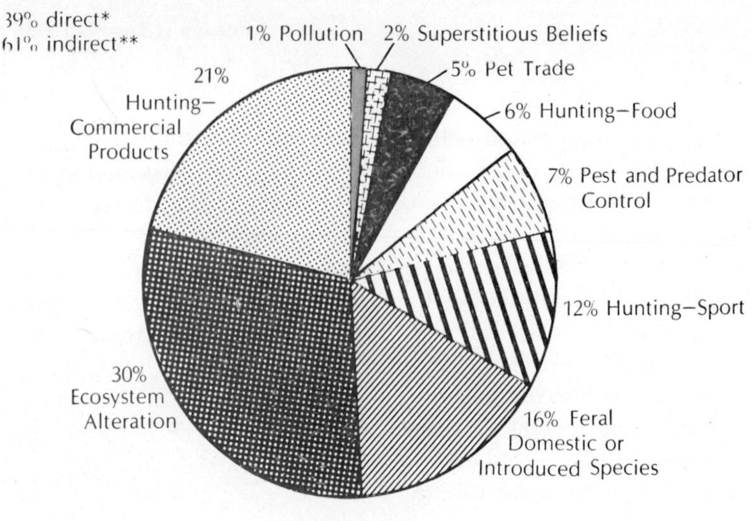

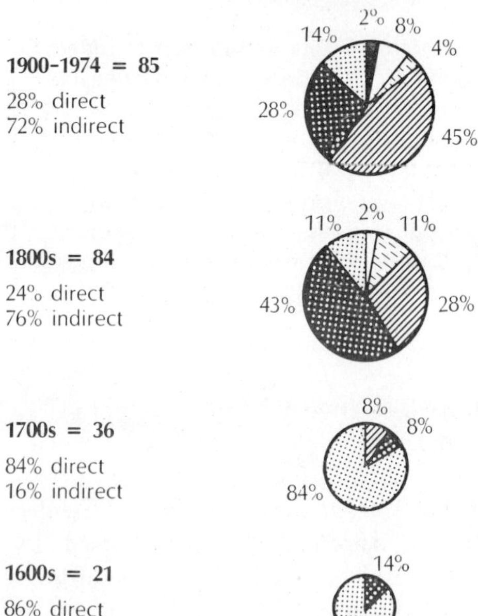

Values given are the percentages of extinct or endangered animals in each time period whose extinction was or is due to ecosystem alteration, hunting, and so on. The lower four pie graphs concern animals already extinct, while the uppermost graph concerns animals presently on the endangered species list. Size of pie graphs is proportionate to the number of species and subspecies that became extinct or endangered in each time period.

*Such as hunting.

**Such as habitat destruction.

Source: Data in first pie graph are from the IUCN, *Red Data Books*, Lausanne, Switzerland: In lower four graphs from Davis, J. A., New York Zoological Society, the Zoological Park, Bronx Park, N.Y., Jan. 1972.

FIGURE 1 *Extinction due to civilization-related causes*

with other factors, habitat destruction or modification, as a result of forest cutting, marsh drainage, agriculture, and development of housing and roads, threatens hundreds of animals that are presently endangered. When wildlife habitats are destroyed or modified, animals must move to other habitats, adapt to a changed environment, or die. The speed with which habitats are destroyed often leaves only the last option. But there are even more subtle and far-reaching consequences for animal species when ecosystems are destroyed or altered by human activities. As people encroach on nature, patches of natural habitat become smaller and more distant from each other. Depending on the size of an animal, and its position on a food chain, it may require a sizable area to provide food and shelter. Even though habitats are not totally eliminated, there may not be enough to allow a species to survive, as in the case of many large carnivores. Fragmentation of habitats into small units isolates individuals and makes maintenance of numbers difficult. Habitat destruction by itself accounts for 30 percent of presently endangered species, and in combination with other factors probably affects over two-thirds of all the species on the endangered list.

Hunting for food, once a major cause of extinction, is now restricted to animals sought as gourmet items or as pet food for cats and dogs in the U.S. and Europe. The passenger pigeon was considered a gourmet item in the mid-nineteenth century, as were buffalo tongues. Several whale species currently endangered are used for pet food. Occasionally, protection of endangered species is usurped by political upheaval, as in the recent Congolese war, when any kind of animal was fair game to a starving population.

A serious menace to many species is the animal products industry. Animals are killed for furs, hides, tusks, horns, and feathers. The harvesting practices associated with such activities make ecological management practically impossible. Selectively slaughtering the young (seals), females (tigers), or pre-reproductive animals (crocodiles) greatly reduces the possibility of a species population regaining lost numbers. Animal poaching in game preserves for illicit sale of skins and illegal, selective slaughtering of particular groups of rare animals, such as the big cats, is another example of human pressure on animals. In 1972 and 1973, the U.S. Fish and Wildlife Service uncovered a multimillion dollar international smuggling ring that illegally processed 86,167 spotted cats from Africa and South America over a seventeen-month period. This number included a large share of the remaining world populations of leopards and cheetahs. Fourteen fur companies from many countries were involved. All pleaded guilty and were fined. However, Fish and Wildlife Service investigators believe the ring constituted only *one-half* of the illegal skin trade, and that still other companies were involved.[3]

Game hunting has been, and still is, a threat to numerous species, such as the oryx, Bengal tiger, and rhinoceros. While most hunting in the U.S. today is tied to conservation measures, this has not always been the case. In the 1870s and 1880s, 25 to 40 million American bison were killed. Bison hunters and the U.S. Army were largely responsible for the near extinction of this species, as they sought to open grasslands for cattle grazing and eliminate the food supply of the Plains Indians. The passenger pigeon, heath hen, and others, were not so fortunate, for hunters so depleted their numbers that they shortly became extinct. Endangered species protection can also be sidestepped by military corruption, as in the recent case in which military personnel were caught surreptitiously hunting in game preserves in Thailand, Cambodia, and Vietnam.[4]

Another form of hunting, the popular nineteenth century pastime of oology (egg collecting or the study of eggs), nearly eliminated the whooping crane, the California condor, and other rare birds. Eggs were put in attractive displays, and high prices were paid for eggs of rare birds. What more efficient way can there be to eliminate a species?

An indirect result of human activity is the introduction of many species of animals into new habitats, or the release of domestic species which become feral. Introduced species affect wild animals by preying on them (cats, dogs, and mongooses), competing with them for food (cattle and rabbits), or destroying their habitat (goats and pigs). Island faunas are especially sensitive to extinction by this means, but continental areas are not immune. The demise of the American elm, for example, was caused by the accidental introduction, into North America, of a disease carried by a European beetle.

Government-sponsored predator and pest control programs in the U.S. cost taxpayers millions of dollars, and sometimes result in the poisoning of innumerable nontarget predators (such as eagles and kit foxes) as well as the predators they are intended to kill. Other species—poorly understood predators or ungainly, frightening creatures, such as spiders and snakes—are killed out of fear or habit.

The capture of wild animals for entertainment, display, and scientific study often results in particular species becoming endangered or extinct, even though their habitat remains intact.

A surprising variety of endangered wild animals (gila monsters, several species of parakeets, parrots, numerous lizards, fish) are sold as pets, especially in the U.S. In 1967, the U.S. Fish and Wildlife Service reported that, among imports to the U.S., were 74,304 mammals; 203,189 birds (not including parrot-family birds and canaries); 405,134 reptiles; 137,697 amphibians; and 27,759,332 fish.[5] Considering that other pet-loving nations, such as Great Britain and Germany, also import large numbers of animals, it is possible to understand how species are endangered by the pet trade. The import figures for wild animals show only a fraction of the impact of this trade because most, and sometimes all, wild animals die during capture, during shipment, or after sale in pet stores. Few pet buyers could hope to meet the specialized food and environmental requirements of the exotic animals they purchase, and most people give up their pets when the novelty wears off.

Since 1967, most of the large zoological parks and aquariums have taken self-regulatory action to control international traffic in endangered species although this will not undo damage already brought about by previous, unrestrained collecting. Smaller, and less scrupulous zoos, however, still continue to bargain for wild animals, some of which are endangered. For every tapir sold to a zoo, 40 to 60 are killed.[6] As animal species become endangered, or rare, their value as specimens or zoo animals increases; higher prices are paid for their capture; and the problem worsens.

Medical research, regardless of its relevance, is a serious threat to certain endangered forms, such as the orangutan, the chimpanzee, and other primates. In recent years, between 56,000 and 200,000 wild primates have been imported annually into the U.S. for medical research.[7] Medical researchers are often wasteful when working with wild animals, probably due to habits developed in working with inexhaustible supplies of laboratory mice and rats. For example, hundreds of animals may be killed at one time for a particular organ (for example, the removal of kidneys from chimps), and hundreds more of the same species killed at a later time for a different organ.[5] Large areas of lowland Venezuela and Colombia are now virtually devoid of primates because they have been collected by Indians for sale to animal traders. India currently exports 65,000 rhesus monkeys, half of which are used in medical research in the United States.[8]

There are numerous species of animals already extinct, or in imminent danger of extinction, because of demand for horns, glands, and other products as aphrodisiacs, magic charms, and medicines. The near-extinct rhinos of Java, Sumatra, Southeast Asia, and India are still shot and poached for their horns, which, when powdered, are believed to make an effective aphrodisiac. A little-known fact is that at the turn of this century, the California seals and sea lions were almost completely exterminated by American sealers seeking oil, and by Chinese who sought their whiskers for cleaning opium pipes and their gonads for aphrodisiac purposes.[9] Thanks to protection laws passed at the urging of concerned citizens and legislators, the seals, sea otters, and gray whales not only escaped extinction but now occur in greater numbers than at any time in the past 100 years.[10]

Chemical pollution, a relatively new, and potentially devastating, means of extinction is unleashed as a by-product of our technological civilization. Chlorinated hydrocarbon pesticides, such as DDT, when consumed by birds of prey result in eggshell thinning and reproductive failure. The peregrine falcon, brown pelican, osprey, and bald eagle have suffered losses due to this type of indirect extermination. By this means too, the California brown pelican and Channel Islands bald eagles have become virtually extinct over the past decade. Some species of fish, for example, the humpbacked chub, have been threatened by increased toxins and by water temperature changes, and many more are undergoing drastic reduction in population for the same reasons.

Few animals are driven to extinction from a single cause. Many species threatened with extinction feel the combined weight of habitat destruction, hunting, pollution, and other factors operating together.

INTERNAL FACTORS

So far, only external environmental factors that influence survival of species have been discussed. It is also important to look at the characteristics of individual species, or groups of animals, that make them prone to extinction, or conversely, resistant to extinction. Means of extinction do not equally affect all species. The number of endangered species varies between animal groups, and major survival differences can be seen. For example, the critical population size (the level necessary for continued existence), is different for each species and should be kept in mind when one attempts to assess the status of a declining animal species, even if it is not officially endangered.

Ehrenfeld has used qualitative factors to construct a hypothetical example of what a "most endangered" species might be like: A large animal, which can only survive in a restricted area, has a long gestation period and few young per litter. It is hunted for some sort of natural product or for sport, but is not subject to efficient game management. It has a restricted distribution but, because it is migratory, travels across international boundaries. It is intolerant of man, reproduces and travels in groups, and has specialized behavioral patterns which make it vulnerable to a changing environment. There is no such animal, but the model closely approximates a polar bear. Conversely, one might take the characteristics of a nonendangered species and come up with a composite picture of the typical animal of the twenty-first century—the house sparrow, the gray squirrel, the Virginia opposum, and the Norway rat—species that exploit habitats created by human activities.[11]

IMPLICATION FOR MANKIND

When endangered species become extinct, the world loses an animal, surely, but an ecosystem somewhere loses a functioning part. It is obvious, however, that not all ecosystems or natural communities are the same, nor will the elimination of a single species, or several species, affect all natural communities in the same way. Some communities are far more diverse than others (they have a greater variety of plant and animal species), and relationships among community members can be extremely complex. Eliminating a species from any natural community changes the relationships among community members, and the system no longer works in the same way. The seriousness of the change depends on the role of the eliminated species in the functioning of the community and the degree of the dependence of the community on its biological complexity. In some cases it is likely that little change would be evident, but in others the loss of several species, or a single key species could have serious consequences. As the impoverishment of communities in terms of species composition continues, the web of interrelationships may be weakened, perhaps to the point of unraveling.

Numerous examples illustrate how this may happen, at least on a small scale. In agriculture, replacement of natural vegetation by single crop species lowers community diversity. The monoculture is ecologically simple, and requires huge energy subsidies from man to maintain it. When insecticides are applied to control pest insects, beneficial insect predators and parasites are sometimes killed off, leaving an unchecked pest population to rebound in greater numbers than before. When crops have been grown experimentally within diverse vegetation, no such pest outbreaks occur.[12] When stream environments are modified by sewage pollution, or channelization, a few species, capable of withstanding the disturbance or exploiting extreme environments (such as mosquitoes, for instance), may occur in great numbers and become a pest problem.[13] In research on intertidal marine invertebrates, Robert T. Paine found that removal of a key predatory species from the rocky intertidal community resulted in one of its prey species monopolizing most of the space.[14] With no check on its numbers, it crowded out several other species, destroying the diversity of the system. Paine and others have suggested that predators act to increase species diversity by directing most of their predatory attention to the most abundant species in a community—thus preventing any one species from monopolizing the community resources.

Although these examples are somewhat simplistic, they point to the potentially critical situation that could arise in many of the world's remaining ecosystems if more species become extinct. Some ecosystems depend on "key" species; for example, predators, which may play a major role in the maintenance of species diversity or in the control of prey populations, and decomposer organisms, which help to recycle nutrients. The salt marsh ribbed mussel, for example, is a seemingly insignificant member of the muddy bottom community, yet is responsible to a large extent for the availability of phosphorus, a critical nutrient.[15]

The capacity of ecosystems to rebound from human disturbance is limited by the amount of time between disturbances, the size of the area, and the distance from populations that could recolonize a disturbed area. As forests, prairies, marshes, and other natural ecosystems become fragmented and smaller in size, chances for extinction of animals become greater. The possibility that an ecosystem might be irreparably damaged by the removal of a single species may then increase as well.

If there is a lesson we should have learned from our past and present ecological crises, it is that living things and their environments are interrelated. The ecological consequences of pollution, ecosystem alteration, or extinction of animals may not be immediate or obvious. Damage to biological systems, however, may be irreparable, and may not be discovered until it is too late. Reducing the diversity of animal species by extinction means permanently removing the components from complex processes in many ecosystems. Unless stopped, it can only result in an ecologically simple world whose future would be unstable at best. Even efforts to preserve rare species in parks, zoos, and preserves, while commendable, may prove futile. Such species may no longer reproduce successfully when removed from their natural habitat. Small, captive populations may lack the genetic diversity necessary for survival when reintroduced to natural conditions.

There appears to be an increasing concern for endangered animals as a result of the environmental movement and the outstanding work of wildlife conservation organizations all over the world. These national and international groups have made some progress in international protection for whales, great cats, birds of prey and other animals.[16] In the U.S., the Endangered Species Act was signed into law in 1973, and provides at least partial protection for domestic, endangered species. The U.S. Department of the Interior and many state conservation departments sponsor research on

> **Cynics rightly say that the world could struggle along without the pelican or the whooping crane; they do not often note that it could as easily do without man. Quite possibly the unecological tendencies of *Homo sapiens* may already have marked him for the extinction he has so freely meted out to other species.**
> LYNTON K. CALDWELL

endangered species and attempt to manage and restore their populations in wildlife refuges and national parks.[17] The magnitude of the extinction problem, however, may well dwarf these, and other efforts, to save endangered animals on an individual basis.

We suggest that the ultimate survival of our world's fauna depends on the conservation of wilderness and natural areas all over the globe. The struggle for the Alaskan tundra, the Amazon rainforest, and other smaller patches of natural ecosystems is a struggle to preserve habitats that are essential to animals and to the smooth functioning of the biosphere—and to mankind's well-being. Should we fail to save them, we cannot decelerate the worldwide extinction process, and human survival itself may be at stake. Mankind probably cannot survive in a biosphere devoid of diversity for very long.

NOTES

1 Data provided by J. A. Davis, New York Zoological Society, the Zoological Park, Bronx Park, New York, Jan. 1972.

2 International Union for the Conservation of Nature and Natural Resources (IUCN), *Red Data Books,* Lausanne, Switzerland. "Critically endangered" classification designates species where survival is unlikely without intervention by man; "endangered and/or rare" includes species in imminent danger of extinction.

3 R. Cahn, "Good News for Ocelots, Pumas, Jaguars," *Christian Science Monitor,* Feb. 5, 1973.

4 *Newsweek,* May 14, 1973.

5 W. G. Conway, "The Consumption of Wildlife by Man," *Animal Kingdom,* June 1968.

6 M. Burton, *Systematic Dictionary of Mammals of the World,* 1962.

7 J. L. Marx, "Shortage of Primates," *Science,* 181: 334, 1973.

8 C. H. Southwick, M. R. Siddigi, and M. F. Siddigi, "Primate Populations and Biomedical Research," *Science,* 170: 1051–1054, 1970.

9 P. Bonnot, "The Sea Lions, Seals and Sea Otter of the California Coast," *Calif. Fish and Game,* 37: 371–389, 1951.

10 P. Bonnot, "The Sea Lions of California," *Calif. Fish and Game,* 14: 1–16, 1928.

11 D. W. Ehrenfeld, *Biological Conservation,* Holt, Rinehart and Winston, New York, 1970, p. 96.

12 D. Pimentel, "Species Diversity and Insect Population Outbreaks," *Ann. Ent. Soc. Amer.,* 54: 76–85, 1961. R. B. Root, "Organization of a Plant-Arthropod Association in Simple and Diverse Habitats: The Fauna of Collards (*Brassica olleracea*)," *Ecol. Monogr.,* 43: 95–124, 1973.

13 R. E. Coker, *Streams, Lakes, Ponds,* Chap. 10, Harper and Row, New York, 1968.

14 R. T. Paine, "Food Web Complexity and Species Diversity," *Amer. Natur.,* 100: 65–75, 1966; "The *Piaster-Tegula* Interaction: Prey Patches, Predator, Food Preference, and Intertidal Community Structure," *Ecology,* 50: 950–961, 1969.

15 E. J. Kuenzler, "Structure and Energy Flow of a Mussel Population in a Georgia Salt Marsh," *Limnol. and Oceanogr.,* 6: 191–204, 1961.

16 R. Gillette, "Endangered Species: Diplomacy Tries Building an Ark," *Science,* 179: 777–779, 1973.

17 *National Wildlife,* Apr.–May 1974.

7 Man's Effects on Island Ecosystems

F. R. Fosberg

F. R. Fosberg is special adviser in tropical biology at the U.S. National Museum, Smithsonian Institution, and affiliate professor of botany at the University of Hawaii. He is the founder and editor of *Atoll Research Bulletin.*

Islands have proven to be very vulnerable to disturbance resulting from human activities. They are unique and unusually interesting scientifically. Island biotas and soils are particularly susceptible to the ravages of introduced animals. This results in the destruction of vegetation cover and the extinction of endemic plants and animals. Exotic plants tend to replace native species on islands, especially after disturbances by man or introduced animals. Fire, used in hunting and in slash-and-burn agriculture, has a profoundly degrading effect on vegetation and ultimately leads to disastrous accelerated erosion and loss of soil. Overhunting quickly eliminates useful animals. Accelerated erosion, from whatever cause, brings about degradation of the entire ecosystem. Plantation agriculture, though efficient, tends both to be vulnerable to pests and diseases and to deterioration of the quality of life for the population. Military activities, both preparatory and belligerent, are among the most destructive of all influences on islands, and the effects are long-persistent. They set the dynamic status of the system back toward the pioneer condition, and they may leave dangerous residues in the form of unexploded ammunition. Mining for phosphate and other products, if extensive, usually leaves an island in totally unusable condition for the native people. Such activity sacrifices long-term habitability to short-term gain. Medical and public health activities, while undoubtedly beneficial to the people, if unaccompanied by any instruction in methods of population regulation result in the rapid increase of human populations which soon tend to exceed the carrying capacities of small island ecosystems. Imposition of alien cultural systems, evolved in continental ecosystems, tend to bring about serious maladjustment and social problems in small islands, as well as material demands that their limited resources cannot satisfy. There seems little reason for optimism about the future of island ecosystems, so long as population growth and pressure for modernization continue. Islands present a unique ecological situation because of their nature and the treatment they have received. Few parts of the world are as vulnerable to mismanagement, and few have suffered as grievously at the hand of man. I am going to depart from the topic of the effects of technology on islands to the extent of substituting "human activity" for "technology." Then I will be able to discuss more fully what man has done to these remarkable microcosms scattered over the seas of the world. To keep the subject manageable I will largely confine my remarks to small, mostly oceanic islands, but this does not mean that there is nothing to say about the large islands. Madagascar, by itself, would be a good case study in destruction if I were familiar enough with the details of what has happened to tell it correctly.

When I say that islands are unique, I mean more than the truism that any spot on the face of the earth is unique. Although many islands resemble each other rather closely in physical geography, islands, with the exception of coral atolls, tend to be biologically more distinct from each other and to have more unique qualities than do comparable continental areas. Almost all oceanic islands higher than sea-level atolls have some, often many, endemic plant and animal species, and relatively small biota, mostly brought together by accidents or dispersal, followed by more or less evolution. Thus, no two have the same complement of species, nor can they have identical combinations of species even locally, such as occur over and over in continental situations and form the basis of the plant association concept in phytosociology.

This uniqueness means that when an island is degraded and its biota largely destroyed, the world has lost something irreplaceable. There is no question of locating another example, as can usually be done for a continental ecosystem. It is, for example, still possible to find remnant examples of the various sorts of mid-U.S. prairie, in spite of the completeness of cultivation of the prairie regions. Not so with an island: when an insular biota is destroyed, the destruction is final. One can find another somewhat similar island, perhaps, but it will be different. A species that is lost can never be relocated. We know it existed only because of a dried fragment in an herbarium or a pickled corpse in a museum jar.

In discussing the results of man's exploitation of islands, I will depart from the currently popular idea of economic values. I cannot prove that the extinct dodo, *Clermontia halakalensis,* or Wake Island flightless rail ever had any economic value. Probably they did not, though we will never know. I am going to assume that because they existed they had value. They were a part of the fascinating diversity that makes life such an interesting and rewarding experience, and provides the basis for the remarkable stability, or homeostasis, of ecosystems that has so far protected man from completely destroying his habitat. I am going to accept as basic that the destruction of

Islands are apt by their seclusion to offer doorways to the unexpected, rents in the living web, opportunities presented to stragglers who might be carrying concealed genetic novelty in their bodies—novelty that might have remained suppressed in a more drastic competitive environment. . . .

Islands can be regarded as something thrust up into recent time out of a primordial past. In a sense, they belong to different times: a crab time or a turtle time, or even a lemur time. . . . It is possible to conceive an island that could contain a future time—something not quite in simultaneous relationship with the rest of the world. . . .

LOREN EISELEY

an island ecosystem causes "life to lose some of its savor" (L. H. Bailey, oral communication). The examples of destruction that I shall describe are matters that should be of serious concern to a generation that finds the world in its present condition to have little enough savor, indeed, and to promise even less in the near future.

My concept of the development of an ecosystem, island or otherwise, is that of a process of building energy into the system, reducing its randomness or entropy, and increasing order, complexity, and integration. Man's usual role has been to reverse this process and to increase very rapidly the entropy in the system. I accept as axiomatic that this role is bad, though to some extent necessary, in order that man may produce food for his outrageous numbers to eat. Even this need, however, is no excuse for his going about it in a completely thoughtless and irresponsible manner, the results of which are only now beginning to become apparent, and to worry us. . . .

I do not want to convey the idea, however, that the consequences of the activities I will discuss have no practical importance. If the soil of an island is washed out to sea, it cannot be cultivated. If the water storage capacity of an island is impaired, the people may suffer water shortages. If a wild species of banana becomes extinct, its disease resistance or other breeding qualities cannot be utilized. But I simply do not intend to detail all the practical consequences, many of which we do not yet know. It is sufficient to say that some of the islands that have been most mistreated are now uninhabited and, unless much energy and ingenuity are directed toward their rehabilitation, they will remain uninhabitable.

EFFECTS OF ANIMAL INTRODUCTION

The sorts of human activity that have had destructive effects on the island ecosystem are many—it is practical to discuss only a few of the more obvious. One of the worst practices has been the introduction of four-footed animals, both herbivorous and carnivorous, for various reasons—as food supplies, as agricultural and pastoral animals, as pets, and even, as in the case of rats, for no purpose at all. The resultant animal populations have been extraordinarily destructive because the vegetation of oceanic islands, and most of the species that make it up, evolved in the complete absence of such animals. These species developed no protective mechanisms at all against the browsing, bark gnawing, rooting, and even the trampling, that are the normal behavior of herbivores. The native animals—birds, reptiles, and invertebrates—on the islands, and previously subjected to predation by cats and dogs, had no fear of them. Many had even lost the ability to fly or to run. The dodo, for example, had no chance at all against dogs.

Introduced animals found no effective enemies or diseases on the islands, so they increased exponentially until, as in the case of the rabbits on Laysan Island, they were controlled by starvation. When nothing remains for a rabbit or a goat to eat, the vegetation of an island is in bad shape indeed. And not only the vegetation. When the vegetation is gone, water is not readily absorbed; it runs off, taking the soil with it.

The sufferings of shipwrecked sailors were so real that it became standard practice to introduce goats onto any island discovered. This was done so early, and the goats were so effective, that we will never know what the vegetation of many islands was like. In the late eighteenth century, the governor of St. Helena wrote to the British Admiralty that the goats were destroying the forests and asked permission to remove them. The reply was that the goats were of more value than the forests and that nothing should be done. Today we would like to know what those forests were really like.

Of course, not all introductions were so successful. Guanaco were said to have been introduced onto Tinian in the Marianas by the Spanish in the seventeenth century. They disappeared without leaving a trace, except perhaps to help various other animals destroy the forests of Tinian.

Sheep and cattle were brought into many islands for ranching. Whether they were kept under control, herded and harvested or became feral, the result was more or less the same. The native forests were destroyed and, at best, the land was converted to palatable grassland—in worse cases to weedy shrubland, and at worst to bare eroding subsoil and rocks. Richard Warner (1960) has vividly described this process in "A Forest Dies on Mauna Kea" (Hawaii). Kahoolawe Island, in the Hawaiian group, is an example of this and other destructive processes. It is now barren and uninhabited. Easter Island, after the introduction of sheep, does not have a single living tree, at least of native species, and has few native plants of any kind. It is now nothing but a bare, wind-swept sheep ranch. Once it supported a thriving population, capable of the megalithic stone art for which the island is famous. The population is now small and the culture destroyed (Skottsberg, 1953).

EFFECTS OF PLANT INTRODUCTION

Exotic plants in many, if not most, islands have almost completely replaced the native flora, except in high mountains or cliffs. In the Hawaiian Islands, for example, a visitor unescorted by botanists, might spend three months without ever seeing a native Hawaiian plant. In the lowland populated areas, he would see gorgeous flame trees from Madagascar, frangipani trees from tropical America, shower trees from the Caribbean and the southeast Asian region, mangoes from Borneo, banyans from India and southeast Asia, samans from the Caribbean, hybrid hibiscus developed locally from largely imported parents, allamanda from Brazil, and many other memorable ornamental and food plants. In the country, he would see forests of algarroba from Peru, guava from Mexico, koa haole from Mexico, and christmas berry from South America, as well as great plantations of sugar cane and pineapples, probably from New Guinea and Paraguay respectively. Papayas from tropical America and bananas from Indonesia furnish fruit for every table. The native forests of strange island trees, without English names, that originally covered these lowlands and slopes are gone except in remote or protected pockets. The visitor does not even learn that the coconut, pineapple and lantana are not native.

The Hawaiians brought with them a considerable number of plants, only one or two of which ever occupied much more land than was specially prepared for them. But when Europeans arrived with a continuing flow of useful plants, ornamentals, and weeds, these plants took over with surprising vigor and replaced much of the native vegetation. They are now the dominant species in most habitats. This success is seldom due to any innate superiority over native species, but rather because introduced animals opened up the vegetation and allowed the newcomers to establish themselves, which they were seldom able to do alone against a closed native vegetation. Once the closed cover was broken up by browsing and trampling, the newcomers found niches that they were able to occupy more effectively, in many cases, than their rather few native competitors.

Only time will show what plants will ultimately prevail, but this must be time without continued disturbance, which in most islands now seems unlikely. Certain observations indicate that the native vegetation may, given a chance, eventually regain some ground from the exotic assemblages now prevalent. For example, in the first two decades of this century, Joseph Rock, C. N. Forbes, and others (oral communication) reported that the native wet forests of the Hawaiian island of Lanai were reduced to a few acres on top of the highest mountain. When I was on this island in 1935, George C. Munro, for years manager of the island, had long been carrying on a relentless campaign to eliminate goats, sheep, and deer and had brought the wild cattle into the domestic herds and kept them under control. There were then very few large feral herbivores on the island. The wet mountain forest had extended itself to cover hundreds of acres of the higher parts of the island (Fosberg, 1936). After the upper parts of the Waianae Range of Oahu were fenced as a water reserve in the 1920's, a few native plants began to make headway against the lantana that had dominated these mountains after it was introduced as an ornamental and spread by introduced mynah birds and perhaps doves into forest openings made by introduced cattle and goats. Its fleshy fruits are eaten by these birds and the hard seed deposited in their droppings. It forms dense thickets of a chaparral-like scrub that resists invasion by most other plants and is not eaten by herbivorous quadrupeds.

In certain cases, exotic animals and plants effect a kind of cooperation in crowding the native plants, as when livestock so effectively scattered the seeds of the introduced kiawe or mesquite tree (*Prosopis pallida*) in their droppings that this tree now dominates the lowlands in the Hawaiian Islands. Here also the introduced mynah scatters the seeds of the introduced guavas (*Psidium guajava* and *Psidium cattleianum*), the introduced lantana (*Lantana camara*) and the introduced blackberry (*Rubus penetrans*), so that these plants dominate large areas at the expense of the native plants.

The questions may be asked: Who introduced these plants and animals, when, and why? For most islands most introductions are undocumented. The Polynesians brought the pig, dog, and chicken, as well as the rat to Hawaii. Captain Vancouver introduced cattle, goats, and sheep to Hawaii in 1790. An early Spanish settler, Don Marin, brought many exotic plants, some of which are now so familiar as to be frequently thought of as Hawaiian. A priest brought the kiawe in about 1830. A Scot, homesick for Scotland, was said to have introduced gorse on Maui. The Territorial Board of Commissioners of Agriculture and Forestry introduced innumerable plants, many of which now occupy large areas (Bryan, 1947). Dr. William Hillebrand, German physician, botanist, and advisor to the Hawaiian kings in the nineteenth century, brought the mynah bird, as well as the banyan tree and many other desirable and undesirable plants. Dr. Willis Pope of the Hawaii Agricultural Experiment Station brought the blackberry. Numerous private citizens, mostly unrecorded, have made their contributions to the exotic fauna and flora, and the process still goes on throughout the island world.

EFFECTS OF FIRE AND HUNTING

Fire, set by lightning or volcanic activity, was doubtless a natural phenomenon in islands, as it was elsewhere. However, its incidence was vastly stepped up when man arrived on the scene. After game was introduced and established, fire was used in hunting. Where upland agriculture was practiced, it was usually of a slash-and-burn type, with fires spreading into vegetation not intended for destruction. Such carelessness accounts for disastrous fires on islands, as elsewhere: and deliberately set fires are common, as on Raïatéa and Guam, for example. Islands like Mangareva and large areas in Guam and Palau that were doubtless once forested, are now reduced to coarse grassland, conducive to continued burning. Not only are the forest and other vegetation destroyed, but the soil is exposed to erosion.

Arid southeast Oahu is now a grassy and scrubby expanse of exposed lava ridges; once it had a fairly deep soil, traces of which were located by Dr. Frank Egler during his investigations in 1937. It will certainly take many thousands of years to re-form soil on this sort of substratum, even if it is revegetated and protected from further burning. At present the area is too bare to burn very much.

Shifting agriculture or other slash-and-burn agricultural practices have a disastrous effect on islands. This sort of agriculture is defended by some ethnologists and agricultural economists as the only way to exploit soil on steep tropical slopes. Clearly it is the way that many primitive agriculturists do exploit these slopes. The practice might have been defensible so long as the population was small, only favorable spots were cultivated, and the period of rotation was long enough for vegetation and soil to recover somewhat. But populations do not remain small, especially when encouraged by modern medicine and hygiene and the elimination of tribal warfare (Taeuber, 1963). The cycles of cultivation and fallow become shorter and shorter, and clearing occurs on poorer and more marginal land. Erosion becomes catastrophic, and the island is destroyed as a favorable place to live.

Where edible or otherwise valuable animals occur—for example, birds, turtles, tortoises, or seals—they are usually overharvested. Green turtles have been almost eliminated from the Hawaiian and Bonin islands and many others where they were formerly plentiful. The Caribbean monk seal is gone. The Galápagos tortoises were hunted to dangerously low numbers and some island populations eliminated altogether. Giant tortoises were exterminated on all the western Indian Ocean islands except Aldabra.

EROSION

In most continental regions accelerated erosion is a gradual enough process that it may not be recognized until the results are catastrophic—dust bowls, dendritic mazes of gullies, and finally badlands. On many islands, the terrain is steep and the soil thin and very erodible if not held by vegetation. Destruction can become irreversible in a few years. One only has to fly over islands like Lanai and Kahoolawe in the Hawaiian group after a rainy spell to see where the soil from the extensive bare red areas goes. There are vivid red plumes in the bright blue sea down-current from every stream mouth. Domestic and feral animals, as recounted above, have, in the last 150 years, reduced these islands to deserts, and even the subsoil is rapidly being removed by water and wind. Fortunately for the present owners of Lanai, the pineapple plant needs only enough soil to anchor itself. This soil can be very sterile, since the plants are funnel-shaped to collect rain water and may be fed by a small squirt of nutrient solution from time to time. Most other plants are not so accommodating. A large area of Lanai is a solid pineapple plantation. The rest is sad-looking. The history of Lanai would have been most instructive to a student of land utilization if it had been written down while George C. Munro, former manager of the island, still lived, as he was fond of relating it from personal remembrance.

The problem of loss of soil from mountainous areas, on continents as well as islands, is an old one. Agricultural techniques have evolved to derive some advantage from silt carried away by streams. Marsh cultures of taro and rice were developed to a high degree of productivity in valley bottoms where the silt tended to be deposited. It is no accident that wet-land taro was the staff of life on islands such as Rapa and the Hawaiian group. This is an example of a people who have evolved a technique to live with a bad situation that they may have created or at least aggravated. Rapa is a perfect example of an island "skeletonized" by erosion. This situation on Réunion has been studied and ably summarized by Gourou (1963).

MONOCULTURE

The most favorable sites on most islands have been pre-empted by plantation agriculture, producing coconuts, sugar cane, pineapples, and a few other crops. This practice has the advantage of efficiency; but its efficiency largely depends on the monoculture technique, which is extremely vulnerable to insect pests and plant diseases and puts the people completely at the mercy of market fluctuations. The diet of the people changes when they shift from a subsistence to a cash economy. Witness the trend to canned meat and fish, wheat bread, and rice in French Polynesia, and the subsequent deterioration of the natives' teeth. The change usually seems to be for the worse, though this perhaps is not wholly the fault of the system. Most people will choose a bad diet if it tastes better, or if it is merely fashionable, even though they could afford a good one. In addition to these obvious disadvantages, the plantation system also reduces the landscape to a deadly monotony, as on the island of Barbados, completely covered by sugar cane; few plants and animals are tolerated if they are even imagined to be detrimental to the crop. Those that are not detrimental are usually eliminated with those that are, as the pesticides used are seldom specific to one or a few kinds of organism. A plantation, though it may be very productive, tends to be made a "biological desert." And on islands, as noted above, this is more serious than on continents, because unique plants and animals that previously occupied these favorable habitats may not have existed elsewhere and may now be irretrievably lost.

MILITARY ACTIVITIES

Especially in relatively recent years military activities have been among the most destructive of all human interferences on many islands. Fortifications and other construction, even in periods of non-belligerency, consume and drastically alter huge areas (Hall, 1944). Most of Guam, for example, has been subject to military construction, as were certain of the Bonins, Iwo Jima, Dublon in the Truk Islands, Kwajalein, Eniwetok, and many others. During actual hostilities enormous quantities of high-explosive shells and bombs were dropped on most of these same islands. The terms "Mitcher haircut" and "Spruance haircut" vividly describe the results of these treatments (Heinl and Crown, 1954, and other Marine Corps Monographs; Richard, 1957). One of the most fought-over islands during World War II was Saipan in the Marianas. Before the war, the arable land on the island was in sugar plantations. Various types of forest, some native, others planted, occupied most of the rougher areas, with some sterile soils occupied by savanna. At the onset of war, the Japanese built extensive fortifications, including caves and underground installations. The attack by the American forces was preceded by a naval and aerial bombardment of incredible intensity, and after an assault landing the island was fought over for a month. Much of the vegetation was burned or cleared and replaced by Quonset huts and other military construction. These soon disintegrated when they were abandoned after military use lessened. In 1948, an ammunition storage area exploded, throwing shells all over the northern third of the island. The unexploded shells make that part of the island dangerous even today. The sugar cane has seeded, giving rise to brakes of wild cane filling old plantations and other clearings. The forests are now mostly rather degraded second-growth of introduced species choked with weedy vines. The native upland forest, with its several endemic species, is mostly gone, though some of the species persist as occasional individuals. Native vegetation largely persists only on cliffs and in strand situations, where it readily recovered from destructive treatment. In general, the dynamic status of the vegetation has been set back toward a pioneer status and native plants have given way to exotics. Most native animals have become scarce with the reductions in their habitats.

Even areas where there was little war activity have not escaped. Some islands, such as Kahoolawe in the Hawaiian group, have been used for bombing or shelling practice and are littered with dud ammunition.

Several islands, such as Bikini, Eniwetok, Christmas Island, Mururoa, Fangatau, and Montebello have been used as test areas for atomic explosives. The results were all that might have been expected. Interestingly enough, since the vegetation of atolls, especially the drier ones, is already of a pioneer type, that is, capable of growing on new or disturbed habitats, it seems to have recovered rather promptly after the earlier of these tests, and there are now active preparations afoot for sending the Bikini people back to their atoll. Fortunately there are few, if any, local endemic species on atolls, so not too much will have been lost as a result of the tests.

MINING

One of the widespread forms of technological destruction on islands is phosphate mining, and in some cases, bauxite mining. As the result of ages of guano deposits by fish-eating birds and subsequent geochemical processes, most of the elevated atolls had large phosphate deposits, and even many low atolls had enough to repay exploitation. Especially on the raised limestone islands the mining removes all loose and semi-consolidated materials, leaving a desolate "badlands" type of pinnacled surface, utterly useless, not even supporting much vegetation, particularly where the rainfall is low or undependable. Banaba (Ocean Island) in the central Pacific was recently abandoned by its inhabitants after it had been worked over for phosphate. Most of the soil had been removed, so that agriculture became difficult without elaborate procedures to convert coral rock to soil. Their water situation, always a marginal one, deteriorated. Water that would have been held by vegetation tended to run off. Evaporation increased from the bare surfaces. The native vegetation, on which they had depended for many essentials, was mostly eliminated. Life became so difficult that, in spite of the traditional islander's love for his home island, it became the easiest course for them to use their saved up share of the phosphate profits to buy an island elsewhere and move to it.

The people of nearby Nauru, a similar large, but almost worked-out, elevated atoll are looking into the possibility of rehabilitating their home.

Fortunately they have large funds held in trust for them—their share of the phosphate profits. It seems likely that a substantial part of these savings, perhaps all, would be used if the island were to be properly restored to a productive condition (United Nations, 1966; Anonymous, 1966, n.d.).

The phosphate islands of the western Indian Ocean are now scarcely capable of supporting anyone, except by their marine resources. A number of local species seem to have disappeared on some of them. On Christmas Island, in the eastern Indian Ocean, the last remaining colony of *Sula abbottii* (Abbott's booby) is now threatened by a stepped-up program of phosphate mining. Fortunately smaller phosphate deposits are no longer profitable because of the opening up of large continental phosphorite mines. The larger ones, however, still seem able to compete. Recently the unwelcome news was published that commercially valuable phosphate occurs on Bellona Island near the Solomons, scarcely explored botanically and the site of a most interesting primitive Polynesian or proto-Polynesian culture. Although there is every reason for leaving Bellona and nearby Rennell untouched, as has previously been the policy of the British Solomon Islands Administration, it seems a safe prediction that this discovery will provide an excuse for abandoning this intelligent policy. Thus another interesting Stone Age culture will be destroyed, and an island biota, indeed, a functioning and unique ecosystem, will be degraded.

One might ask, how is this ecosystem unique and why is it important? In addition to the consideration previously mentioned, Bellona and Rennell islands represent a special class of islands—the elevated atoll type. Elevated atolls are scattered in some numbers through tropical seas; Banaba and Nauru, mentioned above, are examples, as is Makatéa, another phosphate island in French Polynesia. However, because of the prevalence of phosphate in commercial quantities on these islands, only a very few of them remain in anything like an undisturbed condition. Henderson and Aldabra are two which are not inhabited and have not been exploited because insufficient phosphate has been found. In comparison with these uninhabited ones, Rennell and Bellona remain with intact, functioning aboriginal populations still in balance with their habitats. Understanding of these, by comparison with their uninhabited counterparts, might well be one of the most important aims of modern ecological science, since such situations could be considered as microcosms to illustrate some of the principles on which larger balanced ecosystems function. But if they are destroyed for a few million dollars' worth of phosphate, this opportunity will be lost. Some of the most beautiful scenery in Palau, both above and under water, and the hope of a tourist industry for Micronesia are now threatened by phosphate prospecting.

MEDICINE

A more insidious chain of events, but one leading to certain and serious deterioration of most islands, is set in motion by the introduction of modern hygiene and medical practice. Although few would deplore the provisions of health benefits to the charming and attractive island peoples, those responsible may certainly be severely criticized for not, at the same time, introducing effective birth control measures and educating the people as to their use. This idea is not foreign to some of these cultures, but during the recent acculturation following contact with European civilization and the accompanying reduction of population from foreign disease most such lore was either forgotten or abandoned. With effective public health measures populations are now rising at an alarming rate on many islands. Although officials point to this with pride in many cases, they have yet to come up with an effective plan to increase the resources of the islands correspondingly. And on an island the carrying capacity may be very soon exceeded by the numbers of people. Already, on a people-per-square-mile basis some of the smaller islands have an alarming population. Utilization of the resources of the productive reefs of atolls, however, greatly increases their carrying capacity and makes possible the support of present large populations. However, rapid degradation of the resource base, including the reefs, will inevitably follow any significant increase beyond the carrying capacity of any island. The case of Réunion Island studied and described by Pierre Gourou (1963) is an excellent example. The carrying capacity seems to be rapidly becoming less as the population grows. And these people have no place to go (Taeuber, 1963).

IMPOSITION OF ALIEN VALUES

Finally, we may mention something that is belatedly becoming understood about the impact of foreign cultural systems, especially the Western technological culture, on the cultures of small islands. It would seem almost a truism that any culture that has evolved on a small island or that has existed on an island long enough to reach any sort of balance, has evolved to fit the size of the environment that it occupies, as well as the other environmental characteristics. This may be readily seen in the few small island cultures that still remain relatively uninfluenced by outside cultures, for example, Satawal, Ifalik, and Woleai. The pace of life and the daily round of activities are adjusted to the dimensions of the world in which they are carried on. The system of values of such peoples is suited to the resources available and the scope for activity. They do not hurry unnecessarily and they do not accumulate wealth. Capital is not a preoccupation for most islanders.

When Western culture is thrust upon such a system, the result is in almost all cases a crumbling of the social structure and a sense of deep confusion and maladjustment. The values and demands that characterize the impinging culture cannot be satisfied by the limited environment that is available. In other words, a culture evolved in a continental situation is simply unsuited for a small island. The troubles experienced by the U.S. administration of the Trust Territory of the Pacific Islands are a good example. With the best motives in the world, with fantastic financial resources,

Nothing is ever done until everyone is convinced that it ought to be done, and has been convinced for so long that it is now time to do something else.
F. M. CORNFORD

a $50,000,000 U.S. budget in 1970, and with a relatively immense staff, the results can only be described as a failure. The people of these far-flung islands are not happy. Many of them look back on the stern Japanese administration as "the good old days." They are being taught the virtues of a free enterprise democratic system, and their own social systems, evolved to meet the conditions prevailing in their island environments, are crumbling before the impact of a dominant system that is being forced on them by education and political pressure. An alarming proportion of the people, especially of the class that should be leaders, are rapidly becoming alcoholics. It makes one sad and depressed to see these peoples today and compare them with what they were even twenty years ago. They have acquired desires and needs with little or no chance of satisfying them, at least on any sort of longterm basis, as the resource base of their islands simply will not sustain the new levels of exploitation.

In short, almost every human activity, from those of the first arrivals to the islands, to those of contemporary Western culture, has altered or is altering the island ecosystem, and in almost all cases, altering it for the worse in the long run. And the world is becoming more impoverished thereby.

Can we arrest this process? Probably not. But if we can learn to use some self-restraint in our headlong race to seize every opportunity for resource exploitation and development, if we can preserve at least representative samples of what we are destroying, our descendants may still retain the option to restore, in some measure, these fascinating and beautiful environments, after their intangible human values are better understood and appreciated.

CONCLUSION

That there is little foundation for any optimistic predictions is indicated by my final example. Axis deer were introduced on the island of Molokai in the Hawaiian Islands at least sixty years ago. It is not easy to sort out the effects of deer from those of cattle and sheep. However, the result of the grazing, browsing, and trampling of all three herbivores has resulted in a steady deterioration of the native forests on this island. The dry lowland forest was almost gone in 1932, and by 1942 there was one tree left in the patch that had been still wooded in 1932. *Prosopis* had occupied most of the lowlands except on the plateau and windward side of west Molokai, where wind erosion was, and is, having serious consequences.

On this island, the upland wet forests were protected against cattle and sheep grazing for their watershed value. In most areas, however, the forest has been opened up by deer, allowing the entry of serious weeds—*pamakani* (*Eupatorium adenophorum*) and *Lantana camara.* Little of the rich montane rain forest and cloud forest is left intact, and the deer were, at last report, continuing their inroads (Lamoureux, personal communication, 1967). Ecologically, the island is a shambles.

This would be just another of many similar accounts, except that for the last few years there has been steady pressure by hunters to introduce these same axis deer to the island of Hawaii, where they have not previously been and where there are still a few substantial areas of native, relatively undisturbed vegetation. This pressure has been fought by all who have any interest in conservation, scientists and ranchers alike—although, with the example of Molokai, one would think that it would not even be necessary to fight. However, the facts have been misrepresented by the very officials who should know them best. It has been claimed that the deer will not enter wet forests or rough lava. The agitation has been continuous, aided and abetted by fish and game officials who feel they are in the employ of those who buy hunting licenses. Finally, there has been a decision to bring a herd of deer to be kept in a three-hundred-acre sheep pen, to observe how they reacted to this new environment. Anyone who has seen axis deer jump fences on Molokai must realize that this is merely a means of quieting the opposition until they can be presented with a *fait accompli.* If we cannot even profit by an obvious example to avoid such an unnecessary mistake, what hope is there to hand any viable island ecosystems on to our descendants? Or any other ecosystems, for that matter?

REFERENCES

Anonymous. "Background to Nauru." Prepared for the Legislative Council of Nauru (Australia). 1966. 16 pp.

——. Report to the General Assembly of the United Nations. Administration of the Territory of Nauru, 1st July 1964 to 30th June, 1965. Commonwealth of Australia. N.d.

Bryan, L. W. "Twenty-five Years of Forestry Work on the Island of Hawaii." *Hawaiian Planters Rec.*, 51 (1947), 1–80.

Carlquist, Sherwin. *Island Life: A Natural History of the Islands of the World.* New York: Natural History Press, 1965. 451 pp.

Egler, F. E. "Arid Southeast Oahu vegetation, Hawaii." *Ecol. Monogr.*, 17 (1947), 383–435.

—— "Unrecognized Arid Hawaiian Soil Erosion." *Science*, 94 (1941), 513–14.

Fosberg, F. R., ed. *Man's Place in the Island Ecosystem.* Honolulu: Bishop Museum, 1963. 264 pp.

Fosberg, F. R. "Plant Collecting on Lanai, 1935." *Mid-Pacific Mag.*, April–June 1936, pp. 119–23.

Gourou, Pierre. "Pressure on Island Environment." In *Man's Place in the Island Ecosystem*, F. R. Fosberg, ed. Honolulu: Bishop Museum, 1963. Pp. 207–25.

Hall, J. N. *Lost Island.* Boston: Little, Brown, 1944, 212 pp.

Heinl, R. D., and Crown, J. A. *The Marshalls: Increasing the Tempo.* Washington, D.C.: Government Printing Office, U.S. Marine Corps (Marine Corps Monographs Series). 1954. 188 pp.

Richard, Dorothy E. *U.S. Naval Administration of the Trust Territory of the Pacific Islands.* 3 vols. Washington, D.C.: Government Printing Office, Office of Chief of Naval Operations, 1957.

Rock, J. *Indigenous Trees of the Hawaiian Islands.* Honolulu: published by Rock under patronage, 1913. 518 pp.

Skottsberg, C. *Natural History of Juan Fernandez and Easter Island.* Stockholm, 1953.

Stoddart, D. R. "Ecology of Aldabra Atoll, Indian Ocean." *Atoll Res. Bull.*, 118 (1967), 1–141.

——."Catastrophic Human Interference with Coral Atoll Ecosystems." *Geography*, 53 (1968), 25–40.

Taeuber, Irene B. "Demographic Instabilities in Island Ecosystems." In *Man's Place in the Island Ecosystem,* F. R. Fosberg, ed. Honolulu: Bishop Museum, 1963. Pp. 226-52.

United Nations. United Nations Trusteeship Council Document No. T/PV. 1285, 33rd Session. New York, July 11, 1966. Mimeo.

Warner, R. E. "A Forest Dies on Mauna Kea." *Pacific Discovery,* 13 (2) (1960), 6-14.

8

Ecological Effects of Current Development Processes in Less Developed Countries

Kenneth A. Dahlberg

Kenneth A. Dahlberg is professor of political science at Western Michigan University.

Reprinted from *Human Ecology and World Development,* edited by Anthony Vann and Paul Rogers, by permission of the Plenum Press.

INTRODUCTION

It would be intriguing to take a position rather like that of Gandhi and argue that it is the rich Western countries that are really the "less developed" ones as they have not been able to come to grips with their masses of material goods either spiritually or ecologically. The temptation is the greater because many ecologists are doing much the same thing: they are suggesting that we, the rich Western countries, must change our basic attitudes towards nature, our modes of production, many of our institutions, and many of our comfortable habits. The challenges which these "subversive scientists"[1] present to our traditional ways are strikingly similar to the challenges which conventional economists have presented to the non-Western world in urging them to "develop." It is only when we think about the degree and difficulty of change that is being asked for by the ecologists (as well as the social and institutional resistance to such changes) that we can begin to appreciate the ways in which conventional development processes challenge the poorer countries of the world.[2] Such a line of argument would also more clearly stress that it is the rich countries that are the greater sinners in terms of pursuing anti-ecological policies and technologies and that, given the wide-ranging impact of the rich upon the poor, it is really in the rich countries that the major changes must come if we are going to learn to live within the limits of our planet.

While the above "switch" in definitions will not be pursued, it will be argued that conventional understandings of "development" must be modified to include the recognition that what is really involved is a process of transferring or, more precisely, imposing Western values, concepts, technologies, and institutions upon non-Western cultures and environments.[3] It is only when we recognize that development is neither a neutral nor a universal process that we can begin to truly assess the ecological costs involved and begin to suggest some alternative approaches. If, as usually seems to be the case, Western values, concepts, and technologies become even more anti-ecological when they are transferred to or imposed upon cultures and environments to which they are not adapted, then any appropriate action by Western countries towards the poor countries would seem to involve either major changes in approach and/or high degrees of self-restraint in limiting the further spread of anti-ecological processes. However, since the Western countries—through colonialism—are responsible for many of the ecological maladies afflicting poor countries, such self-restraint cannot be equated with non-action; rather, it suggests quite different aid, development, and regulatory policies. . . .

ECOLOGICAL COSTS OF CURRENT DEVELOPMENT APPROACHES

Any attempt to estimate ecological costs—in whatever field—suffers from several fundamental difficulties. There is the inherent complexity of physical and biological processes, the understanding of which is made even more difficult by the fragmented and specialized nature of the academic disciplines. In addition, the conceptions and methodologies of those who specialize in costs—the economists—are in many ways fundamentally anti-ecological.[4] Given these difficulties, the best that can be attempted here is a descriptive survey. Let us first look at some specific projects and their costs.

Specific Projects

It is important to include at least one historical project because many of the difficulties that flow from current development approaches relate to the indifference of most development economists to historical, much less ecological, factors. A recent careful study of public works projects in 19th century India,[5] reads very much like many contemporary environmental "horror" stories,[6] except that the historical remove delineates much more sharply the contrasting cultural conceptions and the very clear Western biases in what were then thought to be universal scientific principles.

The introduction of major public works programmes—in the form of canals, railroads, and roads—was expected to benefit both the peasants of India and the investors in England. Equally, the financial, administrative, and judicial reforms introduced after the demise of the East India Company were expected to remove the barriers to the natural workings of *laissez-faire* economics. The impact of these imported technologies, concepts, and institutions was, of course, quite different than expected. There were two serious effects. First, because the canals and railroads were designed with little concern for proper drainage, some 4,000 to 5,000 square miles of agricultural land was lost to salinity by 1891.[7] Second, while irrigation did increase production, it did so primarily for the spring export crops and at the expense of the autumn crops of millets and pulses upon which most of the population depended for their food and fodder. In addition to these "natural" costs, the canals also had the effect of increasing the wealth and power of those already well off. This was because the British left the construction and control of the minor channels and distributaries to the local rulers and landlords—who naturally used this power to increase their own wealth.

Large-scale railroad and road construction compounded many of these difficulties. Drainage was reduced even more through the building up of these new embankments and barriers. Soil erosion and flooding were speeded up by the heavy demand for timber both for railway sleepers and fuel. Also, the deforestation of many areas meant that poor families now had to use cow dung for cooking fuel rather than as a fertilizer for their crops.[8] The introduction of British legal concepts of "private property" and a series of courts to enforce its attendant rights, while not having an immediate impact upon the environment, did weaken traditional land tenure patterns by encouraging a spate of litigation over titles and debts. Thus, this 19th century attempt at development—albeit phrased in a different rhetoric, but with much the same mixture of idealism and interest as today—led to a serious deterioration of the natural environment, to severe social and economic distortions, and to an impoverishment of a large part of the peasant population.

Contemporary development projects are often planned by international civil servants rather than the colonial administrators of old. However, the Western conceptions and approaches of the latter have been carried over virtually unchanged. This is most apparent in health, irrigation, and agricultural projects, where the approaches of the WHO and FAO—as well as those of various aid donating countries—have been basically technological. That is to say that various "universal" technologies were promoted in an attempt to avoid the many specific cultural and political problems found in each region. For example, the use of DDT to attack malaria typifies Western ideas regarding efficient, inexpensive technological measures to solve a problem. The results show all of the dangers of specialized thinking, dependence on technological measures, and an attempt to avoid dealing with specific peoples in specific environments. One study shows that the levels of malarial infection in Guatemala are now approaching pre-DDT levels while the mosquitos in the area are becoming more and more resistant to hard pesticides.[9] While the technocrat might argue that we have gained 20 years of protection, this is not the case. What has been lost is 20 years in trying to learn how to deal with malaria through biological/cultural means, while at the same time the tremendous long-term ecological costs of persistent pesticides have been imposed upon many tropical regions. What is now needed (and was equally feasible 20 years ago) are programmes that approach such matters from the village level. Various forms of vegetation management, the introduction of larvae-eating fish, the draining of shallow areas, the screening of huts, plus the use of selective biological measures such as the sterile male technique would go a long way towards reducing the dangers of malaria with little attendant environmental cost, while at the same time encouraging the sorts of village activities that most would include as part of any meaningful development programme. The barriers to such an approach lie not only in the vested interests of the urban elites in the developing countries, but in the lack of imagination on the part of national and international aid personnel. Also, such an approach requires not only imagination, but a great deal of *specific* information regarding the local environment, local customs, local patterns of political influence etc., so that the various possible programmes and technologies can be adapted to the requirements of the local situation.

Large scale dams, such as the Aswan Dam in Egypt, the Akosombo Dam in Ghana, and the various dams suggested in the Colombo Plan for the Mekong Basin, offer another example of large-scale projects, justified in the name of development, which have been conceived with very little concern for the larger environmental and ecological effects that they produce. The Aswan Dam, for example, while more than doubling Egypt's irrigated farm land, has also greatly reduced the fisheries in the eastern end of the Mediterranean (the rich nutrients which formerly fed the fish now producing a rapid silting behind the dam and a profusion of plant growth in Lake Nasser). Ironically, the increased food production has been barely able to keep pace with the growing population—which itself is in large part a result of cheap public health technologies such as water and sewage treatment, DDT spraying, and immunization—which are much easier to introduce than birth control techniques, since they encounter fewer cultural barriers. In addition, the expansion of irrigation through canals had led to great increases in the debilitating disease of bilharzia (schistosomiasis)—a parasitic flat-worm that is carried by microscopic snails living in the canals which enter the bare feet of those working in the canals.[10]

The Akosombo Dam has created many similar environmental problems. In addition, there were disruptions caused by the dam flooding a number of tribal lands. While the government made various studies in an attempt to minimize the difficulties, and even offered to build new villages for those displaced, this meant little to those tribes concerned because their cosmologies were based on their living and dying, like their ancestors, on that specific piece of earth. Not having our cultures so deeply rooted in specific environments, it is easy for Westerns to suggest that such displacement is a small price to pay for progress—although the resistance of London homeowners to the building of new motorways suggests that it is really the planners and engineers who are insensitive to the real impacts of projects, rather than those directly affected. It should also be pointed out that while the Akosombo Dam does offer relatively cheap electricity for urban elites and manufacturers, the major beneficiary of the dam is clearly the consortium of British and American aluminium companies which obtained a guaranteed long-term quota of electricity at fixed cheap rates with which to process bauxite. While they did provide some of the financing of the dam, they were also able to obtain significant financial support from international agencies. . . .

GENERAL TRENDS AND INTERACTIONS

The above discussion of several specific projects suggests that if environmental dimensions are disregarded there will be not only unexpected ecological costs, but additional social and economic costs—which generally tend to fall most heavily upon the poor. A number of systematic dimensions are also hinted at and should now be made more explicit. The theoretical basis for this is found in the analysis of linkages between more and less organized subsystems made by the ecologist Ramon Margalef.[11] In discussing the links between a number of such subsystems, specifically including those between agrarian communities and industrial societies, he says that the latter subsystem

> . . . experiences more predictable changes through time. In so doing it stores information better and is a more efficient information channel. The first subsystem is subject to a stronger energy flow and, in fact, the second system feeds on the surplus of such energy. It is a basic property of nature, from the point of view of cybernetics, that any exchange between two systems of different information content does not result in a partition or equalizing of the information, but increases the difference. The system with more accumulated information becomes still richer from the exchange. Broadly speaking, the same principle is valid for persons and human organizations. . . . Such relations are compounded in an hierarchical organization and are reflected at every level.[12]

These relationships appear to be valid both at the national and international levels. Nationally, as industrialization has progressed, the farmers have generally become relatively poorer (on average—though there are similarly increasing gaps between rich and poor farmers). Internationally, the gap between the rich industrial countries and the poor agrarian countries continues to increase. While it is important to recognize such natural tendencies, in human affairs they are not to be accepted fatalistically. The basic purpose of welfare and development policies is presumably to place a limit on the size of the gap and to establish the levels below which society will not permit the poor to slide. Unfortunately, an unawareness of class biases at the national level and of general Western biases at the international level make many well-meaning policies counterproductive. In addition, many national and international programmes that are basically exploitative in nature are cloaked in the morally respectable rhetoric of welfare or development.

In terms of post-World War II development approaches, two broad trends have been visible. First, there was the expectation that technical assistance and capital financing for industrial projects would be sufficient to launch the poor countries on the paths of modernization. This approach soon ran into serious cultural barriers—though it took development economists many years to recognize this. While the goals for the Second UN Development Decade include references to the need to deal with population pressures, to consider the employment consequences of new technologies, and to pay attention to pollution, the major thrust is still towards industrialization as the best approach to development. The development decades would appear to be a clear embodiment of the attempt to project Western values and technologies upon the rest of the world.[13] And if the ecologists are correct regarding the degree of change that will be required within the

All economic forces operate to promote and hasten annihilation; none operate against it.
KENNETH E. F. WATT

Western world, the development decades will tend to result in the promotion of obsolete approaches and technologies—that is to say, the West will be exporting its most anti-ecological products to the rest of the world at a time when it is changing its own approaches.

The second major trend, which is now gaining momentum, is the shift in thinking regarding agriculture that has occurred with the so-called green revolution. The biological idea behind the development of the new varieties of wheat and rice was that if tropical agriculture was to be improved, seeds specifically adapted to tropical conditions would have to be developed. Unfortunately, this partial ecological insight was vitiated by specialized and technological thinking.

Specialization led to a neglect of any consideration of the social and economic dimensions of peasant farming. Technological thinking led to the typical reaction once better seeds were produced: that they should be promoted universally, and more importantly, that peasant cultures should change—or be changed—to meet the requirements of this new and "superior" technology. Peasants are thus expected to change cropping patterns, invest any capital they have in tube wells, fertilizers, and pesticides, and generally shift from local, barter-oriented cultures to larger, market-based systems. In short, they are expected to shift from what in most cases is an ecologically sound traditional agriculture to a form of modern industrial agriculture—with all its attendant ecological costs.

There are several ways of trying to estimate the ecological costs of such a shift. One would be to estimate the systematic impact of introducing high levels of both artificial fertilizers and pesticides. This has been attempted in at least three major—and quite different—surveys.[14] Each of these makes clear in its own way that industrial agriculture in the West is overdue for a thorough re-thinking and re-structuring. Another approach is to analyse what such a shift means in terms of global energy reserves (recognizing that our terrestrial stock of low entropy resources is limited):

> . . . the ultimate and the most important result is a shift of the low entropy input from the solar to the terrestrial source. The ox or the water buffalo—which derive their mechanical power from the solar radiation caught by chlorophyll in photosynthesis—is replaced by the tractor—which is produced and operated with the aid of terrestrial low entropy. And the same goes for the shift from manure to artificial fertilizers. The upshot is that the mechanization of agriculture is a solution which, though inevitable in the present impasse, is anti-economical in the long run.[15]

A third approach is to consider the risks involved in planting huge acreages of crops of the same genetic stock. The spatial extension of single crops over large areas reduces natural variety and protection mechanisms—thus greatly increasing the risk from pests and plant diseases. Using crops with the same genetic base (hybrids) risks losing a large part of the crop if some new crop disease or mutation appears to which the crop is susceptible. This occurred recently in the United States when one-fifth of the maize crop was lost to a new variety of maize blight. Historically, the most disastrous example of dependence upon a single crop was the Irish potato famine.[16]

A fundamental misconception regarding the green revolution that clouds most discussions relating it to development is that it is seen as the only approach which offers sufficient increased production to offer a "breathing space" in which to get population under control: an approach which—so the reasoning goes—must be adopted even with its great ecological risks. In fact, there are a number of alternative strategies, many of them much more ecologically sound, which offer the prospect of even *greater net production* than the green revolution. Recent research has suggested two important points: (a) local improved varieties, for example in Iran, give better results than the "miracle seeds" if they are given the same inputs;[17] (b) net production per unit of land on small farms—whether using traditional or new varieties—is greater than the equivalent crop (on similar soil) on large farms.[18] If this is so, one might then ask why the green revolution has been adopted rather than other approaches. One main reason is that in the real world of peasant farming the "miracle seeds" are clearly "landlord biased." That is to say, only the large landlords have the various resources—information, capital and/or borrowing power, and political power—to successfully adopt the new seeds and all of their requirements.[19]

Another main reason, which combines with the first, is that given the fact that in most developing countries a small number of people own most of the agricultural land, large increases in production result when these few landowners adopt a new technique that is more productive. What is argued here—and can be seen in practice to a large extent in Taiwan—is that the same amount of land redistributed among a large number of small farmers would be much more productive.[20] Equally, such an approach would be labour intensive—drawing upon one easily available resource of the developing countries—and would not involve the degree of social and economic disruption that has occurred as the green revolution has increased gaps in wealth, led to greater unemployment, and often reduced the local high protein crops the poor depend upon.

It is instructive to consider the sort of agricultural research that would be required—in addition to land redistribution—to make peasant agriculture more productive while maintaining its ecological soundness. The research would have to be self-consciously designed to be "peasant biased," that is to say, it would have to consider how to improve local food crops, how to improve varieties that are grown on rain fed, rather than irrigated land, and how to reduce the losses to local pests, etc.[21] Such research would involve a massive research investment in each country—because to be ecologically sound, new varieties, husbandry practices, protective measures, etc, would all have to be adapted to the local ecological and cultural conditions. And, of course, the research would have to be interdisciplinary to include the social as well as the physical dimensions of agriculture.

Such a research and aid programme—which would require a major shift in attitudes on the part of both government and international agencies—

would tend to reduce another major risk in the green revolution. That is, that as the multinational corporations become more involved in the commercial aspects of the green revolution—the selling of seeds, fertilizers, pesticides, tractors, irrigation equipment, etc.—they will tend to want to standardize (and limit) the kinds of seed available, the number of different mixtures of fertilizers, the size of tractors, and so on. What this really means is that the peasant becomes much more dependent upon outsiders. Decisions as to what the appropriate seeds and fertilizers for him will be are more likely to be made on the basis of ease of production and commercial profit than upon ecological grounds that relate to his region and his farm. At the same time the new cultivation practices force him to abandon many traditional ways of dealing with pests, etc.—with the clear danger that his localized knowledge will soon be lost. These points show the great importance of including the actions of private corporations and investors when considering the "developmental" impact of Western agencies upon non-Western peoples.

CONCLUSIONS

The above discussion hopefully makes clear thc need for new conceptual and organizational approaches to understanding and dealing with "developmental" and environmental dilemmas. Conceptually, three easily stated, but most difficult to realize changes are needed. First, there is a need for much less "universal" thinking and categorizing (which as mentioned often contains large admixtures of Western values) and much more "contextual" thinking—thinking that relates to specific groups in their real natural and historical environments. Next, and related to the first point, is the need to recognize that technologies are not neutral, but reflect (as well as shape) the values and environments in which they were developed. Once this is recognized, then perhaps we can go about the more important task of learning how one *adapts* technologies to fit different cultures and environments (this would suggest that we should ban the phrase "transfer of technology"—since such "transfer" can only occur through imposition). Finally, it is clear that conventional economic theories and measures are inadequate, if not positively misleading, in developing ecologically sound policies—which as has been stressed, often are more productive in real economic and social terms. When all the above points are taken into account, much of the supposed "conflict" between environmental and developmental goals disappears.

The sorts of organizational changes which are suggested include the following. To the degree that international development programmes are encouraged, they should be conceived in terms of counteracting and placing limits on the natural tendency for resources and information to be extracted by the richer, more organized subsystems from the poorer, less organized ones. This, of course, would require a thorough re-thinking of current approaches to aid, trade, and the activities of multi-national corporations. In terms of specific projects there would appear to be a general need to research and develop low-level, peasant-biased technologies that are adapted to the particular culture and environment concerned. While some of this might be done on a fairly broad basis, there is also a great need to have specific projects—such as dams, canal systems, etc.—planned from the beginning by interdisciplinary teams that include some of real participation by the various groups affected by the project. The fact that such planning and participation is seldom realized (and usually feared) even in most Western countries suggests both the arrogance of most "experts" and the great distance we have to go to learn how to adapt and humanize our own technologies, much less reduce their harmful impacts on other cultures.

NOTES

1 P. Shepard and D. McKinley (eds.), *The Subversive Science: Essays Toward an Ecology of Man,* Boston: Houghton Mifflin, 1969.

2 The parallels are not complete, for the ecologists represent an *internal* challenge, rather than one with external origins. Also, the goals are rather different: the economists suggesting quantitative goals relating to material progress, the ecologists stressing qualitative goals—which, however, must be consistent with, as well as encourage, survival and adaptation.

3 In these terms, the USSR and most of Eastern Europe are included as Western countries. For a full discussion of the ways in which Western values are intermixed with modern science and technology, see Kenneth A. Dahlberg, "The Technological Ethic and the Spirit of International Relations," *International Studies Quarterly,* Vol. 17, No. 1 (March 1973), pp. 55–88.

4 They are fundamentally anti-ecological in the following ways: first, there is no attempt to conceptually link the economic system with real natural processes, say for example, energy transfers; next, economic theorizing is usually a-historical, attempting to ignore the evolution and real space-time context of both natural and social systems; finally, current economic thought contains a large number of value assumptions—especially those relating to growth—that are anti-ecological. For a detailed discussion and critique, see Nicholas Georgescu-Roegen, *The Entropy Law and the Economic Process,* Cambridge: Harvard University Press, 1971.

5 Elizabeth Whitcombe, *Agrarian Conditions in Northern India: The United Provinces under British Rule 1860–1900,* London: University of California Press, 1972.

6 For perhaps the most complete catalogue to date, see F. T. Farvar and J. P. Milton (eds.), *The Careless Technology,* Garden City, N.Y.: Natural History Press, 1972.

7 Whitcombe, *op. cit.,* p. 11.

8 *Ibid.,* pp. 93–94.

9 F. T. Farvar, mimeo. paper presented to the AAAS Meeting, December 1971.

10 For a discussion of these and other health problems, both in Egypt and especially in the Mekong basin, see J. P. Milton, "Pollution, Public Health and Nutritional Effects of Mekong Basin Hydro-Development," US AID mimeo. report (n.d.).

11 Ramon Margalef, *Perspectives in Ecological Theory,* Chicago: University of Chicago Press, 1968.

12 *Ibid.,* pp. 16–17.

13 "The assumption that development is a *generalizable* concept must be seen in this context. It is far more potent than the crude instruments of 'neo-colonialism.' It is the last and brilliant effort of the white northern world to maintain its cultural dominance in perpetuity, against history, by the pretence that there is no alternative." Address by John White, "What Is Development and for Whom?," given at the Quaker Conference on "Motive Force in Development," April 1971, p. 4. (My italics.)

14 W. H. Mathews, F. E. Smith, and E. D. Goldberg (eds.), *Man's Impact on Terrestrial and Oceanic Ecosystems,* London: MIT Press, 1971; The Institute of Ecology, *Man in the Living Environment,* Madison: University of Wisconsin Press, 1971; and D. H. Meadows *et al.*, *The Limits to Growth,* New York: Universe Books, 1972.

15 Nicholas Georgescu-Roegen, "Economics and Entropy," *The Ecologist,* Vol. 2, No. 7 (July 1972), p. 17. Another dimension of this shift is that it involves moving from a highly localized form of production where there are high levels of re-cycling, to a process where many elements—particularly the fertilizers—are imported from outside, put through the crop, and "exported" to rivers or ground waters, thus increasing the risks of eutrophication or contamination.

16 Between 1700 and 1846, Ireland shifted from a grain-based agriculture to a potato-based one, with the population rising from 2 million to 8 million. During the years of famine, some 2 million starved to death and another 2 million emigrated. The inappropriateness of conventional scientific thinking at the time, the insensitivity of the British Government to the realities of the Irish situation and the actual aggravation of the crisis caused by bureaucratic centralization of relief measures and the dominance of *laissez-faire* economic thinking are all portrayed in detail in Cecil Woodham-Smith's *The Great Hunger,* London: Hamish Hamilton, 1962.

17 Ingrid Palmer, *Science and Agricultural Production,* UNRISD Report No. 72.8, Geneva, 1972, pp. 6–7. This would reduce genetic risks, if not those related to the use of irrigation and artificial fertilizers and pesticides.

18 Keith Griffin, *The Green Revolution: An Economic Analysis,* UNRISD Report No. 72.6, Geneva, 1972, pp. 31–38.

19 See *ibid.,* pp. 46–48, for full discussion.

20 What often makes the large farms appear more productive than they are is the tendency to measure productivity in terms of production per man hour or in capital terms rather than in terms of crop productivity per land unit.

21 Given the fact that crop losses in the tropics average 25–45% of the crop (from the field through milling), Ingrid Palmer, *op. cit.,* p. 72, asks what might have been the result had research since World War II been directed at reducing these losses, rather than concentrating on increased production through hybrid seeds.

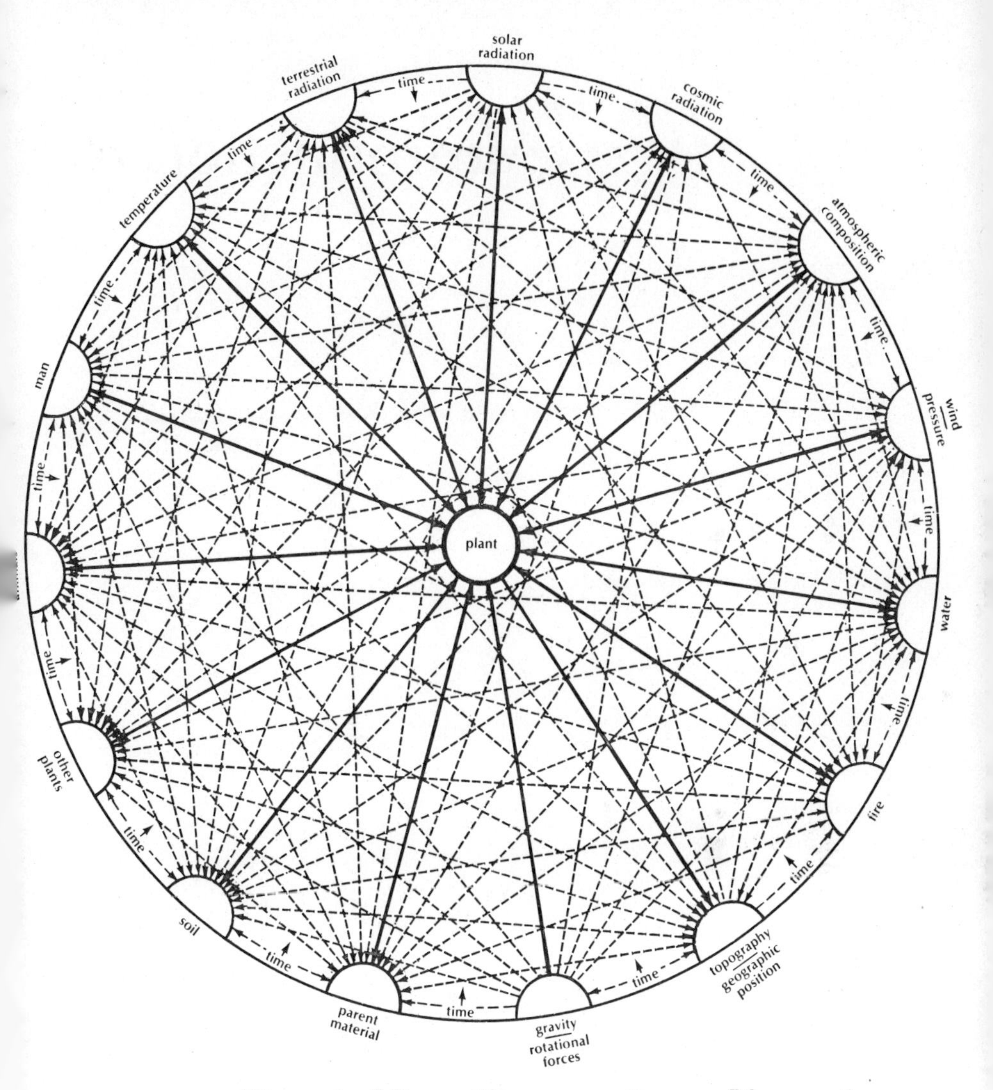

Part II Case Studies in Environmental Disruption and Eco-Management: A Global Survey

Introduction to

Part II

Throughout most of the world the prevailing ideologies, be they political, cultural, or economic, are wholly inconsistent with ecological realities. Even in North America and Europe, where an ecological orientation is at least perceptible, the rudimentary principles of environmental protection are understood and responded to by only a few. Many of the developing countries, cued by the industrialized nations, seem dedicated to unbridled economic development whatever the ecological costs. The developed world is agonizingly shedding a few of its cherished notions about growth and progress, and the developing world is picking up the discards. It is unreasonable to expect the Third World peoples to be disabused of these notions while the industrial behemoths continue their unecological rampage.

Each country and each continent has countless and singular environmental problems. When these are viewed from a global perspective, the complexity and immensity of the task of biospheric protection are staggering to contemplate. The task would be gigantic even if there were a steady flow of good will and hearty agreement among nations.

Science and technology have only recently been applied to the protection rather than merely the exploitation of the environment. Yet, as the selections in this part illustrate, the technical problems, formidable as they are, are overshadowed by the social, political, and psychological barriers to the protection of the biosphere. So great are these that, as Lynton Caldwell sees it, "What is required of the peoples and nations of the Earth is that they transcend their own histories."[1]

What is particularly nagging to political leaders the world over is that an environmental crisis will not respond to the usual panaceas. It will not respond to what Kenneth Keating has called "the Washington reflex": you discover a problem, throw money at it, and hope that somehow it will go away. Nor will it be placated by good intentions, hope, subterfuge, short-term fixes, or political rhetoric. It will respond only to the last thing that most people seem willing to do: change their aspirations, priorities, and values to accommodate the ecological facts of life.

Among the many impediments to controlling the causes of environmental deterioration in any nation is the tunnel vision of many in the professional world and in government agencies. Erik Eckholm observes that "when reading the analyses of economists, foresters, engineers, agronomists, and ecologists, it is sometimes hard to believe that all are attempting to describe the same country."[2]

Admittedly, all this sounds unrelievedly pessimistic. But, as somebody once said, "Pessimism has no survival value"—except to point out the problems. And as the selections in this part show, not *all* the news is bad.

Two selections are devoted to the United States, a nation in which the bulk of goods and services contribute to the exhaustion of the life-support system; a nation in which each citizen, on the average, accounts for more toxic wastes discharged into rivers and oceans than 1,000 Asians; a nation whose most unappreciated inhabitants are the desperately poor who, with their outside privies and their inability to wallow in wastefulness, lead the most ecologically sound lives. In Selection 9, Jon Margolis breezily reveals the underpinnings of the American environmental movement. In Selection 10 John Mitchell laces his lyricism with more than a dollop of cynicism as he tells of the inroads the insatiable proponents of development are making in south Florida, where a wilderness like none other on earth lies in jeopardy.

The Canadian Arctic, which is rich in resources but poor in species diversity, is the subject of Selection 11. As the 1970s began, man had wrought more changes there in five years than at any time since life on earth began. M. J. Dunbar, a Canadian, describes the ecology of this unique area and ponders the impact that large-scale development and settlement could have not only on its biota but on its people, the Indians and Eskimos, who share an uncommon reverence for life and the land. As another Canadian has asked, "If Canada—an industrialized and affluent country with a high-material standard of living and natural resources far in excess of immediate needs—cannot 'afford' to slow down development in one of its frontier areas, what country on earth can?"[3]

Latin America, whose population growth is the most rapid in the world, is a continent increasingly burdened with ecological disasters and dilemmas (see Selection 3). In Selection 12, Gerardo Budowski reviews the developmental history of Middle America, emphasizing the incalculable aesthetic, material, and scientific values that have been lost through mismanaged development. His outline of a new approach to land use in wet forest areas focuses on forestry as the most promising means of benefiting man and ensuring long-term productivity of these ecosystems. Using a Costa Rican example, Darryl Cole in Selection 13 sadly shatters the myth of the fertility of virgin tropical lands.

Selection 14 is a hard-hitting report on Brazil, whose government is calling for unimpeded expansion of industry and population alike. Remarkably unattuned to ecological considerations, Brazil is cutting its rainforest (see Selection 5)—the largest remaining forest in the world—at breakneck speed. The author warns of the immense harm Brazil could do to both its own and the global environment if its plan of action is fulfilled.

Darwin's "treasure islands," the Galápagos, are the exposed tips of submarine volcanoes that erupted about a million years ago. The Bishop of Panama, who first discovered them in 1535 (300 years before Darwin visited them), wrote that they looked "as though God had caused it to rain stones." It is this forbidding landscape, plus the almost complete lack of fresh-water sources, that has so long protected the archipelago from widespread and ruthless exploitation. Scattered over 150 square miles, the islands offer a unique "natural laboratory of ecology and evolution," and in Selection 15 Charles Gilbert discusses their vulnerability, the unsavory history of human impact, and the urgent need for and movement toward conservation. The government of Ecuador today is wisely following a policy of *un*development in order to "develop" a natural ecosystem.

In the most highly industrialized countries of Western Europe, some environmental problems have become so insufferable that they are finally, if ponderously, being tackled; in the less industrialized nations, the citizens tend to regard the spread of ecological abominations as unpleasant but necessary side effects of development. Selection 16 reviews the past and recent destruction of the West European terrain, the present comprehensive planning of the landscape, and the results of rehabilitation efforts. S. J. Holt, in Selection 17, discusses the myriad sources of ecological damage to the Mediterranean Sea, and offers a generally encouraging account of what can be and is being done to save it.

Thanks to recent international cooperation, there is hope, too, for the Baltic Sea, as Selection 18 indicates. One of the most polluted and most investigated areas of the world, the Baltic has become a stagnant sea. Fifteen times larger than Lake Erie—another notorious industrial-urban cesspool, which is gradually being restored to health—the Baltic is plagued with a pollution threat much like that which the U.S. lake has endured.[4]

Selection 19 surveys the rapid pace of environmental disruption in East-Central Europe and notes the general insouciance with which East Europeans regard ecological hazards. As John Kramer shows in Selection 20, the U.S.S.R. has environmental problems that vie in severity and extensiveness with those of the United States, and there is a sharp divergence of Soviet theory and practice in the realm of environmental protection. In Selection 21, Rudolph Chelminski reports on Lake Baikal less dourly than Kramer does, pointing out how, paradoxically, a cellulose plant may have saved the extraordinary lake from the clutches of industrial developers.* It should not be surprising that the United States and the U.S.S.R. have similar ecological crises, for both are devoted to maximizing production and resource exploitation. In fact, there is an ecological détente between the two nations: a recent U.S.-Soviet Environmental Agreement encompasses 11 major areas for collaboration, including the prevention of air-and-water pollution, the study of the biological effects of pollution, and the prediction of earthquakes.[5]

Japan is the world's foremost example of how a nation hell-bent on industrial development can turn an economic dream into an ecological nightmare. The authors of Selection 22 first explain why Japan, whose 110 million people are crammed onto a small and exquisite archipelago, may represent the nearest approach yet seen to a human-ecological climax situation, and then discuss the primary geographical concentration of industry, the urban congestion, pollution and its hideous effects, and the despoliation of scenery. Only recently have the Japanese begun to reconsider a Western-style future, and in their disenchantment many are now desperately attempting to reclaim their unique cultural legacy, which embodies an indifference to material progress and a reverence for nature.[6]

The Chinese have no word for "ecology" in common usage but they have shown a respect for the principles of ecology that is probably unparalleled in any other major nation. In striking contrast with the Western notion of man's need to conquer nature or to "straighten it out," China has a long tradition of living in harmony with nature and of honoring man's "duties" toward the earth. Still somewhat under the sway of this tradition, Communist China tends

---

*Since the U.S.S.R. spans both Europe and Asia, Selection 20, which deals primarily with the heavily industrialized European area, has been placed in the "Europe" section, while Selection 21, which concerns only Lake Baikal, appears under "Asia."

to formulate environmental policies as part of the social and economic development of the country, an alternative to Western methods that other developing countries might contemplate. The Chinese environment has not been victimized by a value system that condones waste as salutary in any way; nonwaste is therefore not regarded as simply a necessity but as a virtue. Leo Orleans, in his comprehensive report on China's environmental policies and programs (Selection 23), does not picture China as an environmental paradise. But he shows why China—through a combination of wisdom and luck—clearly has good reason to be proud of its efforts in fighting human and industrial pollution.

The monstrous ecological effects of America's assault on the land of Vietnam—the most massive warfare against the environment in the history of the world—are analyzed by Malcolm Somerville in Selection 24. Somerville's article was written while the war was still raging, so the devastation wrought by the end was, of course, greater than he could indicate. Between 1965 and 1973 the total weight of U.S. high-explosive munitions used in South Vietnam was more than 7 million tons, or the equivalent of one Hiroshima-sized bomb every five days. Some 100 million pounds of assorted herbicides were sprayed over 5 to 6 million acres, an area roughly the size of Massachusetts. Thousands upon thousands of acres of forest land, fruit orchards, and fields were cleared by the Rome plows. The long-term effects of all this on the land and its inhabitants remain unknown. Perhaps the land will prove as resilient as its people. Meanwhile the principle of environmental warfare lives on. The U.S. military-appropriations requests for the fiscal year 1974 included an item on the financing of research into ways of making military herbicides even faster-acting; and the feasibility of geophysical weapons and techniques of weather modification is being seriously considered.[7] "Military intelligence," as Groucho Marx once remarked, "is a contradiction in terms."

The millions of wretchedly poor in India are not concerned about pollution or Malthusian theories or national parks. Prime Minister Indira Gandhi, though she spoke for these millions, could have been speaking for billions of other people when she asked, "How can we speak to those who live in villages and in slums about keeping the oceans, the rivers and the air clean when their own lives are contaminated at the source? Are not poverty and need the greatest polluters?"[8]

The population of India represents over one-seventh of the human race and, at the present rate of growth, could soar to over 1 billion by the year 2000. An Indian diplomat once commented that whereas the national sport of the

United States is baseball, the national sport of India is sex. Perhaps apocryphal, this statement does reflect an underlying cause of high procreation in India, namely, the tenacious tradition of having "an heir and a spare," or at least two sons. To provide enough food for India's enormous population is a well-nigh impossible task; in fact, the country was for a time receiving one-third of its food supply from the United States—until U.S. policymakers declared "farm surpluses" a thing of the past. Yet even today, rats, which outnumber humans by a ratio of 8 to 1, eat or destroy nearly half the Indian grain consumed each year—about 100 million tons. Moreover, in its urge to industrialize, India has tended to neglect its agricultural and population problems, and recently even resorted to setting aside areas where famine was officially expunged by a fiat denying its existence. In Selection 25, M. S. Swaminathan earnestly examines the agricultural assets and liabilities of his country. While pointing out India's considerable agricultural capability, he acknowledges that the improper use and management of resources compounds the problems of population growth in a country that is somehow going to have to produce "more and more food from less and less land."

Economic need, unsound land use, and destructive nationalism constantly weave a pattern of death for wildernesses throughout most of the world. Selection 26 illustrates in three examples—the Great Thar Desert, Kashmir, and the Middle East—how a living wilderness can be irresponsibly and rapidly turned into a desert. In a variation on this theme, Selection 27 suggests methods of combating environmental degradation in vast areas of the Near and Middle East that, arid by nature, threaten to become wastelands.

The ecology of development in Africa is the subject of two selections. In Selection 28, ecologist D. F. Owen, focusing on tropical Africa, explains why he believes that the prospects for economic development there are depressing. Not only is sustained economic growth inhibited by a phenomenally rapid rise in population but it is further impeded by Africa's role as a producer of raw materials for consumption mainly in the developed countries. Owen also sees little value in current aid programs or the advice of outside experts (see also Selection 8). An important point he makes elsewhere is that "with the possible exception of man himself, insects are by far the biggest overall threat to economic development in tropical Africa." This is borne out by a 1976 report that in Kenya an estimated 75 percent of the crops is lost to insects, and in Tanzania insects destroy 25 percent of the country's crops after harvesting.

In Selection 29, geographer Simeon Ominde discusses ecodevelopment in the whole of Africa, analyzing the obstacles to, and possibilities for, integrating economic and social development with a rational management of the environment. Like Owen he emphasizes the predicaments engendered by the population boom, but he also examines the tremendous variety of other ecological problems and the varied strategies needed for integrated development in Africa's distinctive urban and rural settings. (See also Selection 3.)

In Selection 30, Anthony Netboy describes the wondrous animals, their conservation, and the impact of tourism in the world's last great treasury of wildlife—the national parks and reserves of East Africa. Although Netboy sanctions the cropping of surplus numbers, this recourse is opposed by some. For instance, ranchers would like to raise certain game animals (such as the buffalo, oryx, and eland) just like domestic animals, thereby controlling their numbers, producing meat and hides, and eliminating the widespread and highly lucrative practice of poaching. Some conservationists oppose both cropping and "natural" elimination (allowing the animals to destroy their own food supply and habitat) and prefer opening up more land to the animals to forestall their extermination.[9] On the other hand, many Africans believe that the size of the parks should be reduced in order to provide more farm and grazing land. Peter Beard, a passionate champion of the elephant, despairs for its future (as well as that of most wildlife), convinced that any conservation measures taken now will be too late, too misguided to do any good, and merely a way of easing our guilt. "We are killing them with kindness," he says, and they are dying excruciating deaths because of insufficient food for the large populations in the parks; even some young elephants have advanced states of heart disease apparently from the stress of competing for food.[10]

At the turn of the century artist Paul Gauguin, writing of Papeete in Tahiti, complained about the "absurdities of civilization" that had infiltrated the town. Oceania has long been a victim of these "absurdities" and of man's biological meddling (see Selection 7). For the most part the islands are hardly models of ecological excellence today, and their fast-growing populations promise future crises. But the peoples of the Pacific region are gradually making restitution for many of the ecological calamities of the past. For instance, the eight-nation South Pacific Commission reviews the ecological problems of 4.5 million people on the islands garlanding more than 7 million square miles of ocean; New Zealand has clean hydroelectric power, is not overpopulated, and maintains fine national parks, while Australia has saved its kangaroos and is working to combat pollution of every type.[11] Selection 31 presents an eco-development plan for Nissan Island, Bougainville, New Guinea; it was written by the two men who were commissioned by an enlightened government to formulate the plan.

As they point out, the problems of Nissan epitomize the problems not only of Pacific islanders but of peoples all over the world. The plan outlined here could well serve as a model for working with nature and encouraging self-reliance.

The last selection in this part is devoted to Antarctica, "the last continent"—an awesome ice sheet that covers five and one-half million square miles, averages more than 7,000 feet in thickness, and contains over 90 percent of the world's snow and ice. In this coldest place on earth are forms of life that have adapted to the most severe conditions, for they exist in the middle of an ice age. Antarctica has been called the most spectacular natural phenomenon on our planet. Spectacular, too, is that it is the focus of a unique international effort. Research work on the continent spearheaded a 12-nation treaty in 1961 to preserve the continent for scientific and other peaceful uses. The subsequent Agreed Measures on Conservation (not to mention Antarctica's exceedingly inhospitable clime) offer hope that on one continent, at least, ecological insults may not be inevitable and may, in fact, be forestalled. In Selection 32, M. W. Holdgate discusses Antarctic conservation objectives, the state of the continent's ecosystems, the marine ecosystems, conservation measures in Antarctic and subantarctic land areas, and the conservation of the southern seas.

Dare we hope that the international harmony and the wise management attained in Antarctica will one day generously grace other parts of the planet?

NOTES

1 Lynton K. Caldwell, *In Defense of Earth: International Protection of the Biosphere* (Bloomington: Indiana University Press, 1972), p. 230.

2 Erik P. Eckholm, "Losing Ground: Impending Ecological Disaster," *The Humanist,* March/April 1976, p. 7.

3 James Woodford, *The Violated Vision: The Rape of Canada's North* (Toronto: McClelland & Stewart, 1972), p. 118. For a look at what is happening to a vast subarctic region of northern Quebec and its inhabitants, see Boyce Richardson, *Strangers Devour the Land* (New York: Alfred A. Knopf, 1975).

4 For an excellent analysis of the stagnation of the Baltic, see Stig H. Fonselius, "Stagnant Sea," *Environment,* July/August 1970, pp. 2–11, 40–48. For more details about Lake Erie, see W. T. Edmondson, "Fresh Water Pollution," in William W. Murdoch, ed., *Environment: Resources, Pollution & Society* (Stamford, Conn.: Sinauer Associates, 1971), esp. pp. 224–26; and Alfred M. Beeton, "Eutrophication of the Great Lakes," *Limnology and Oceanography,* 10(2), 1965, pp. 240–54.

5 See also Marshall I. Goldman, "The Convergence of Environmental Disruption," *Science,* 170 (October 2, 1970), pp. 37–42; Marshall I. Goldman, *The Spoils of Progress:*

Environmental Pollution in the Soviet Union (Cambridge, Mass.: MIT Press, 1972); and Philip R. Pryde, *Conservation in the Soviet Union* (New York: Cambridge University Press, 1972).

6 A detailed analysis of Japan's ecological plight is given in Norie Huddle and Michael Reich with Nahum Stiskin, *Island of Dreams: Environmental Crisis in Japan* (New York and Tokyo: Autumn Press, 1972).

7 See also J. B. Nielands et al., *Harvest of Death: Chemical Warfare in Vietnam and Cambodia* (New York: Free Press, 1972), esp. pp. 150–75; J. S. Bethel et al., "Military Defoliation of Vietnam Forests," *American Forests,* January 1975, pp. 27–30, 60–61; and Frank Barnaby, "Towards Environmental Warfare," *New Scientist,* January 1, 1976, pp. 6–8.

8 Quoted in Philip W. Quigg, "Environment: The Global Issues," *Headline Series,* no. 217 (October 1973), p. 10.

9 See Iain and Oria Douglas-Hamilton, *Among the Elephants* (New York: Viking Press, 1975), for the story of the compassionate research that led to this conservation method for the elephants of Manyara National Park in Tanzania. And for an excellent account of wildlife conservation on Tanzania's Serengeti Plains, see Harold T. P. Hayes, "A Reporter at Large: The Last Place," *The New Yorker,* December 6, 1976, pp. 52–133.

10 Wilborn Hampton, "African Dilemma: Reason an Endangered Species," *Los Angeles Times,* January 25, 1976, pt. V, p. 17.

11 See Mary and Laurance S. Rockefeller, "Problems in Paradise," *National Geographic,* December 1974, pp. 782–93.

NORTH AMERICA

9 Our Country 'Tis of Thee, Land of Ecology

Jon Margolis

Jon Margolis is a journalist.

Senator Ted Stevens first noticed them last September. Stevens is an Alaska Republican; as such he supports the right of every able man to get rich by building things. When he saw late last summer that this faith too had its heretics, Stevens could not keep silence. "All of a sudden all these conservationists are coming out of the woodwork to tell us how to save Alaska," he said. He did not approve.

The statement did wonders for Stevens politically. Alaskans don't like conservationists. Nearly everyone else does, though, and in the past few years they have been coming out of the woodwork more often. Going right back in, too, and staying out of sight once they have made their point. Conservation may be the first revolution led by unknowns.

And make no mistake, it is a revolution, very likely the most contentious and the most important of the rest of the century. The radicals know this now; the politics of confrontation has been used on conservation disputes, and the politics of ecology is now a frequent term in the underground press. If only the conservationists knew it. Conservation is in revolt against its own past as much as against the country's present. Like the unfettered technology it fears, the conservation movement may contain within itself the seeds of its own destruction.

There is little doubt that conservation is quite the vogue. Hardly a county lacks some sort of citizens' conservation group. Usually, they are formed not to protect the environment in general but to battle a specific threat to it, often a threat close to home. With increasing frequency, the battles are being won. Consolidated Edison has not been able to build its power plant on Storm King Mountain in New York State; Dade County and the U.S. Transportation Department may yet be prevented from building the world's largest jetport in the Big Cypress Swamp of Florida, and there are more nuclear power plants on the drawing boards than will ever be on the waterside killing fish.

So successful have the protectors been that the developers have begun to fight back. In 1967, naturalists found a covey of ivory-billed woodpeckers, then considered extinct, in the heart of the Texas Big Thicket. But they didn't say precisely where for fear that timber and real-estate interests would kill the lovely birds. The theory is that the fewer natural wonders there are in any plot of land, the less reason Congress will have to preserve it. Similar tactics were used by the California lumber companies which tried to cut down the best of the redwoods while a bill to protect them was being debated.

Not that business is unaware of the growing public concern with conservation. Advertising and public-relations departments across the country are hard at work telling us how much corporations care about the environment they are befouling. In full-color, two-page ads in fancy magazines, oil companies inform how the fish actually like detonations for underwater wells, or that their brand of gasoline will pollute the air just a bit less. Magazines and television networks are devoting unprecedented time and space to conservation, and even newspapers, usually the last to know what the public really cares about, are beginning to cover the subject. Possibly they noticed the appointment of a hard-line law-and-order attorney general did not create nearly as much fuss as the naming of an interior secretary who dared deride conservation for its own sake.

Conservation crosses party lines and even ideologies. John Birchers and Communists, *Ramparts* and *Reader's Digest,* The New York *Times* and the Los Angeles *Free Press,* Max Lerner and James Kilpatrick, Barry Goldwater and George McGovern. This may not last, for the New Conservationists are about to enter the political arena full force, perhaps even borrowing from older revolutions such tactics as the sit-in, the boycott, and, who knows, can the Molotov cocktail be far behind? This will make some enemies. Even now, not everyone is a conservationist, though opposition also is bipartisan. Ronald Reagan, Richard Daley and Nelson Rockefeller are among those uncommitted to the cause.

Nonetheless there are places where a man can get elected by being a good conservationist. Westchester County, New York is full of old-fashioned Republicans, and Richard Ottinger is a liberal young Democrat. But because he has pledged to save the Hudson River, those Republicans keep sending him to Congress. Richard McCarthy, another liberal New York congressman, comes from a working-class backlash district in Buffalo. When he goes home he doesn't talk about race and welfare; he talks about how dirty Lake Erie is. Union men out collecting money for the Committee on Political Education likewise have learned to get a dollar out of right-wing workers by telling them the money will support candidates committed to clean streams and lakes. Some local unions have even taken to opposing industrial expansion which would increase their membership. The current members, lacking the money to move to the suburbs, have begun to notice something foul in the air they breathe, and they don't want it any dirtier. Rich or poor, most people are lining up on the conservationist side of any given dispute.

Yet few of these newly enlisted troops know who their generals are, or even their sergeants. Nearly everyone knows which senators are prominently for or against the war, civil rights, and unions. But how many know that were it not for Gaylord Nelson of Wisconsin every drop of fresh water in America would be too sudsy to drink? Nelson, a liberal, is the leading conservationist in the Senate. John Saylor of Pennsylvania, a friend of the American Medical Association and the Chamber of Commerce, is the leading conservationist in the House.

One reason conservationists remain unknown while their cause is embraced is that many of them are Westerners, while the people who make people famous are in the East, meaning of Northeast. Conservation, after all, means the wise use or preservation of natural resources. There are young women in New York who talk about conservation every evening but aren't quite sure what a natural resource is. Of course the East is very conservation-conscious now, but it is a recent concern. Westerners have been involved in conservation for years, especially the kind which subsidizes farmers, ranchers, miners, and lumber entrepreneurs, which is certainly not what the women in New York have in mind. The typical informed Westerner, for instance, may know that Boyd L. Rasmussen is director of the Bureau of Land Management. The typical informed Easterner has never heard of the Bureau, though it controls twenty percent of his country's land.

Yet East and West alike are probably ignorant of such names as Dan Poole, Stewart Brandborg, Elvis Stahr, or Paul M. Dunn. These are some of the men who are not in government but who are at the center of the Conservation Establishment, an establishment which has been left behind by its cause. Which is not necessarily bad at this point for both cause and establishment. If some of the young intellectual liberals who have lately embraced conservation found out what some of the leading conservationists thought about the world in general, they might go back to the peace or civil-rights movements.

As an organized movement, conservation still deals in specifics—getting a national park established, preserving a wild river or a species of bird. But the cause, the unorganized but ever more popular movement, has become holistic without quite realizing it. When conservation started in the last century, the basic concept was wise use. Later it was beauty. Now it is ecology, and if you doubt the potency of conservation consider that three years ago you probably didn't know what that word meant; now you surely do. In ecology, by definition, all of a nature is connected, all relates to everything else.

Yet if the North American Wildlife Foundation and the National Audubon Society spent all their time worrying about ecology, fewer woods and rivers would have been saved in the last few years. Sometime soon the New Conservationists must define their beloved ecological conscience and figure out how best to organize to put it to use. But the concept is too new and the organizations are in their infancy. Meanwhile the bulldozers roll on, and someone has to stop them, lest there be no life left to interrelate.

There are really two Conservation Establishments. The first, centered in Washington, is dominated by the "user" groups, or tools of the interests as children of another revolution might call them. It is made up of foresters, fishery managers, and state fish-and-game officials. Its friends in government are in such agencies as the Bureau of Commercial Fisheries or the Bureau of Reclamation. Its friend in Congress is Wayne Aspinall, head of the House Interior Committee, who makes sure that every time some wild land is set aside there remains a way for someone to make money from it. It was Aspinall who included in the Wilderness Bill an amendment allowing all mining until 1984, a loophole which may create in the northern Cascades a pit which can be seen from the moon.

This traditional approach to conservation is the one most often found in the Interior Department, whose task is less to preserve the environment than to subsidize nature's users. Interior's public relations are conservation-oriented—its annual report features colored pictures of natural wonders and inspirational words about saving them—but its budget is not. In fiscal 1968, half of the department's $1,800,000,000 went to "water resources," most of which was for "reclamation," meaning flood-control and irrigation projects which reclaim rivers from running free and reclaim salt marshes from being breeding grounds for fish and nesting spots for birds. Flood control is needed because people decided, before we knew as much as we do today, to live along flood plains. Actually, houses are still being built on flood plains, though everyone knows they will necessitate a new dam somewhere. But then, building homes is often the most profitable use for flood-plain land, and once people need flood control, the government is sure to provide it. Floods also occur downstream because of reckless deforestation upstream. Among their other values, trees hold water. When they are removed, the waters run into the streams in greater quantity than the stream beds can accommodate. The solution to that is to build a dam, upsetting the river basin's ecology, and possibly reducing the number of fish.

The second Conservation Establishment, the one everybody is talking about, is not really headquartered anywhere, and it isn't very established. Nor is it united. It consists of perhaps twenty national organizations, most of them as ineffective as they are unknown, and an uncountable number of local groups, a few of which are very tough indeed. Conservation has all the disadvantages of citizen-oriented politics. Most of the official leaders, especially of the national groups, are part-timers, people who do other tasks, usually well-paid ones, most of the time. For instance, Dr. Edgar Wayburn of the Sierra Club is a successful San Francisco physician; Warren M. Lemmon, the head of the Nature Conservancy, manages extensive California real-estate holdings; Robert Winthrop of the North American Wildlife Foundation is on the board of directors of the First National City Bank. The opening bankroll for the John Muir Institute, the educational and tax-exempt arm of Friends of the Earth, came from Robert O. Anderson, chairman of the board of Atlantic Richfield; and Laurance Rockefeller runs the American Conservation Association. It is not a poor-man's movement. The day-to-day operations are left to small professional staffs made up in large part of unknown men. The citizen leaders like it that way. When David Brower began to act as though he ran the Sierra Club, the paragons of San Francisco society who controlled his executive board got rid of him. And when he challenged them in an election among the club members he had recruited, they beat him badly.

Because the movement is so loosely organized, strong personalities can have an unusually strong effect. Allen Morgan is credited with making the Massachusetts Audubon Society perhaps the most effective conservation organization in the country, and the West Virginia Highlands Conservancy is a force in that little state primarily because of Grover C. Little. Local conservation groups usually start on an ad hoc level, coming into existence to fight one threat to the local environment, but they often stay around to look for more trouble. In Hawaii, a group called Save Diamond Head saved Diamond Head, and it's aiming now to save all of Hawaii. The women of Morris County, New Jersey, did not disband after they had stopped the airline industry from filling in the Great Swamp to make a jetport; now they are fighting power companies and the Army Engineers elsewhere in northern New Jersey. Nearby, the Scenic Hudson Preservation Conference formed to fight Con Ed's Storm King Plant, but when the fight ends there will be more. The Hudson is a big river.

Despite the increasingly political character of the movement, there remain several national conservation organizations which stay out of politics and lobbying in favor of more restricted tasks. The American Conservation Association gives away Laurance Rockefeller's money to select research projects. The Conservation Foundation studies public policy's effects on the environment. The North American Wildlife Foundation and the Wildlife Management Institute, really one outfit supported by the gun industry, have undertaken the research which has brought wildlife management a long way from the days of indiscriminate predator control and un-ecological stocking, though some of that still goes on. Perhaps the organization closest to the heart of the matter is the Nature Conservancy, which with a pittance from private sources does what the government won't do enough of—buy land valuable to nature and about to be lost to it. The Conservancy thus far has been able to save 150,000 acres of the American Earth.

But the Conservation Establishment has been taken over by the militants, some of them so militant they don't even call themselves conservationists, but preservationists. The militant preservationist is a radical. Well, about nature he's a radical; about people he may be quite conservative. This is not a new irony. Since the movement began before the turn of the century, it has been split between those who wanted to save and manage nature for people—all the people—and those who wanted to leave it just as it was for nobody at all, or perhaps for those with the time, money, and culture to appreciate true wilderness. At the beginning, this schism was personalized by Gifford Pinchot, head of the Forest Service under Teddy Roosevelt, and a populist, and John Muir, founder of the Sierra Club, and a mystic. They were friends for a while, camping out together in the Grand Canyon, but they became estranged, first over the fight about sheep grazing on the public lands, then over the Hetch Hetchy Dam, which violated Yosemite National Park but brought low-cost public power to San Francisco. The preservationists fought the dam bitterly, making common cause with the private-utility industry. Then or later, they seemed not at all embarrassed about the bedfellow.

Today, within the militant conservation movement, united opposition to dams can usually be counted on unless the dam will also create a big lake for motorboating and stocked fishing. Which it always will, the power companies and the Army Engineers and the Reclamation Bureau knowing they then have a good chance of splitting the recreationists—the "parks are for people" crowd and the sportsmen—from the preservationsists.

Within the Conservation Establishment, these arguments rage within as well as between organizations. There are Sierra Club members who think saving Mineral King Valley not worth the effort, and there are ardent preservationists in the National Wildlife Federation. But in general the recreationist cause is taken up by the hunters, who far outnumber everyone else in organized conservation. Theoretically, they are hunters and fishermen both, but fishermen per se have no clout. Fishing rods are cheap. The only militant and effective sportsmen's group is the National Wildlife Federation, which is run by the state councils of rod-and-gun clubs, thus representing local groups totaling about 2,500,000 people. Many of these members are not really conservationists, and the Federation gets most of its money from conservationists who are not members, but who subscribe to *National Wildlife Magazine* or purchase the wildlife stamps the Federation sends out without mentioning who controls the beneficiary. Nationally, the Federation is not as strong as some other conservation groups, but on the state level its councils are often the only working conservation lobby. Non-sportsman conservationists have only begun to lobby in state legislatures in California, Oregon, Maine, and New York, but the Wildlife Federation has close ties with legislators and state fish-and-game departments. At present, the Federation is supporting an anti-conservation proposal backed by the Western state fish-and-game bureaucracies which would give state departments total control over all wildlife in their states, a move which could open the national parks to hunting and effectively erase federal protection now given eagles and migratory birds.

If the recreationists are most likely to come from the sportsmen's groups, the extreme preservationists are most apt to be found in the Wilderness Society, a single-minded collection of 55,000 persons to whom a well-managed campground is as hideous as an oil refinery on a promontory would be to other conservationists. Wilderness Society members want parks which are not for people other than themselves. They do not seem motivated primarily by ecology, but speak of preserving the pristine for the sake of "our physical and spiritual regeneration." Among other effects, such statements give aid and comfort to the enemy, be they mining interests, the

Wilderness is a bench mark, a touchstone. In wilderness we can see where we have come from, where we are going, how far we've gone. In wilderness is the only unsullied earth sample of the forces generally at work in the universe.
KENNETH BOWER

Agricultural Department's Forest Service, which manages the national forests, wherein are most potential wilderness areas, or the power-boat manufacturers who wish every mountain stream to be turned into a huge lake courtesy of the Army Engineers.

Hard-core preservationists are also found in large numbers in the National Parks Association and Defenders of Wildlife. The former spends half its time fighting the National Park Service, which in recent years has suffered the delusion that it is a federation of highway departments commissioned to build roads through all the national parks. The Association's staff head is Anthony Wayne Smith, an abrasive man whose background offers a good defense against charges of elitism. Smith was once conservation director for the C.I.O., where he apparently found a hero. A few years ago, he got angry at the Natural Resources Council of America, a forum for national conservation groups, and, in the manner of John L. Lewis, he disaffiliated.

Defenders of Wildlife began as a group of women who were upset about inhumane trapping and roadside zoos, and expressed their anger in a mimeographed newsletter. As such causes are wont to do, this one attracted the attention of several persons with excess money, and the newsletter became a slick if virtually unedited quarterly which combines valuable facts on conservation with little cutenesses about squirrels. As fierce as any species, Defenders of Wildlife is less nimble than most, hence not very effective.

The bridge between the straight conservationists and the sportsmen is the Izaak Walton League, which is a bit of both. The national Izaak Walton League is a loose collection of chapters which maintain virtual autonomy. In states like Montana, where conservation sentiment is minimal, the Walton League is strictly a sportsmen's group, delving into social issues only to pass a yearly resolution against gun-control laws. In crowded Indiana, the Walton chapter pays no mind to fish and game, concentrating instead of preserving what remains of nature there. At times this local autonomy can be embarrassing. For three consecutive years, the League's national convention was rent by the emotional, and successful, effort to override the California chapter's opposition to the 80,000-acre Redwood National Park, opposed by most sportsmen because hunting is now allowed in national parks. More recently, the refusal of Florida chapter president James Redford—again the importance of personality—actively to oppose the Everglades jetport left the Walton League the only national conservation group outside the anti-jetport coalition. Coalition is the League's hallmark though, and its conservation director, Joe Penfold, until recently headed the Natural Resources Council, essentially a coalition of warring groups.

Actually, they get along rather well, considering basic differences in outlook, because there remain enough specific problems for them to unite on: the lumber industry's raid on the national forests, the jetport that would ruin the Everglades, the dam which would have created in Alaska a lake the size of New Jersey to provide more power than Alaska needs, the possibility that the trans-Alaska oil pipelines will ruin the tundra ecology, the danger that a little-known federal commission will recommend this summer that the public lands be parceled out to the highest bidder. If these fights are won, they will be won by cooperation and old-fashioned lobbying with the conservationist friends in government. The Old Conservation has its work cut out for it while the New Conservation figures out just what its work is.

To the New Conservation, rivers, mountains, and wildlife are no more important than cities and suburbs. The New Conservationist's concern is the total environment, and Penfold thinks it's in trouble. The rallying cry of the old preservation was poetry, Thoreau's: "In wildness is the Preservation of the World." The New Conservationist quotes not poets, but scientists such as Paul Ehrlich, Eugene P. Odum, and Barry Commoner, who say, in effect, that in unfettered technology is the destruction of the world. In place of the whooping crane, the roseate spoonbill, and the California condor, the New Conservation has interposed another endangered species—the two-legged predator, man.

This is not an assumption on which to base moderation. New Conservationists have adopted some of the rhetoric and even a few of the methods of the Left; and the Left, especially the psychedelic Left, has discovered conservation. In Berkeley, something called Ecology Action, which participated in the People's Park campaign and conducts symbolic destruction of automobiles, grew out of a splinter wing of the Peace and Freedom Party. Elsewhere in California, ecological revolutionaries have burned cars as hateful artifacts, lain down in front of trucks carrying what once were redwoods, and pulled out the surveyors' stakes marking the path of a road through Sequoia National Park. The underground press, especially the Los Angeles *Free Press,* devotes considerable space to conservation matters. Despising science, such of the young that meet the turned-on label adore ecology, a science. Logic is not their strong suit.

But militant new Conservation is not solely, even mainly, the province of the young and radical. In Patchogue, Long Island, an outwardly ordinary lawyer named Victor Yannacone has formed the Environmental Defense Fund, which plans to take every polluter to the Supreme Court if need be. In Colorado, the Environmental Protection League threatened to disrupt recent underground nuclear tests. In St. Louis, a group of scientists concerned about nuclear pollution formed the Committee for Environmental Information, now prepared to fight all environmental threats.

Thus the radical and the technical. But political movements succeed by being careful and emotional, and if the New Conservation succeeds, it will be through its mainstream—the Sierra Club, the Audubon Society, and now the Friends of the Earth. The first two are the giants of the movement. It was the Sierra Club which stopped the dams proposed for the Colorado River, dams which would have filled in part of the Grand Canyon, and it is the Audubon Society which is leading the fight against DDT and other hard pesticides. Each has more than 80,000 members, and both have enough knowledge of and connections in media to gain the support of many thousands more whenever needed. Charles Callison, executive vice-president of the Audubon Society, is often on the phone to John Oakes at The New York *Times*, who is never unwilling to rise to any conservation cause. The power of both Audubon and the Sierra Club is acknowledged by the powerful, which is a form of power itself. They complement each other nicely. The Sierra Club, hitting hard and fast, constitutes the shock troops; Audubon, choosing its targets more deliberately, using its well-written, handsome magazine to recruit support, provides the bulwark of the line. The Sierra Club, still West-oriented though half its members now are not from California, leads the fight on saving the public lands and the redwoods, preserving wild rivers, and Alaska. Audubon, strong in the East, concentrates on wetlands and wildlife.

The bitter Sierra Club election which ousted Brower did not weaken the club, other than to deprive it of his services. But neither did it weaken Brower. A committed, even demonic conservationist, Brower was out of a job but not a vocation. Like all conservationists, his emotions may tend him toward the remote and pristine past, but his talents are contemporary. He is an editor, a writer, a public-relations wizard. In 1966, he correctly read the mood of the country by gambling that if only the people knew of it, they would oppose the Colorado River dams. Now he bets the country is ready for an avowedly political conservation group. Friends of the Earth will openly espouse candidates and raise money for them. More, Brower wants conservation to go on the attack, to go back over where men have trod, often unwisely, and reclaim the land in a way the Reclamation Bureau never dreamed. Reclaim shopping centers, not swamps, cities, not mountains, polluted rivers, not free-flowing streams. More open space and less development, more wild animals and fewer people are the political aims of Friends of the Earth, and Brower intends to use money and political muscle to get them. He feels he has to.

For basic to the ecological conscience is the belief—no, the scientific fact—that the earth is finite, and the opinion that we are approaching its limits, that every assault on its nature starts a chain of certainly unknown and possibly cataclysmic events, which could make life for all of us uncomfortable, unsafe, even impossible. No longer is the militant conservationist worried that the whooping cranes might die out, or that the lovely land will disappear, or that the oil will run out. He is worried that the oxygen will run out.

Nor is this a needless panic; the supply of oxygen is not infinite and it is diminishing. Every time a tree is cut down, there is just a touch less, and we cut down a million acres of trees a year. Of course, that's not the source of most oxygen, most of it comes from the phytoplankton in the oceans, oceans now so polluted that the phytoplankton is endangered. Only half-facetiously, Ehrlich has predicted the death of the oceans in 1979, and when the oceans die, we all die. When all the water is salty or polluted we die also, and there are those who say that at current rates we have thirty years of water left. Then there is the weather. Suppose the sun's rays can't get through because the air around the Earth is full of glop. Would the glaciers march south again? Or suppose, once here, the heat could not filter away. Then would the polar ice caps melt, raising the seas and flooding the coasts?

To be more mundane, take southern Florida, and what could be more mundane.* By nature, men ought to live only along the coasts there, where the land is high enough and dry enough. What is in the middle is really not land at all, but a unique river which does not flow so much as seep through the saw grass from Lake Okeechobee to Florida Bay. It is no place for men to live, but years ago some pioneers saw it was a good place for men to make money. So, like most pioneers, they went to the government and got an elaborate and expensive system of canals and reservoirs to drain the river of grass, leaving rich black soil, and some roads to cross the newly created farmland. (They also got laws restricting foreign competition and no laws regulating wages and hours, but that is another matter.)

Now the glades produce vegetables in quantity, adding to the national surplus, but the water is not seeping south the way it used to, and the southern Everglades, including the national park, is dryer than it should be. Because the downward flow of fresh water has declined, Florida Bay may be getting too salty and too warm, so that while southern Florida may continue to produce plenty of tomatoes, it may soon cease to produce menhaden and black mullet and spotted sea trout. Tomatoes we can grow in New Jersey, or at least we could until we paved it over. Menhaden and black mullet in such size and numbers come nowhere else. Perhaps the New

---

**Editor's note* For another view, see Selection 10.

Conservationists are right. Perhaps nature knows best. Perhaps we should not have messed with it.

As if they don't realize that messing around with nature is both What Made America Great and the basis of Western Civilization. The trouble with society, to a New Conservationist, is not that it is unjust or immoral, but that it is . . . anthropocentric. And it is. We were, after all, given dominion. And now we have established it. We were all brought up to admire the pioneer who cleared the forest and blazed trails though the wilderness. We think Boulder Dam was a triumph of mankind, that swampland filled and built upon is useless land reclaimed, that it is virtuous to build million-dollar ditches to make fallow land fertile. Now who are these people come to tell us different?

America is the only nation in history which, miraculously, has gone directly from barbarism to degeneration without the usual interval of civilization.
GEORGES CLEMENCEAU

They are among the most handsome beneficiaries of dams and irrigation projects and dirty air and water which lie, let's face it, at the foundation of our prosperity. Most of those in the forefront of the fight for clean lakes and rivers belong to country clubs with chlorinated swimming pools. The New Jersey women who saved the Great Swamp had, in addition to fervor, money. Those committed to preserving the Hudson Valley live along it; it is not a low-income neighborhood. When a road is to pass through loveliness, the denizens of the lovely arise to fight it, and if the engineers retreat to the point of rerouting the highway through the ghetto, well, for some, that's victory enough.

Everyone wants to preserve Alaska save the Alaskans, who want to make a passel of money tapping their natural resources. The Sierra Club saved the Red River Gorge from inundation by a proposed dam fought by everyone except the small farmers who live along the river's often-flooded banks. Northern New England is full of impoverished families living in near secrecy off back roads and paying absurd prices for electric power and fuel oil. The prices could come down if the Dickey-Lincoln School Dam were built and an oil-import complex constructed at Machiasport, Maine. And prosperity could come to the hamlets of the North if factories were lured there. But such steps are strongly fought by a lot of $40,000-a-year executives from Boston and New York, who point out that the dam might block free-flowing rivers, the oil pollute the water and deface a lovely harbor, the industry and resultant housing and shopping-center development would dirty the air and spoil the region's character, a character quite irrelevant to many of the region's residents.

The New Conservationists are telling us to stop, and from their perspective, they are quite right. If the power companies are to be prevented from ruining more hills and warming more rivers, we must stop needing more electricity. If we are not to be buried by mountains of bottles and cans, we must use fewer of them, or use them over again. If we are not to pave over all the open land, we must have fewer cars. If we are not to run out of food, or out of open space because we need all the land for farms, we must stop adding to the population. If we are to have clean air and water and decent places to live, perhaps we need less manufacturing. Brower says we must "do more with less." Odum says we must stop being "consumptive" and "learn how to recycle and reuse." Either way, they are saying that this vaunted economy, this sainted Gross National Product, must stop growing.

Stop growing? But growing is the secret of our success. We have mass affluence, to the extent we have it, not because we took from the rich and gave to the poor but because we became—we *grew*—so much richer that even most of the poor live tolerably. They still get the short end of the stick, but the stick is so long now that one can get at least a fingerhold on that end. To stop growing is to stop elongating the stick, and since most people are still clinging to that short end, this presents some problems.

Because the conservationists are not on that end. They are not steelworkers or assembly-line workers or small farmers or hotel clerks. They are Wall Street lawyers and junior faculty and editors and writers and corporate vice-presidents. One does not become a conservationist until one has had the time and learning to care about whether there are eagles or Everglades. Searching for their hundred-fifty-year-old Vermont farmhouses, conservationists wonder how people can actually want to live in a new, $25,000 split-level in the suburbs, apparently never thinking that for most people the alternative is a three-room walk-up in the downtown smog. The suburbs are open to them, as Vermont is to the more affluent, because of technology, because draining swamps and dirtying streams and damming rivers and polluting the air gave them high-paying jobs. Shouting about the environmental catastrophe, urging an end to growth, the conservationists are $20,000-a-year men telling all the $7500-a-year men simply to stay where they are so we can all survive. Ethics aside, there is a serious tactical problem here; there are more $7500-a-year men and they are likely to say no. True, money would go farther in a good environment. True, as Ian McHarg said, ecological planning can give any given area more high-paying jobs and more profits plus good environment. But for the nation as a whole, for the economy, the conservationist's dichotomy remains, and he has not faced up to it: if we do not stop expanding, we ruin the environment; if we do, we condemn the lower-middle classes to their present fate.

Unless. Unless of course we did redistribute the profits of affluence by legislative fiat. Unless we planned where industries could locate and how much they could produce and where people ought to live in what numbers, and where, ecologically, no one ought to live, or drive, or even walk. Unless we instituted such extensive public regulation over use of the land, water, air, and people, that hundreds of enterprises, perhaps most of them, could not operate profitably, especially if they couldn't grow, so that perhaps they would have to be operated on a basis other than profit. There is a name for such a system. And can you see Laurance Rockefeller financing a feasibility study on that, and can you see all those $40,000 executives endorsing it?

Well, maybe. Huey Long supposedly said that if fascism came to America it would come from the working class. Now we may have come to the point where if socialism comes to America, it will come from Wall Street lawyers concerned less about the welfare of people than the survival of spoonbills. Maybe. But days after Dave Brower announced formation of Friends of the Earth, he invited to Aspen, Colorado, a select group of scientists and professors and businessmen to discuss "progress in a living environment," and because he is concerned about conservation's upper-class, all-white constituency, he invited also one Ted Watkins, Negro, chairman of the Watts Labor Community Action Committee. For two days, amidst Aspen's beauty, Watkins listened to the learned, concerned men, and then he spoke:

"What are you going to sacrifice to do the kind of conserving we want to do? Which teacher is going to give up his nice two-story home? What doctor, what architect, is going to sacrifice some of his practice to help the cause along? What advertising man is going out and campaign to raise funds without a twenty-five- or thirty-percent fee? Who of you in this room is ready to make a sacrifice to do what you say you are going to do?"

There were no takers.

10 The Wings of Man

John G. Mitchell

John G. Mitchell is a free-lance writer and a contributing editor of *Audubon.* A former newspaperman, he has been the editor-in-chief of Sierra Club Books, the editor of *Open Space Action,* and the science editor (1966–68) of *Newsweek.* A forthcoming book is on North American wildlife.

In the early winter of each year, about the time the first snow turns the color and texture of mushroom soup in the gutters of northern cities, and even as frost sidles across the higher hills of Dixie, there is an urge among certain large warm-blooded creatures to seek the golden sunshine of South Florida. I do not mean the birds migrating down the ancient flyways, though they undoubtedly experience a similar yearning. I mean that portion of the human species which is airborne, rising above the clouds on the Wings of Man, riding the thermals in Yellowbirds generic to Boeing, strapped in three abreast, a pale face at every window. These people are flying to Florida to fulfill their version of the American Dream. They are dreaming of two weeks in the sun, white sand beaches, fresh orange juice, Royal palms, calypso music, dog races, and a full moon over Miami. For some of the passengers the expectations are greater, transcending the brief delights of tourism. One sees the faces lined with age and guesses that for these it is a one-way trip to the social insecurity of retirement. The old people sit uneasily as the plane dips toward Miami International Airport. Possibly they wonder now if they have made the right decision after all, moving so far from home to take up residence in some trailer court or bungalow village, with inflation waiting to unravel the flesh from the skeleton of their life savings. Well, no matter. It is cold and drear back home. The kids have flown the coop. Old friends are gone. It is better to face death under blue skies than under gray. . . .

Florida—or, rather, the southern half, which is where most of the tourists and immigrants go—is divided into two parts, one part for people, the other for birds and alligators and cypress and saw grass and similar wild things that function best where humankind functions least. The two parts are more or less mutually exclusive, and there are quite a few of the one kind who still believe that there is no longer any room for the other. Given Florida's rate of growth, they are probably right, or will be soon enough. Of course, a few individuals understand that should the wilder half of South Florida ever disappear, it will be curtains for the tamer half. They understand it and, of late, have been talking bravely of limiting growth. But so far no one has quite figured out how to do that. "It's a nice idea," says one Dade County planner, "except for one thing. Limiting growth keeps coming up against the Bill of Rights."

In the late 1960s it was the fashion for out-of-state journalists to write South Florida down the drain. Only the travel writers dissented, as was to be expected. Jim Dooley, a pitch man for one of the airlines, was constantly on television exhorting his viewers to "come on down." Invariably, Dooley would add: "The water's fine." As it turned out, the water was anything but fine. It was disastrously scarce, low in the aquifers supplying the cities, spread thin over the conservation pools of the Central and Southern Florida Flood Control District, choked with algae in the sloughs and 'gator holes of Everglades National Park and the Big Cypress Swamp in Collier and Monroe counties. Park Service rangers and Audubon wardens, fisheries biologists, apostate poachers, and almost everyone else in a position to know the backcountry well enough to care, all predicted apocalypse.

It had not quite come to that, but it was an effective way to draw some general attention to a local problem—sort of like screaming "Rape!" in response to a raunchy bar-room proposal. In Florida, however, a certain amount of carnal knowledge was already demonstrable. Levees completed in the early 1960s by the Army Corps of Engineers had blocked the historic overland flow of water from Lake Okeechobee to the mangrove wilderness of Florida Bay. What little water did manage to reach Everglades National Park carried pesticides and fertilizers from farmlands upstream. North and west of the park, in Collier County, corporate land hustlers promising "new worlds for a better tomorrow" moved into the cypress swamps with draglines and dynamite, opened canals to the Gulf to suck the land dry, and used limestone spoil from the dredging to build access roads to their mail-order lots. Finally, as if this were not enough, the Dade County Port Authority announced the raunchiest proposal of all: a plan to establish the world's largest jetport in the Big Cypress just six miles north of Everglades National Park. And the press, knowing that bad news is good copy, descended on

Florida like locusts. The promoters of flood control and jungle jetports couldn't understand what the fuss was about. "What's more important," they asked, "alligators or people?" One Miami transportation official even suggested that it was the responsibility of all men to "exercise dominion over the land, sea and air above us as the higher order of man intends." Hearing such rhetoric, and seeing the environmental havoc such rhetoric had already unleashed, one was hard pressed to remain journalistically objective. For my own part, as time went on, I gave up trying.

I was a stranger to Florida when the first reports of havoc began to percolate through the conservation press into the national prints. My editor had sent me south not to count birds but to cover the countdown and launch of a Gemini mission, manned and orbital, from Cape Kennedy; and as the auspicious event approached, I grew increasingly anxious to be somewhere else, anywhere else, away from the neon of Cocoa Beach, the plastic motel

The secret of the Everglades, as far as man will ever be able to fathom it, is that they seem to adapt their image to each individual's expectation of what he is about to see. It is a land of nothing, a dream world, where a wanderer . . . loses his way along the highroad of time and reality. . . .

JAN DE HARTOG

bars named after missiles, and the sterility of the NASA briefing rooms. It was not that I lacked respect for the venture; I was bored by it. I was out of my element in a situation defined by the mysteries of physics, chemistry, metallurgy, and Einsteinian math.

On the eve of the scheduled launch, word was passed down from Mission Control that a defect had been detected in the launch rocket's propulsion system. . . . The man from Hearst stalked to a telephone to beg permission to fly home. Me, I managed to conceal my own delight, and in the morning NASA made it official. Human error somewhere in the drafting rooms or on an assembly line had given us five days' freedom. I called my editor. "There's a good story in the Everglades," I said, "They're drying up." He was a moon-and-missile man himself, and not much impressed. But he was generous. I went.

I drove out across the Tamiami Trail, U.S. 41, west from Miami under cumulus clouds taller than mountains, along the levee the Corps had built; past the Blacks and Cubans fishing canals with cane poles, the airboat concessions, the thatched shops of the Miccosukees, the billboards offering acreage for almost nothing down. I noticed these things by the side of the road, but what I saw was the country beyond them. As far as the eye could see, the saw grass rolled away yellow-green in the sunlight and gray under cloud, a "river of grass," as Marjorie Stoneman Douglas described it in her epic history of the Everglades, grass rooted in ancient beds of peat and washed by the imperceptible southward flow of water—southward, that is, until it reached the levee and the four huge water gates clamped down over the Shark River and Taylor Slough. I tried not to think much about the gates and the levee, for without them it was strange and wild country, full of mysteries infinitely more awesome than the calculated ones of Cocoa Beach and Cape Kennedy.

Out near Forty Mile Bend, the oceanic horizon of grass slipped away into a tassled gray wilderness of cypress trees, and at the slough crossing now the open-prairie look of things gave way to the tangled appearance of tropical jungle. At a place called Monroe Station, I stopped for gas. Fantastic country, I said to the man at the pump.

"There ain't none better," he said. He was an old man with white hair and furrows like sloughs meandering over his sunbronzed face.

"Been out here long?"

"Fifty years," he said. "On and off." And for a while we talked, since there was not much business that afternoon at the pumps, and he told me how it had been before the Tamiami and the tourists and souvenir stands, before the engineers had come in battalions to alter the flow of the water. There had been a simple dirt track west from Miami, he said, and where it ended you parked your car and walked. The old man admitted he didn't know much in the beginning, being soft and green, but soon he learned the swamps from the Indians he met along the way and the oldtimers who had been there when the profit was in plumes. He would be gone for weeks at a time, traveling light with a tin cup, a poncho, a sack of oatmeal and dried fruit, and a sawed-off .22-caliber rifle. "You didn't need for more," he said. "There was plenty of eating out there, and more than enough water." And then his eyes left me and he looked across the Tamiami Trail at the low gray façade of the Big Cypress Swamp. "It's still there," he said. "Most of it, anyway."

I asked him how long he thought it might last.

"Not long enough."
And what was his reason for saying that?
"People," he said. "It's as simple as that."

The Big Cypress Swamp is so named not because of the size of its trees but because of the extent of its territory, which covers just about every available acre south of Lake Okeechobee that isn't already occupied by wet prairie (the true Everglades), pine ridges along the coasts, mangrove estuaries, and assorted human places that no longer qualify as landforms. Strangers who mistakenly interpret "big" as modifying the tree instead of the swamp are invariably disappointed, for most of the trees are of the dwarf pond-cypress variety—short and scraggly. The bald cypress can be something else. It *was* something else, despite intermittent logging, until the tough water-resistant qualities of its wood were discovered by Naval procurement officers early in World War II, after which many of the true bald cypress giants sailed off to the South Pacific in the form of PT boats. Only a few big trees are left, mainly in the Fahkahatchee Strand near Ochopee and the Corkscrew Swamp northeast of Naples. And Corkscrew was where I was heading.

At the time, the situation at Corkscrew was something of a *cause célèbre.* Its 10,000 acres had recently been designated a National Natural Landmark, in recognition of its fine stand of bald cypress ancients, some of which, it is said, were already one hundred feet tall and festooned with moss when Columbus blundered into the New World. Corkscrew, moreover, was and is a National Audubon Society sanctuary, and its purpose as such is quite specific: the tall trees along the Corkscrew slough provide what is possibly the most significant rookery remaining in North America for the endangered wood stork. Bald cypress and wood storks are especially dependent on water—neither too much water, nor too little, but a cyclical natural flow just sufficient to protect the peat-bedded cypress roots and to concentrate fish in convenient pools for the foraging storks. But now the cycle had been broken by the promoters of new worlds for a better tomorrow. The Gulf-American Land Corporation, boasting that it was "literally changing the face of Florida," had invaded the country downstream from Corkscrew and, with a multimillion-dollar fleet of bulldozers, drag-lines, and 55-ton "Tree Crushers," was draining 200 square miles of the Big Cypress (and, because of the nature of water and gravity, Corkscrew as well) down puckered canals to the Gulf of Mexico. Just a few months before I arrived there in the summer of 1966, Gulf-Am had poked its main 60-foot drainage canal to within a mile of Corkscrew's south boundary. "It was like pulling a plug in a bathtub," said Phillip Owens, the sanctuary director. "We were left with soup."

Copious summer rains had since diluted that soup and raised the water level of the slough, and as I followed Owens along the boardwalk trail into the big trees it was difficult to reconcile the healthy appearance of the place with the director's dour remembrance of things past. It was early evening now; the sun was low. Thin shafts of light, penetrating the green canopy and silver-gray shrouds of Spanish moss, slanted across the strand as if through stained-glass windows, and the broad tapering cypress trunks rose from the dark water like columns in a cathedral. The air was still and heavy. We heard the mournful shriek of a limpkin, the splash of a cooter in the water. We saw an alligator in a bed of water lettuce, a wild turkey on a cypress branch, otter scat on the boardwalk. The storks? "They're up north till December," said Owens. "Then they start drifting in, just a few at first, sort of looking things over. By January, we'll have three to four thousand pairs."

"What about next winter?"

"They'll be back," he said. "But how well they'll do depends on how much rain we get between now and then. That, and on how much goes down those canals to the Gulf. If it keeps getting worse, we'll have to build dikes to slow down the runoff."

For the next three days I moseyed around Collier County looking at canals and ditches and dying pond-cypress, and talking to people in Naples who weren't terribly happy about what was happening up in the boondocks. Not that all of them shared Owens's concern for the wood stork—some of them just didn't fancy submerged cypress logs coming down the canals and into Naples harbor to bust the propellers on their fancy yachts. And I spent one morning at Gulf-Am's sales headquarters, observing the unctuous hustle. Already Gulf-Am's eager salesmen, buoyed by a nationwide advertising campaign, had committed 20,000 buyers to the dotted line. Thus, for a pittance down and so-much a month, could one exchange his good judgment for the privilege of someday owning a modest piece of the Sunshine State, complete with resident water moccasins and no utilities.

On the fourth day I returned to Cocoa Beach. . . . In the morning I dragged myself to the press observation area in the dunes behind the launch pad and sat in a canvas chair in the sun, watching the crew from CBS set up their cameras. Cables were spilling from the network's trailer like spaghetti. Directly in front of me, in line of sight with the Atlas-Agena rocket that was to propel a pair of astronauts into an orbital docking exercise with some hardware from a second rocket, I saw two of the network men setting up chicken wire around a patch of beach grass through which, from time to time since my arrival, various heavy-footed people had been tromping. I had no idea why they were doing this, nor did I particularly care. I assumed that chicken wire must have some technical relevance to the business of broadcasting from a sand dune, and that whatever its purpose, the whole thing was undoubtedly beyond my comprehension. Or so I thought until the men went away and I saw that inside the corral of chicken wire, protected now from the careless feet of cable stringers and other such technicians, was a bird nest with young in it, camouflaged against the dun landscape of sand and grass.

The countdown was into its last minute when the parent bird returned, saw the wire, circled it, and came down for a landing in the beach grass nearby. I am no birder, and was even less of one then, so I cannot cite the species. A bittern perhaps. It doesn't matter. What mattered was that here was this wild bird bewildered by the wire strung around its nest to spare the fledglings a ghastly stomping, and here were two brawny men who, in the midst of this vastly complex and anthropocentric adventure, had concluded that birds did matter after all. And then the cables of the distant gantry cranes fell from the poised rocket, and we saw the flash of ignition at the rocket's tail, and the puff, and the slow grudging lift, and with the sound of it came the first waves of shock, like gusts of hot wind, and tremors from the ground rattling the inside of your belly. . . . And as the gleaming rocket vaulted on an orange plume to the sky, I watched the other bird fly in terror and haste across the dunes toward the ocean, and wondered then how the storks would fare next winter in the cypress strand at Corkscrew Swamp.

I wonder still. Not only about Corkscrew and its birds, but about Florida and its people and how well or badly they are relating to the land and the water and the wild things they need to have around if they want to go on living there with some prospect of a decent environment. I wonder because nothing seems to get resolved in Florida, not that things generally get resolved anywhere else. But for Florida one keeps hoping. I mean if men can go out of their way to string chicken wire around a vulnerable bird nest, if Phil Owens can raise dikes to save a slough and others can open dikes to spare a national park from the crushing burden of artificial drought, if good and courageous people can stand up and say what ought to be done, if ill-sited jetports can be blocked—well, then. There is hope. There is the promise of something better coming. Or is there? Since that first visit, I have been back to South Florida four or five times looking into such things, and each time there is the promise of something better coming, a solution to this, or a halt to that. Yet the very next time I am down that way, something better isn't happening. Something bad is getting worse.

Like water. Historically, since people began living there in sizable numbers and mucking around with the hydrosystem, water has been such a great and vexing problem in South Florida that from afar one might think the region at times was totally arid, a misplaced Mojave appended to the underbelly of the U.S. of A. Actually, South Florida is among the wettest regions in the Lower Forty-Eight, second only, I believe, to Western Washington. South of Orlando the average annual rainfall is figured at about sixty inches. That is five feet of water, and if it all fell at once you could drown in it. As a matter of fact, quite a few people did drown in it back in the days of the great Okeechobee floods; which is why all the dikes, levees, and canals were built to pool the water and send it safely to the sea. Well, not all of it—a quarter of it, about sixteen inches, about three billion gallons a day down the drain to the Gulf of Mexico and the Atlantic Ocean. This still leaves forty-four inches, figuring annual rainfall. But now comes the tropical sun, and green plants with their roots in the spongy peats and soils; and evapotranspiration soon takes its cut, which is the largest by far—some forty-two inches right off the top. And what's left? Two inches. Two inches of rain to percolate into the wells and aquifers that are supposed to sustain the human herd of South Florida for better or for worse, in sickness and in health, from this day forward. Two inches on a sustained-yield basis might do just fine for the present population of South Florida. The trouble is you can't count on it (the taps of Miami almost ran dry in the spring drought of 1971), and the herd keeps growing. Despite recession, inflation, and the energy pinch, the so-called Gold Coast from Coral Gables north to Palm Beach is growing at the rate of 3,000 new residents a week. And until tight money put the squeeze to housing starts, Florida as a whole was accounting for more than a quarter of the national increase in building permits.

From time to time in recent years, schemes have been trotted out by people determined to deprive the sky and the ocean of their full share of the water, thereby making more of it available for increasing urban and agricultural growth. One ambitious, if not altogether brilliant, plan called for coating parts of Lake Okeechobee with a paraffin-like emulsion to suppress evaporation. The hope was that this might cut the losses from the surface of the lake—and substantial losses at that, up to an inch or more a week in the hottest weather. But experimenters soon discovered that their suppressant was inhibiting something else besides evaporation, namely the emergence of insect larvae essential as food to the lake's commercially valuable catfish and bream. At the same time, another plan was advanced to reduce the profligate waste of water pouring down canals to the sea. Here, the idea was to install pumping stations that would intercept and backpump the water into the conservation pools of the Flood Control District in Dade and Broward counties. But this plan was flawed as well, for the water coming down many of the canals was spiced with the runoff from citrus groves and melon fields and cow pastures, a witch's brew heavy on such things as nitrogen fertilizers and, until recently, chlorinated-hydrocarbon pesticides. Now this might have been acceptable to the farmers. Like the penurious character in a contemporary novel who refused to rinse his toothbrush after brushing, on the theory that the residue saved him the equivalent of one tube of toothpaste every two years, the backpumped farmers might have calculated certain savings on their chemical bills. But urban consumers were not overly delighted with the plan, and neither were the administrators of Everglades National Park, which depends on the conservation pools for much of its overland flow of water.

Since its dedication in 1947, the park has figured prominently in the water woes of South Florida, usually as the reluctant—and innocent—adversary of such other consumers as commerce and industry. In the horrid droughts of the 1960s, some worm-brained Floridians were incredulous that the Park Service should expect any water at all from the Flood District. And while most of these critics were content to pit rhetorical alligators against people, others were carried a step beyond, even to claiming that wetting the park would be tantamount to condemning the children of Palm Beach to the agonies of death by dehydration. Fortunately, wiser heads prevailed. By law, the park is now guaranteed a minimum annual release of 315,000 acre-feet of water from the Flood District, through the gates under the Tamiami Trail to the Shark River and Taylor Slough. So the park is sitting pretty again, at least by most official accounts. But there are rangers who say otherwise; who, knowing the glades and how they work moving water to the mangrove estuaries, say the guaranteed release is not properly distributed; who say more culverts under the Trail are needed to correct this situation;

and who report that in recent years the west side of the park, the side watered not by the Flood District but by the Big Cypress Swamp, has begun to dry up.

But now there is the promise of something better coming: The Big Cypress National Preserve authorized by Congress in 1974. Conceived as a deterrent to the kind of drainage and development pioneered by the Gulf-American Land Corporation, among others, the Cypress Preserve covers some 570,000 acres—or will, when and if the federal government appropriates its share of the acquisition funds (the State of Florida already having put aside $40 million as its contribution). One property in the Big Cypress, however, is not for sale. This is the forty-square-mile site of the proposed Everglades Jetport, which everyone thought had been laid to rest in 1969 when former Governor Reuben Askew and the Nixon Administration

Save a piece of country . . . intact and it does not matter in the slightest that only a few people every year will go into it. This is precisely its value . . . we simply need that wild country available to us, even if we never do more than drive to the edge and look in. For it can be a means of reassuring ourselves of our sanity as creatures, a part of the geography of hope.

WALLACE STEGNER

ordered a halt to its construction. By agreement between Dade County aviation officials and loftier levels of government, one completed runway and some support structures were allowed to remain, to be used by airlines as a training facility. Aye, the chink in the armor, the foot in the door. Meanwhile, Dade officials went through the motions of searching for an alternate site for their super jetport. And what did they find after more than four years of weighing options? Why, they found that the alternate sites are unacceptable. And now, friends in Florida tell me, the Everglades Jetport is back again on the front burner, stewing away under a lid that is likely to blow at any moment.

To some officials, including many of the original promoters, it seems to make no difference that time and events have greatly altered the rationale for another major jetport in South Florida. It seems not to matter that the need for another jetport is already anachronistic, since air traffic at Miami

International Airport has been declining, as it has everywhere else, because of recessionary incomes and inflationary fares. Or that training flights are now so few and far between at the Everglades site, because of fuel costs and a need for fewer pilots, that the Federal Aviation Administration no longer has much heart for staffing the control tower there. Nor does it seem to matter much that the original arguments against development of this site still prevail, including all the logic assembled by the National Academies of Science and Engineering and the U.S. Geological Survey to show that a jetport of this magnitude, at this location, would surely accelerate the collapse of natural systems in Everglades National Park. It is, of course, foolish to condemn the absence of logic on the other side of the argument. Logic has no place in the pork-barrel ethic.

One encounters curious attitudes in South Florida. I suppose they are no more peculiar than the attitudes of some people almost everywhere in America today, but they are distressing nonetheless. And they drag one screaming to the conclusion, if I may articulate an unabashedly snooty bias, that for all the environmental and cultural awareness we the people are supposed to have acquired in the past decade, we the people are still largely Visigoth at heart. We suffer yet from arrested sensitivity. I mean what can one expect of a society in which a Disney World attracts nearly fifty times more visitors than an Everglades National Park; in which "Monkey Jungles" and "Love Stories in Stone" and caged bears and stuffed serpents and cock fights and 'gator 'rassles draw *ooo*'s and *ahh*'s from the gullible throngs, while the glorious backcountry at best provokes a disdainful yawn? At Everglades Park headquarters in the winter of 1974, a Park Service official told me that the way he figured it, the gasoline shortage could only *help* attendance. "I know it sounds crazy," he said, "and maybe we won't be getting as many visitors from out-of-state. But the locals from Miami will be pouring in here by the droves. You see, most of them have never been here—and now they'll come because it's so close."

I was then heading through the park to Flamingo, and on the way I stopped at the Pahayokee overlook, where the Park Service maintains an elevated observation platform at the edge of a vast saw grass prairie. There was only one car in the parking lot, and it bore Florida plates. The owners of that car were on the platform, a man with a camera and a woman with flaming red hair.

"Harold," the woman was saying, "what in heaven's name are you doing?"

"I'm taking a picture," said the man.

"There's nothing to take a picture of," said the woman.
"There's grass," he said.
The woman laughed. "Oh brother," she said. "*Big* deal."

The wood stork, *Mycteria americana,* is a large white wading bird with a baldish black head and a passion for maximizing mileage on a minimum of fuel. The stork is an energy conserver. By soaring on thermal air currents, it can travel great distances with hardly a flap of its wings. Contrary to archaic opinion in some quarters, the stork is not where human babies come from. The stork is where stork babies come from, and this great bird's struggle to keep enough of them coming out of its nests each year to sustain the species is a matter of some public concern; especially to those who not only appreciate wild diversity for its own sake, but who see the bird as a kind of living barometer of the natural pressures on everything else in South Florida, including *Homo sapiens americanus.*

Though other stork species hang on here and there throughout the world, this is the only one that is native to North America. It breeds almost exclusively in South Florida, where thermals stack up quickly under the hot sun and the fishing is easy, or was, before drainage. For some time now, great numbers of nesting wood storks have convened in the cypress at Corkscrew Swamp, clearly because the large trees provide excellent shelter and, more importantly, one suspects, because Corkscrew is strategically located within striking distance of good fishing holes, such as Catherine Isle, the Fahkahatchee, and the Okaloacooche Slough. The wood stork is an unusual fisher. Unlike its distant cousin, the heron, the stork locates its prey not by sight but by touch; it gropes with its beak in the water, using a sideways raking motion. Thus, to fish successfully, the stork must have its food concentrated in shallow pools. And though I have boasted of its frugal use of energy, it nevertheless requires a great deal of protein. During the breeding season, with young on the way or in the nest, an adult stork must rake up some five to fifteen pounds of fish each day.

In the nearly twenty years since men began counting storks and nests in the cypress at Corkscrew Swamp, the number of nesting adult birds and fledglings produced there has declined. Not steadily, year by year, but in a sort of oscillating downward curve, from a peak of 12,000 adults and 17,000 young in the winter-spring breeding season of 1960–61, to 2,000 adults and about 4,000 young in 1973–74. To be sure, one cannot pin the entire blame on the Gulf-American Land Corporation, which is blessedly defunct, or on its successors in the drainage racket, for the record shows that nature, too, can be ruthless with storks. Cold snaps, freak pre-season hurricanes, high water from heavy rains (instead of low water concentrating the fish during the October-to-May dry-down period) have taken their toll in the rookery, so that as many as three straight years have come and gone with the great flocks producing no young, or worse, with the nests deserted and the young abandoned to crows and vultures. So it goes out there in the screaming wild; and so it has always gone. Nature has its own reasons for mucking around, perhaps reasons that become very logical in human terms as we begin to understand a little about the dynamics of population and food supply, among other things. The wood stork has been confronting such logic for millions of years. Somehow it has coped. But now the stork must confront an additional and unnatural interference from humankind, with its drag-lines and tree crushers and artificial droughts. And somehow, the stork is no longer coping.

Not that it isn't trying to cope. I have before me now a report by Joan Browder of the University of Florida's Center for Wetlands in Gainesville. Ms. Browder is a bright young environmental scientist I first met wading through a cypress swamp at the height of the jetport controversy in 1969. More recently she has been up in the air in a small plane, monitoring the feeding areas of wood storks nesting at Corkscrew Swamp. Ms. Browder's research is part of a larger project called the South Florida Environmental Study. Her part was undertaken on the assumption that the wood stork is an indicator of wetland productivity. In other words, as the stork goes, so goes the land's ability to capture, store, and process energy.

Ms. Browder started monitoring the storks' fishing holes in December of 1973, right after the birds began to convene at Corkscrew Swamp (and right after my last visit there). Flying out of Ft. Myers in a Cessna chartered by the National Audubon Society, Ms. Browder and her pilot crisscrossed the skies of Southwest Florida at intervals over a six-month period. Feeding birds were spotted from altitudes of up to 1,500 feet, and occasionally soaring storks were followed on thermals up to 5,000 feet. As nesting activity progressed at Corkscrew through the winter, the birds gradually shifted their feeding from scattered sites throughout the region to the major sloughs, where the dry-down had not yet affected the level of the pools and fish were still plentiful. Rainfall was far below average during this time, and soon even the sloughs and marshes that normally held water until June were running dry. Yet despite the accelerated dry-down, the storks produced.

Eggs began hatching in late February, and by May young birds were testing their wings above the rookery.

By May something else was happening. South Florida was drying down to its marl underpinnings, and the Weather Bureau at Ft. Myers was calling it the most severe drought in eighty years. Now the storks in their rookery were facing a critical time, for there were 4,000 fast-growing fledglings to feed. But in the Fahkahatchee Strand and the Okaloacooche Slough, the old reliable cupboards were bare.

Until May, it had commonly been assumed by those familiar with the habits of wood storks that the bird's effective range—that is, the maximum distance it will travel out from the rookery to its feeding areas—is about forty miles. And forty miles more or less from Corkscrew covered just about all the sloughs and marshes in which Joan Browder had sighted wood storks before the final bone-dry stages of dry-down. On May 2, 3, and 5, Ms. Browder and her pilot flew sorties south and west from Ft. Myers over the old feeding areas. In her log, she noted: ". . . dry everywhere . . . nothing but mud puddles . . . fires burning in scattered spots over entire area . . . dense haze of smoke over Southwest Florida." And no storks at the traditional fishing holes. But there were storks in the sky, riding the thermals on a coordinate from Corkscrew toward Lake Okeechobee. When the Cessna banked over the northwest shore of Okeechobee, Joan Browder looked down at the vast wet marshes spreading back from the lake along the Kissimmee River and counted two hundred wood storks (and thousands of other waders) feeding there. And this was *sixty* miles from Corkscrew Swamp.

I am awed to think of it—these tenacious birds, brooding like black-cowled monks in the treetops, knowing the fish are dead and rotting in the Fahkahatchee, waiting for the sun to get up and for thermals to rise, then rising themselves and soaring sixty miles to the only place in South Florida where food is still concentrated in shallow pools, and fishing all day and storing the catch in the stomach, to be regurgitated at sunset for the fledglings after sixty more miles on heavy homeward wings. It is inspiring as well as awesome, and one comes away from Joan Browder's report with a new sense of hope, with a vision of wet marshes hunkering forever along the shores of Lake Okeechobee as a fail-safe for foraging wood storks in times of drought. Yes, if the Okeechobee wetlands can permit the storks to conclude a successful year of reproduction, as they did in the season of 1973–74, there is hope all right. And the promise of something better coming. Right?

Wrong. We should have known better. For here comes the Army Corps of Engineers again with a scheme to increase the storage capacity of Okeechobee by raising the levees around it several feet, and no one knows what effect this endeavor might have on the Okeechobee marshes. The best educated guess is that elevated levees will probably muck up the area's hydrological rhythm and turn the marshes into dehydrated flats, fine for sugar cane maybe, but not for storks.

So Florida continues to grow by metes and bounds and breakwaters and canals. And levees. Someday, at the rate it is growing, Florida may be right up there at the top, behind New York and California. It will be a curious sort of place, all patched and emulsified and planted with exotic palms from the Caribbean. And along with the Disney Worlds and the Monkey Jungles and the moon missiles replacing the howitzers in front of city halls, there will be roadside museums filled with plastic animals. Among the specimens, surely, will be the replica of a large white wading bird with a baldish black head. And glass eyes. And a little boy with a pasty face—tans no longer being as prevalent as they once were in the Sunshine State—will point at the big bird and say, "Daddy, what were *those* for?" And Daddy, knowing perfectly well where babies come from, won't have the foggiest idea.

11 The North

M. J. Dunbar

M. J. Dunbar is chairman of the Marine Science Centre, McGill University.

Reprinted from *Environment and Good Sense: An Introduction to Environmental Damage and Control in Canada,* by permission of McGill-Queen's University Press.

Were we the harbingers
of a brighter dawn,
or only messengers
of ill-omen,
portending disaster?
Diamond Jenness

Diamond Jenness wrote those words in 1928, over forty years ago, as the last sentence of his classic study of the Eskimos, *People of the Twilight.* It was an excellent question, and the answer is still in doubt. The invasion of technological man into our Arctic has not been entirely for the common good, nor for the good of the native people. This is a very Canadian problem, for we possess over a million square miles north of the tree-line, a region not strictly comparable to any other in the world. Greenland, Alaska, and the Soviet north obviously have much in common with Arctic Canada, but it is a popular and serious mistake to assume that they are all the same in terms of problems of development. Siberia did not undergo the extensive glaciation that covered the whole of Canada with ice during the Pleistocene, so there is much more soil in northern Russia than in our Arctic. Moreover, the transport of heat by ocean currents is so distributed that the tree-line in Siberia is farther north, which means that the true Arctic is more extensive in Canada than in Russia. A large part of Alaska was not glaciated either. On the other side of the ledger, Greenland is still glaciated, so that only the coastal strip is ice-free. Even within our own Arctic there is considerable differentiation of climate and geomorphology; as a consequence, it is not possible to treat the Arctic as one unit, convenient though that may be.

In terms of environmental damage, this is a critical time in Arctic Canada. Until recently the Arctic has been peopled by the Eskimos (the real owners), the traders of the Hudson's Bay Company, the Missions, the Royal Canadian Mounted Police, and a handful of scientists, government officials, prospectors, and the first few mining-company men. We are talking here of the region north of the tree-line, not the subarctic sparse woods or the conifer belt of the "Middle North," although what is true of the Arctic proper in terms of the vulnerability of the environment is also in part true for the Subarctic. During the war considerable military activity in the Arctic did a great deal to develop air traffic over the north; this method of transport remains important today. For some years after the war the Distant Early Warning stations were strung like a necklace along the 70th parallel. The early developments, some of which go back to the seventeenth century, did not overload the north with people, but even these beginnings did enough damage to the environment to cause serious concern among biologists and anthropologists, as we shall see.

Now suddenly we have oil and a greatly stepped-up search for other minerals, and half-formed plans to cause major environmental changes by altering the course of rivers. The Arctic is faced with a rapid inrush of machines and men armed with highly effective techniques in the exploitation of commercial resources which were designed and perfected in different climates and different environments farther south. Small wonder, therefore, that there are signs of a hurried retreat—behind scientific barricades—against this threatened invasion. The statement of policy concerning the sea channels between the arctic islands and the call for a 100-mile zone outwards from Canadian shores in which Canada should have jurisdiction in the field of pollution control—which aroused such instant reaction from Washington—is an excellent example of this preparation for trouble.

Unfortunately the scientific barricades are not in the best of shape. We have been caught in a state of scientific near-nudity in the particular respect in which we now so urgently need protective covering: namely, knowledge of what the proposed developments will do to the environment, in precise terms, and knowledge of what should be done to conserve and to protect. Scientific research in the north, especially ecological research, does not have a long history. As a young and idealistic student in 1939, I remember standing in an office of the (then) Department of Mines and Resources and hearing an official, one with considerable authority of decision in arctic matters, saying "We don't want any——scientists in our Arctic." We have come a long way since then, but in developing scientific work in the north we had first to deal with that sort of attitude. The war in fact dispelled that attitude once and for all, because military activity in the north suddenly demanded a great deal of scientific work. But it still has not been possible to obtain from government anything like enough money to finance northern research as it should be financed. Perhaps the present crisis will change that.

One important result of the simplicity of arctic systems is that the component species oscillate in abundance over periods of time. In the example given above, the period of oscillation is controlled by the length of life and reproductive capacity of the lemming, and is maintained at from three to five years with quite remarkable regularity. These oscillations are severe in amplitude, so that they give rise frequently to what amounts to local extinction of species; the populations then have to be built up again by immigration from adjacent areas. The upsetting of this already rather shaky equilibrium by man's activity is probably very easy to do, and hence one must suppose that the north is more, rather than less, sensitive to pollutants and other environmental dislocations. This is the sort of thing upon which we need more precise information than we have at present, and which we need time to obtain.

One important ecological factor that may well be dependent both upon food supply and temperature is growth, the rate at which animals reach maturity. This is especially true of the poikilothermal animals and of plants. This means that damage done to populations of animals and plants takes a long time to repair. One may, for instance, come upon a remote lake full of arctic char, or lake trout, and thrill at the prospect of such excellent fishing. This has happened not infrequently in the north. After two years of fishing by Eskimos, or by visitors, the lake appears to be devoid of fish; the reproductive rate and the growth rate of the fish have not come near to making up for the fishing take, and it may in fact require a rest of many decades before the fish population is restored. The arctic char of the Sylvia Grinnell River, at Frobisher Bay in Baffin Island, take twelve years' growth in the female before ripe eggs are produced, and even then each female spawns only every second or third year. Small wonder that such resources are soon fished out and destroyed. In Canada we have been guilty of such overfishing, even under government supervision. In Greenland the Danes have handled this problem rather better, ensuring that the fishery is mobile and able to take a controlled number from each stream each year, to conserve the char population as a whole.

The factors of population oscillation, then, and of slow growth rates, appear to give the northern ecosystems a quality of sensitivity, a knife-edge balance. A third factor is the simplicity of the system itself, for where so few species are involved the extinction of just one must be a serious matter. Yet one cannot at the moment be dogmatic on this point, because the situation has not been experimentally tested; we do not know how much stress the systems will bear and still survive.

The whole of the tundra ecosystem floats on ice; below a foot or so of the surface vegetation, the land and the rock are permanently frozen, summer and winter. This is the famous permafrost, upon which not nearly enough work has yet been done. We know enough not to destroy it if we want to build houses on permafrost terrain; we must lay gravel-platform foundations on which to build, so that the ice beneath is insulated against thawing. And we know that once the vegetation cover—itself an insulating layer—is removed, the progressive melting of the frozen ground produces a terrible mess, a wound in the tundra landscape which heals very slowly if at all. To quote . . . from the Wildlife Federation brief (Passmore 1970) on the subject of permafrost: "Disturbance of surface vegetation destroys its insulating qualities and permits slumping of the local terrain—a process known as 'thermokarst'—which alters drainage patterns, induces soil erosion, forms barriers to the movement of animals and results in serious, probably irreversible change in an ecological balance which has taken thousands of years to develop." Wheel tracks recently observed on Melville Island were made by an exploratory party 125 years ago and remain unhealed today. I have seen a copy of the Bible that had been picked up on an arctic island beach and which seemed to date from the whalers' days, still entirely readable. And on another island was found a barrel of rum which dated to the Franklin days, over a hundred years ago; the barrel was in good shape, but unfortunately the rum was not.

Permafrost becomes an urgent problem now that oil development is upon us in the north. Reports from across the Alaskan border, where an oil rush is in progress in the vicinity of Prudhoe Bay, are not encouraging. Techniques developed in the south cannot safely be applied in the north without great modification, and this will require research. Engineers, geophysicists, and ecologists with arctic experience are apprehensive about the effect in permafrost terrain of pipe lines carrying warm oil; they are as apprehensive as oceanographers and marine biologists are about the prospect of giant tankers in the north. Quite apart from the problems of drilling and site maintenance, the big question is going to be how to get the oil out of the north safely.

Even under the best circumstances, scientific research cannot be hurried. Even had we foreseen, thirty years ago, the present rush to the north, it is doubtful whether we could have reached by now the stage of ecological expertness necessary to meet the crisis. There was an enormous background of basic ecological work to be done, and much remains to be done. By basic ecological work I mean the inventory of species present, their relative abundance, their life-cycles, growth rates, population oscillations, distribution and dispersal, and interrelations. A great deal of this information we now have in hand, thanks to several government agencies on both sides of the border, to the work of university scientists, and to the activities of the Arctic Institute of North America. We must complete this basic work and move on rapidly to the next stage.

This next stage, as I see it, is that of ecosystem research on the one hand and practical experimentation on the other. We have to know precisely how delicate or how tough the arctic ecosystems really are, so as to be able to decide how much stress, or technologically rough treatment, they can stand. We have to know precisely how the tundra, with its permanently frozen subsoil, will react to the laying of pipe lines and to the building of towns; how spilled oil behaves in arctic sea water and beneath the ice and how it can best be cleaned up, and so on. In February 1970, the Canadian Wildlife Federation prepared for the Minister of Indian Affairs and Northern Development, a statement on the "Crisis in the North" which called for a moratorium on oil and other mineral development in the north until scientific research had caught up. I quote from that statement here (Passmore 1970):

> A partial moratorium, lasting until 1974, would do much to bridge the serious knowledge gap which now exists. It would allow time for:
>
> 1. Completion of government-sponsored ecological research now being carried out by Canadian universities.
> 2. Development of new techniques, or modification of existing ones, for exploration, development and production without excessive damage.

3. Ecological research to establish the levels of disturbance which can be tolerated in the different zones to be explored.
4. Testing the feasibility of transport of oil through Arctic waters by tanker or submarine.
5. Conducting research into the effect of oil spills, on land and at sea, under Arctic conditions, and development of techniques for accomplishing the degree of clean-up indicated by that research.
6. Development of stand-by facilities, equipment and staff necessary to ensure adequate clean-up in cases of accidental oil spills.
7. Studying the feasibility of permitting offshore drilling under the conditions of ice, wind speed, and temperature prevailing in the Arctic.
8. Training Indians and Eskimos in the skills used in all phases of exploration, development and production of oil, so that native people may play a significant role in helping to develop their country.

The response to this call for a moratorium, on the part of the government, was less than hoped for and much less than the situation demands, at least in the opinion of the ecologists concerned. It is therefore necessary to keep up the pressure.

The question of oil spills in the sea is a specialized field demanding a whole book in itself. As far as the arctic waters are concerned, there is every probability that the hazards will be very much greater than in warmer seas, and that the damage will be complicated by the presence of the ice. The most toxic elements in crude oil are the most volatile; in more temperate seas they disappear reasonably rapidly, but in the near-freezing-point temperatures of the Arctic they would be expected to dissipate much more slowly. As to the ice, oil very easily forms thick films on solid surfaces, so that the mess it would make on and beneath sea ice can well be imagined. Crude oil has a density between that of sea water and that of ice, so that it would accumulate and spread out just under the ice; this again would prevent the volatization of the lighter parts. Finally, the process of degradation of oil in the sea by microorganisms would probably be greatly slowed in the Arctic; nothing is yet known about how the low temperature would affect the emulsion formation (oil-in-water emulsion and water-in-oil emulsion) which determines the behaviour of oil in temperate waters. Dr. R. E. Warner,

Memorial University of Newfoundland, has been quoted as saying that the sinking or damaging of just one large tanker in the Arctic could have "catastrophic effects" on the coastal environment.

Add to this the enormous size of the proposed tankers and the fact that the *Manhattan* suffered two large punctures in her hull in 1969—a light ice year and at the best season of the year for ice navigation—and it is not surprising that arctic ecologists and arctic administrators are worried. It was reported in 1970 that the Humble Oil and Refining Company, owner of the *Manhattan,* had decided to build two new ice-breaking tankers, without waiting for the results of the 1970 voyage to be complete. These new ships were to be 1,095 feet long and some 180 feet in the beam, with a horsepower of 130,000 and a carrying capacity of over 200,000 tons of oil. The *Manhattan* is a 43,000 h.p. vessel, carrying 115,000 tons of oil. The attitude seems to be

Across the northern reaches of this continent there lies a mighty wedge of treeless plain. . . .

It is a land uncircumscribed, for it has no limits that the eye can find. . . . Brooding, immutable, given over to its own essential mood of desolation, it showed so bleak a face to the white men who came upon its verges that they named it, in awe and fear, the Barrengrounds.

Yet of all the things that it may be, it is not barren.

FARLEY MOWAT

that if you want to beat the arctic ice,. build ever larger and more powerful ships. This seems a somewhat childish and frontier kind of thinking, but I may be wrong. It contravenes the Baconian warning that "nature to be commanded must be obeyed," and one tends to become concerned less with the ice-breaking power of the ship than with the ship-breaking power of the ice. Ships of these dimensions would permit the ice to stress the hull with great leverage. But this is a matter for the marine architects.

Later in the same year (1970) it was announced that the plan to use the sea route to take oil from Alaska through the Northwest Passage had been abandoned in favour of a pipe line south through Alaska. Arctic marine ecologists can therefore breath a little more easily for a while, at least until submarine tankers are given serious consideration. . . .

Arctic ecosystems are simple compared with those in temperate and tropical regions; that is to say, they consist of a comparatively small number of species. There are about 8600 species of birds in the world; of these only some 56 breed in Greenland, and perhaps a little over 80 in Labrador-Ungava. Colombia, on the other hand, has 1395, Venezuela 1150. Of the 3200 species of mammals known in the world, only 9 are found in the high Arctic, on land, and only 23 in the Cape Thompson area of Alaska. The world is full of fish; well over 23,000 are known. But only about 25 live in arctic waters. The same proportions, approximately, are shown in other groups of animals and plants.

As an example of such simple systems: the lemmings (there are two species in the north, but with fairly separate distributions, so that they are seldom found together) form the herbivore link between the mosses and grasses (the primary producers) and the foxes, snowy owls, and weasels. Here we have only one dominant herbivore, three common predators, and a few species of plants: so far only four species of mammals and birds in any one region. In certain areas, add two more predators: the rough-legged hawk and the gyrfalcon; elsewhere, add caribou and ground squirrels, two other herbivores; here and there, a wolf. In more southerly regions of the north another fox, the red fox, is also found; and a few herbivorous and insectivorous birds, perhaps five species. This gives only fifteen species of homotherms or warmblooded animals, and it is rare to find all of them in one "system" or restricted region. To these must be added the invertebrates and the plants, but this is enough to show how simple the pattern is when compared with the variety of birds and mammals found together in temperate parklands, or, even more so, in the tropical rain forest. In arctic lakes the number of species is very small indeed, and in the sea the same general proportion of species numbers is maintained in comparison with lower latitudes. Other similar examples could be given for coastal communities and for islands.

The cause of this simplicity is not the low temperatures themselves, contrary to common belief. Living organisms can adapt very easily to low temperature as such; this is true not only of the warm-blooded forms but of the poikilotherms ("cold-blooded" species) as well. The limiting factor is the ability of the system to produce life in abundance. In the sea, at least, and in lakes, this means that the limiting factor is the supply of inorganic nutrients (fertilizers). . . . On land the limiting factors may be both this lack of nutrients and the long frozen winter when the food supply is very greatly, though not entirely, reduced. In either instance it is food supply rather than low temperature. It is probable that arctic ecosystems are in a slow process of evolving toward greater complexity and greater capital supply of nutrients, but this is not a concern in the present discussion that deals with present problems and with what can be done within the next few years.

It may seem alarmist to be so concerned about damage and potential damage to an area as large as the arctic tundra. It has been estimated by the oil industry that the area which will be affected by oil development amounts to only one-third of one per cent of the total tundra, which expressed thus does not sound like very much. But it amounts in absolute terms to thirty-five hundred square miles. Furthermore, the figure is probably underestimated; once a development starts it is very likely to keep on growing. This appraisal, moreover, does not include the area to be covered by the development of minerals other than oil. And the fear is that damage to this extent, scattered in separate places over the north, is likely to disturb the ecosystem as a whole, or at least over areas very much larger than that directly affected by drilling, building, mining, and transportation of oil and ore. The behaviour of caribou, for instance—a migrant species on a large scale—may well be so disturbed that migration does not in fact take place, which would be disastrous. Thus the concern of the ecologists, and of those involved in the welfare of the Eskimos, is fully justified, and the call for a moratorium to give the necessary research time to catch up seems entirely reasonable; in fact it is an urgent and desperate call.

How the ecosystems of the arctic land will respond to these technological commercial enterprises is one thing. The response of the marine systems is another. The menaces here are: first, spilled oil; second, pollution by industrial chemicals, especially metallic ions. Of these, oil is by far the most immediate threat and also the most serious; other pollution would have much the same effects as elsewhere, . . . except for the factor of slow growth and therefore slower recovery. Oil, however, would hit two groups which dominate the arctic marine scene and which are vitally important to the native population, namely the sea mammals and the sea birds. In any large-scale oil spill, the seals, polar bears, walrus, ducks, murres, terns, and other species would be bound to suffer severely and directly. The rest of the fauna and flora might be less disturbed. Oil is a surface danger, and there is virtually no intertidal invertebrate fauna in the Arctic, because the intertidal zone is scraped and eroded by the ice. The damage to the plankton, however, is problematical; if the light supply were significantly cut down, the effect would be serious.

But the effect on birds and mammals would be decisive and drastic, and cannot be allowed to happen. They are the most valuable production of the arctic seas, and the Eskimos have depended upon them for centuries. Without them, the Eskimos as part of the ecosystem would disappear—a good example of the effect of damaging one part of a simple system of this sort. And the Eskimo, to the rest of Canada, is more than part of the arctic system. He represents a strong and established culture, with an elaborate technology of its own adapted to its peculiar environment. If we destroy that culture we do something irreparable and unforgivable.

The damage already done in the Arctic is by no means negligible. Arctic caribou, feeding predominantly on lichens which obtain their nourishment from the air and therefore absorb air-borne radioactive products rapidly and directly, have acquired a heavy load of radioactivity. Insects which breed in the water are carrying large DDT loads and handing them on to their predators. On human activities of immediate relevance in the north I quote from Dr. Ian McTaggart Cowan, University of British Columbia (1969):

> The disposal of the waste products of our culture is a continuing problem in the Arctic. Human sewage is usually dumped into a local river if one is available. In this the north differs little from more southerly areas. However, rates of decomposition are much slower and we know almost nothing of the impact of such action on Arctic rivers. Research is urgent. The beaches of some of our most remote Arctic Islands are littered with plastic bags of human excrement grounded ashore after drifting miles from some northern outpost of our culture. . . .
>
> Military and industrial activity in the Far North has shown widespread disregard for the environment. Anyone who has visited the sites has been appalled at the litter that is left behind. In almost every instance government involvement or control was intrinsic and it becomes the responsibility of government to insure a clean-up of each such project in the north.
>
> Mining is one of our most productive uses of the natural resources of the Far North. In most instances it is an activity with high waste residues. In land as heavily glaciertilled as much as the western Arctic is, the physical wastes of subsurface mining may be relatively unimportant except where they occur in areas of special value in their original state. The discharge of chemical pollutants, however, is a

different matter and should be carefully studied and controlled where ecological changes may arise.

In the October 1969 Speech from the Throne, the Prime Minister set forth certain points on Canadian policy for the immediate future: the development of the north, national security, the maintenance of Canada's international stature, and the protection of the arctic environment from ecological disaster. It is the latter policy that led to the establishment of the jurisdiction over our arctic sea waters and the 100-mile pollution-control zone, and it is clear that government departments in whose authority the northern wildlife and the native peoples are included are greatly concerned about the effects of this new technological northern crisis. Much less concern seems to have appeared from the side of industry, and there is a high probability that government may have to push hard to insist on the working out of the necessary precautions and on their application, even to the extent of veto on development and wherever it turns out that no other course than veto is possible within the avowed policy. The attitude of industry, in my experience and with certain exceptions, is at present to be willing to listen but unconvinced of the seriousness of the situation. It is quite extraordinary, for instance, that at the Fifth National Northern Development Conference in November 1970, organized by industry, not a single biologist of any kind was on the program and biology had no representative at all on the Advisory Board of the Conference.

REFERENCES

Cowan, I. McT. 1969. Ecology and northern development. *Arctic* 22 (1): 3–12.

Passmore, R. C. 1970. *Crisis in the north.* Canadian Wildlife Federation.

MIDDLE AND SOUTH AMERICA

12 Middle America: The Human Factor

Gerardo Budowski

Gerardo Budowski is director-general of the International Union for the Conservation of Nature and Natural Resources (IUCN). He was previously with UNESCO, where he headed the organization of the Man and the Biosphere Program.

The Land and the People

Middle America is a conglomerate of many countries, most of them cited in Table 1 in alphabetical order together with other demographic characteristics.*

Other islands in the Caribbean and British Honduras would add another 1.8 million, bringing the total population to about 74 million. In the table, the low annual rate of increase of Jamaica and Puerto Rico since 1958 can be explained by the immigration to the U.S. rather than a low birth rate.

---

*According to chart published by the Information Service, Population Reference Bureau, 1755 Massachusetts Ave., N.W. Washington, D.C.; October, 1963.

TABLE 1 *Population and other demographic characteristics of Middle America*

Country	Population est. mid-1963 (millions)	Annual rate of increase since 1958 (per cent)	Population under 15 (per cent)	Life expectancy at birth (years)	Population literate 15 years and over (per cent)
Costa Rica	1.3	4.4	46.4	50–60	75–80
Cuba	7.2	2.1	36.4	45–55	75–80
Dominican Republic	3.3	3.4	44.5	40–50	40–45
El Salvador	2.7	3.6	41.2	35–45	35–40
Guatemala	4.1	3.1	42.3	35–45	25–30
Haiti	4.4	2.2	37.9	30–35	10–15
Honduras	2.1	3.0	48.1	40–50	35–40
Jamaica	1.7	1.6	41.2	60–65	70–75
Mexico	38.3	3.1	44.4	45–55	65–70
Nicaragua	1.6	3.5	44.5	40–50	35–40
Panama	1.1	2.7	43.4	40–50	65–70
Puerto Rico	2.5	1.6	42.6	65–70	80–85
Trinidad and Tobago	0.9	2.9	42.4	60–65	70–75
Total	72.2				

Variation from country to country is the rule and should always be understood when looking at the whole region. This refers to the sizes of the countries and their resources, the racial and cultural background, the political and economic patterns, and many other aspects.

It must also be borne in mind that changes are coming fast to the regions. Some are of the same nature as in the United States or Canada, but many are different. It is the dynamics of these changes, their motivations and results, which need to be carefully evaluated. . . .

A Balance Is Upset

An ecologist's dream is to see man in balance, or rather in harmony, with nature. Naturally it is not a static balance but dynamic, involving, if possible, gradual improvements of the physical, spiritual, economic, and social values. This balance has been violently upset for a large part of Middle America, especially for the last thirty years when technological advances and increases in population have resulted in land-use patterns that have taken a tremendous toll in the use and abuse of natural resources. It should be borne in mind that a large percentage of the active population depends on agriculture varying from 37 per cent for Puerto Rico to as high as 83 per cent for Haiti and Honduras (CIDA, 1963). Farming over most of the area is carried out under very primitive conditions quite often at the level of subsistence. The process has been well described by Bartlett (1955) for tropical areas. Annual income is very low, usually below a hundred dollars, for the very large majority of the farmers. Illiteracy is high among the sector connected with agriculture and animal husbandry.

There are many other often linked factors which can be judged as unfavorable for the improvement of man and his habitat, such as food and health conditions, the prevailing political and social structure, and the general lack of knowledge of the physical environment in which man is living.

With few exceptions, the interaction of all these factors has resulted in poverty for a large sector of the population. True, such poverty is only relative, and for many the present situation can be judged as an improvement on conditions existing in the past over the same area. But there is a great fallacy in such reasoning when it comes to people who have received the benefit of some education. The present array of better communication media has tended to have man compare himself not with former generations as in past centuries but with others of his kind who have a better life which he sees in the movies, reads about in newspapers and magazines, or hears about on the radio or from neighbors. The result is what many have called a revolution of expectations, of which we are witnessing violent outbreaks in the forms of revolutions, general discontent, and other manifestations of disconformity. What is worse, it appears that what many believe to be a violent upset due to present conditions is possibly very mild in comparison with what may come. The result may lead to more and more violent situations unless awareness of the problem by the people who could do something about it and proper solutions understood by a large majority are the next immediate steps. Presently this does not appear to be the case for the great majority of the region.

The Impact on Resources

From the physical and biological angle the most destructive effects have been the tremendous impact on soils, natural vegetation, water regime, and wild animals. Productive areas are being depleted at fast rates and converted to sterile lands. Water resources are being mismanaged so as to make them unusable for the future. Immense genetic reservoirs of valuable plants and animals have been or are being destroyed. Even worse, this loss of biologically indispensable material is often irreparable, since it involves an irreversible trend.

The destructive impact is apparently geometric in its progression, as are so many of the other changes, especially population increase.

For centuries, use of resources of man has been in a geographical area where conditions have been relatively favorable for human occupation and the impact on the resources has been relatively mild. It is a region where precipitation is not excessive or, as it has been outlined by Holdridge (1959), where the ratio of precipitation over actual evapotranspiration is close to 1.

Some of these areas are some of the densest, most heavily settled of the world—such countries as El Salvador, Haiti, Puerto Rico, some of the Lesser Antilles, or some selected areas of other countries, especially the central valley of Mexico, the highlands of Guatemala, and the Meseta Central of Costa Rica. Curiously even in these heavily settled areas natural development schemes have left untouched, at least until recently, what were typically submarginal areas such as those with very steep topography, poor soils, very heavy rainfall or absence of it. In overpopulated countries such as Haiti, the Forêt des Pins, south of Port-au-Prince, an area of very poor soils covered with beautiful pine stands, or in El Salvador the flat and often alkaline *Crescentia alata* savannas, are typically depopulated, as are some of the steep slopes in most other countries—at least until recently. Many other areas were left untouched because they were inaccessible.

The great change in this pattern witnessed in this century, a change presently in full swing, is the impact on those marginal areas, or on what are becoming accessible areas, through the opening of new roads. Sometimes such a change appears justified and the opening of new lands to produce food for the increasing population can be done without detriment to the natural resources. Examples are well-conceived irrigation schemes such as those witnessed in Mexico. Most of the time, however, such "planned" or unplanned encroachment leads to a rapid and often irreversible destruction of the present land cover with incalculable losses in material, esthetic, and scientific values.

In this connection it should be borne in mind that marginal areas are much more sensitive to the effects of disturbance through standard methods of farming. Large areas of fresh volcanic soils have been farmed in Mexico and Central America* for centuries without adding fertilizers or using refined conservation techniques. But it takes only a few years, usually less than five, to produce a great deterioration or complete loss of the topsoil when it comes to the marginal lands not favored by natural fertility or inherent favorable soil structure.

This destruction of areas not suitable for settlement, at least with our present methods or knowledge, confers an alarming character to this upset in the natural balance. It engenders further destruction, like a chain reaction,

Accuse not Nature; she hath done her part; Do thou but thine.
MILTON

leading ultimately not only to an unproductive environment but to an equally negative force affecting adjoining productive areas. No one connected with natural resources can escape the feeling that it is his obligation to help in stopping this process while there is still time.

The Population Explosion and Its Repercussions

Population explosion is possibly the most significant single factor responsible for this change . . . . (see Table 1). It implies a tremendous stress on the land so as to produce more food and hampers economic development. It does not take much training in economics to realize that for a country with such a large part of its population below the working age, the economic growth rate must be of such magnitude that it remains outside of the present potential of most of them. The fact is that this dramatic situation is not understood, or not mentioned for several reasons, by the majority of people who influence policy. A single number such as, for instance, an annual net population increase of 3.5 per cent does not tell the whole story. The average life expectancy still remains much below that of the United States. Although infant mortality is below what it was some twenty years ago, it is still much higher than in the United States and

---

*Middle America and Central America are often erroneously thought to be synonymous. Middle America is the designation for Mexico and Central America and sometimes (as in this selection) for the West Indies as well. Central America extends from the southern boundary of Mexico to the northern boundary of Colombia.

Canada. A sad factor is that the number of births per woman is very high, and so is the number of abortions and deaths, still leaving a relatively high margin of living children. Many of the children who survive are further weakened by malnutrition, parasites, and other diseases. One could go on and on showing the contrast and the waste of energies and resources.

It is of course obvious that there is room for more people and that under planned conditions a much larger population than the present would eventually be able to live in good conditions, at least in many of the countries. This reasoning has often been advocated in favor of a policy of *laissez-faire* in relation to birth rate. The trouble of course is not the ultimate population number to be reached but the rate of increase which should be expected yearly. With an annual high population increase, the strain to achieve economic improvement is so great that it cannot be matched through the growth of productivity as expressed by the gross national income of the countries. Hence stagnation or even a worsening of the human condition may be the unavoidable outcome of a very fast population growth.

Attitude Toward Natural Resources

Indians, when organized in communities with a certain degree of self-government, often had a good knowledge of ecological balance. This was well described by Cook (1909) and more recently by Sears, using historical data (1953). It can still be witnessed today by the elaborate system of terraces for conserving soils and productivity. Good examples can be found north of the steep Cuchumatanes Mountains in Guatemala, or in that same country the preservation of natural forested areas very close to population centers where cuttings or encroachments are regulated by a special council chosen among elderly people. For instance, in the highlands of that country, where a heavy Indian population prevails, one can still find beautiful forests on the Maria Tecum range in central Guatemala, covered with beautiful pine and fir trees, in spite of the vicinity of the rather large Indian town of Totonicapán and the passage of several roads and numerous trails, all with heavy traffic. This area has been densely settled for centuries, even before the arrival of the first Spaniards. The mountain range is of utmost importance for water conservation and for the supply of timber products and fuel to several of the nearby Indian communities.

The above picture is sharply in contrast with other areas where opening of new roads leads invariably to the clearing of the prevailing forest in a relatively short time and often subsequent destruction of the productivity of the land. Some authors dealing with the impact of Indian agriculture on the soil resources have forwarded the hypothesis that the decline of the Mayan culture may be connected to their population increase, leading to exhaustion of soils by the shortening of the rotation of the *milpa* system (Gourou, 1953). Although this assumption has been challenged, it can be concluded that there are reasons to believe that with the smaller population in Pre-Columbian years, impact on nature was relatively small. Moreover, it is uncertain that marginal areas such as those on poor soils or high rainfall were ever densely populated. Most of the areas of dense Indian population were dry or moderately wet. The wet habitat of many of today's small nuclei of Indians should not be interpreted as their most favorable niches for the past centuries. Rather, they have been driven to them because of pressure from other population groups.

In contrast to the controversy on the impact of Indian groups on land resources in the past, there is no doubt that presently they are definitely the cause of much of the destruction, but then they are also exposed to a completely different set of values.

On the whole, however, it can be stated that on the basis of present attitude and past available evidence Indians had and still have a better comprehension of natural ecological balance than their European masters and descendants.

The arrival of Spaniards and their introduction of cattle meant a very large change in the environment. A characteristic attitude of the new landlords was a generally adverse feeling toward the forest. A man who had cleared the forest or literally had "opened the land" was doing good. The Spaniards, of course, used fire even more liberally than the Indians, mainly because it was indispensable to maintaining grasses and avoiding forest encroachment. Forest products were thought to be plentiful, and for a long time it was unthinkable that the immense reservoirs of forests would ever be threatened.

This feeling still prevails today in those countries where forests presently occupy a large portion of the country. Characteristically only the wet forests have escaped this massive destruction, mainly because they would not burn. A recent survey of natural habitats in need of conservation, derived from the published ecological life-zone maps based on the Holdridge system and covering Panama, Costa Rica, Nicaragua, Honduras, El Salvador, and Guatemala, shows that there is very little left of the original lowland dry forest —although this life zone occupies about 100,000 square kilometers or 22 per cent of Central America while the wetter lowland forest areas, which occupy about 181,000 square kilometers or 38 per cent, have been largely untouched (Budowski, 1964). This tendency is also true for the highlands, where the dry forests have practically disappeared because of higher density of population. Here the encroachments on wet forest areas have been much more severe in recent years. The wet forest areas are then the last frontiers of wilderness in most of Middle America but for how long?

Any programs of preservation today must take into account that feelings toward forest protection are practically absent in rural areas. The case for forests as regulators of watercourses is not generally understood. And to demand feelings for esthetic and scientific values or preservation of wildlife is to be completely unrealistic *vis-à-vis* the small farmers or cattlemen.

This may be the reason why in almost every country in Middle America there is so much emphasis on legal aspects to achieve conservation aims. Such emphasis derives its origins from well-educated urban groups that have good intentions but are quite ineffectual on the whole. As a result, many countries have had several forest or conservation laws succeeding one another over short intervals. Even foreign experts, dealing with technical assistance, are often requested to participate in the making of a new law to remedy the situation. It usually does not take long to realize that ignorance, resulting in deeply entrenched adverse feelings toward the forests and the strong motivations for clearing or misusing more land for food production mainly as a result of population increase and poor farming systems, makes those laws completely ineffective.

Altogether rural sociology, which is a relatively young science, has a very long way to go to bridge the gap of understanding between the policy makers and the poor farmers of Middle America, who are presently ignorant of what goes on beyond the immediate surroundings of their dwellings. Lee (1955) has forcefully summarized the situation when he wrote: "Too poor to learn, too ignorant to improve, and too frightened to try, a large mass of the tropical peasantry is seemingly doomed to an endless round of inadequacy. Here and there, a spark may be kindled, now and then an improvement made; but in spite of a growing awareness of the outside world, such people may still present a considerable drag to any progressive force."

HOW IMPROVEMENTS MAY BE ACHIEVED

The Need to Emphasize Family Planning

Scientific studies to ascertain the attitudes of different sectors of the population toward the size of the family, sexual practices, and other factors connected with birth rates are generally lacking. Puerto Rico is a notable exception.

There are good reasons to believe that a large percentage of the children who are born every year are the result of unwanted pregnancies. One cannot escape the conclusion that in order to remedy such a situation a change in the system of values and morals appears to be in order.

Moreover, the general attitude of many influential sectors in connection with this formerly very controversial factor is swiftly moving in a direction where the dangers of a policy of ignorance of the problems are stressed.

What is needed now is a kind of mass education in this respect. Such education needs a policy understood by and acceptable to a large majority of the population and enforcements by the necessary agencies. Such objectives as "a plea that every child born is wanted by its parents" and similar well-understood slogans would undoubtedly help to change the attitude of large, less educated sectors of the population.

The publications distributed by different international and national agencies, written in a language accessible to everyone, are important instruments for achieving this purpose. A . . . booklet published by the Center for International Economic Growth and translated into Spanish is an excellent example (Jones, 1962).

At any rate, ecologists or scientists in general, connected with natural resources, should understand that it is their duty to raise this question. An era of tabu in writing or speaking about this theme is nearing its end and scientists and other persons connected with natural resources should play a much more important role in promoting a sound policy toward family planning. A great majority of those people know the problem and privately will explain their views, almost always favoring family planning. It is the channeling of these opinions in a way that they can be heard and used by governments, and other agencies influencing and promoting programs of action and public opinion, which is the greatest need at present.

A New Approach Toward Land Use

In most of Middle America, man, through a process of trial and error leading to selection, has selected many of the soils and crops and sometimes even the methods of cultivation which were most suitable for the different life zones. Certainly there is still room for plenty of improvement such as the use of irrigation, fertilizers, improved varieties, weed control, and so on. But there are also limitations for these improvements due to social, economic, and ecological factors. In relation to the latter it can be said, as a generalization, that in marginal areas, especially those of heavy rainfall, these improvements if used in a conventional fashion are less effective.

Since man has the tendency to stick to his methods of cultivation when moving from a dry area to a wet one, his expansion into the wet environment is often doomed to failure unless he is willing to change to new crops better adapted to the different environment.

The effects of high temperatures and abundant rainfall hasten such processes as decomposition of organic matter and leaching of nutrients. On the other hand, plant growth is extremely fast, especially that of weeds. There are many detailed studies on these matters (see, for example, Lee, 1957) showing that it is necessary to make the necessary adjustments to meet the environmental factors if the latter are to be used for the best benefit of man.

A pathetic example is the attempt to establish pastures. In areas where a long dry season prevails, maintenance of pastures, while still a problem, is definitely easier than for areas of high rainfall and high temperature. At

Where grows?—where grows it not? If vain our toil,
We ought to blame the culture, not the soil.
ALEXANDER POPE

present, the clearing of natural forests in most of the hot tropical wet lowland areas to make room for pastures is a process which leads to extensive animal husbandry with a very low production per unit of surface. What is worse, most of these artificial grasslands are not permanent and the increased presence of aggressive shrubs and other woody species usually leads to the abandonment of the land after a few years. Notwithstanding the costly destruction of the original forest, of which very little is being utilized, this process is very wasteful when the production is compared to the large labor input.

Consequently a new approach for wet areas must be found and promoted as has been suggested by Holdridge (1959) and more recently by Tosi and Voertman (1964). Essentially a plea is made for a combination of crops that involves the use of tree crops either alone or in combination with annual crops of short duration. Holdridge indicates that this would best simulate the forest environment and its high biomass productivity. It is commonly found in many of the yards or orchards of peasant families, where fruit trees, vines with useful food products, and animals grow together. This would imply intense use of the area while protecting the soil and environment as opposed to pure cultivation of the same annuals or perennials over a long period of time.

Plantations of bananas, cacao, or African oil palms under scientific management systems has been practiced notably by the United Fruit Company in many Central American countries. Although praised by Sears (1953) this is only a limited solution because here, too, good soils with level topography are essential. Besides the economic limitations such as capital investment and ready export markets, there are numerous social limitations which make the plantation system unpalatable to the bulk of the peasant community and politically unstable under the present trends of nationalism and demands for agrarian reform.

Combination of tree crops producing food is a distinct possibility which deserves further exploring. Some pejibayepalm plantations (*Guilielma gasipaes*) have recently been established in the wet lowlands of northern Costa Rica, where the mean annual temperature is about 25°C and the mean annual rainfall about 160 inches or 4000 millimeters, with apparent great success. The fruits have a very high nutritive value for human consumption and they can also be fed to hogs. The pejibaye palm is a well-known food plant and cultivated by many lowland Indian communities from Costa Rica to Peru, and feeding hogs is a common practice in some areas—for instance, among the Indians of northwestern Panama, equally a very wet area.

Altogether, however, it should be clearly understood that with our present knowledge, the possibilities of using the wet lowlands of the tropics for food production are very limited, in spite of the assumption widely held by many sectors (see, for instance, CIDA, 1963) that unpopulated areas could easily be "colonized" by farmers. The few exceptions, such as rice cultivation in Southeast Asia, demand a series of special factors linked with cultural backgrounds (Gourou, 1953) which cannot be duplicated in Middle America, at least within the next decades. Hence, taking again a good look at nature it seems obvious that the most promising use of the wet tropics for the benefit of man is forestry.

The Role of Forestry in the Development of the Wet Tropics

Anyone who is aware of the high biomass production potential of the tropical forests, or simply looking over the vegetation maps of tropical countries and finding out that many countries with a large percentage of their land under forests have to import wood or its derivates, faces one of the greatest paradoxes of our present land-use patterns.

We know that, generally speaking, trees grow much faster in tropical than in temperate countries. True enough, climax species of the tropical forest grow slowly in their environment, but secondary species, including many of the most valuable, have amazing growth rates. Some pines introduced in tropical wet areas and properly inoculated with mycorhiza have done remarkably. Some simple managing practices can achieve very much when markets are nearby, even if most of the species are of limited value. A few years of research can pay off handsome dividends—as has been shown, for instance, in Malaya. In the lowlands of Middle America there is presently very limited forest research except in Trinidad, where the oldest continuous programs are to be found; Puerto Rico, where forestry has a rather limited future; and Costa Rica, where it has been practiced on a small scale. In other countries, research is either very recent or practically absent. Forest-harvesting practices in most countries imply the cutting of a few valuable species out of the virgin forest when the latter becomes accessible, and hence the development of forestry has often been linked with communication. But this of course is not forestry but high-grading and the end result is the gradual impoverishment or disappearance of the forest altogether, since logging roads are later used by farmers practicing shifting agriculture. Silviculture has been mostly planting of native or introduced species, often without sufficient knowledge of their ecological requirements. Many of these trials have shown to be failures, but there have been some notable achievements and this may lead to promising developments.

1. *The combination of agriculture with forestry* (*taungya system*). The use of shifting cultivation to the purpose of establishing plantations of valuable tree crops has long been practiced in India and other Asian countries, but has scarcely been given a test in Middle America, except in Trinidad (Cater, 1941) and most recently in Costa Rica (Aguirre, 1963). It appears a logical way to direct natural land-use systems, notably shifting agriculture toward much more intensive production of the land. Species like teak (*Tectona grandis*), Caribbean pine (*Pinus caribae*), some eucalyptus species (*E. deglupta, E. saligna*), and laurel (*Cordia alliodora*) appear very promising in areas of suitably drained soils and hot temperatures and heavy rainfall, although teak still needs a dry season. Laurel, native to tropical America, is a typical example of such a promising tree. It is an aggressive secondary species of

fast growth and unusually good stem form, of easy propagation. In some wet areas of heavy encroachment by man this species has become a dominant feature of the landscape of Central America. When clearing the land for crops or when weeding the grass from shrubs and undesirable trees, farmers and cattlemen of many Central American countries are prone to leave the valuable laurel trees. An economic rotation of less than thirty years can be achieved for timber production, and there can be several intermediate crops in the form of thinnings, when plantations are established.

The taungya system then appears to be very promising under certain conditions where ecological conditions, patterns of shifting cultivation, pressure on land, and suitable markets combine favorably. It could well play a decisive role in future controlled colonization schemes. If successfully applied, it would gradually transform inefficient farmers into efficient forest workers while the establishment of industries using the products of the plantations would also absorb a good part of the population. As can already be witnessed by the increased importance of newly founded "colonization institutes" or "agrarian reform" programs, it is appropriate to think that for wet areas, successful colonization can be achieved through the application of forestry practices and schemes based on proper research. Taungya is one possibility; there is still another.

2. *The large-scale forest industry as a land-use scheme.* The industrial complex with a variety of different products derived from the forests is not new for the United States, Europe, or Australia, but has witnessed few examples in Latin America, where the problem is complicated by the larger number of tree species. One successful example is located on the Yucatan peninsula and appropriately called Colonia Yucatan. It covers an extensive area on soils not suitable for permanent agriculture. At least six industries producing plywood, veneers, saw boards, fiberboard, compressed boards, frames for windows and doors, as well as other products, have been installed here. The Mexican government supports the industries through the system of *unidades forestales* (forest units) by which provisions are made to manage the forest and to ensure sufficient supply of raw materials for this large industry. A well-planned village, with living conditions much above Mexican rural standards and a sense of dignity and security for the inhabitants, is linked with the industrial complex. Colonia Yucatan is an excellent example of what can be done elsewhere to make the forest land produce. Here again, provided that capital, markets, and technical skills are available, such forest colonization schemes offer much better promise for development and im-

provement of human condition than uncontrolled cutting of the forest for wasteful high-grading of timber, or simply clear-cutting for shifting agriculture.

It Is Time to Stress Scientific, Recreational and Esthetic Values

Clearing large areas of primeval forests to produce only subsistence crops is obviously a tremendous waste of scientific, recreational, and esthetic values. At first it appears that preservation has not a chance when confronted by the demands of an increasing rural population. But the examples of the preserved areas in Puerto Rico and around Mexico City, as well as other national parks throughout the world, offer a clear mentis. With an increasing number of scientists being trained—this curve, too, fortunately appears to be geometric in most countries—as well as more leisure time for city workers and growing pains of city life, pressure toward preservation of natural areas is increasing. A more comprehensive attitude toward wildlife is slowly developing among educated people. Since urban sectors hold the political power, what is mostly needed is to bridge the gap so that urban and rural sectors can achieve a program ultimately benefiting both.

Such action would require an intense program of enlightening of public opinion, leading to the establishment of national parks and other protected areas. In many countries it is true that the desire for having and enjoying national parks must first be created, but working toward improving such desire seems an excellent investment. Because there are at present few leaders and practically no resources available, there is right now a pressing need for external support. These two factors, the lack of leaders and the proper channeling of external support, are of paramount importance if national parks, wildlife sanctuaries, and other preserved areas are to be successfully established.

Training of Leaders

One of the saddest aspects of the countries of Middle America is possibly the lack of sufficient politically powerful leaders with a sound scientific training to deal with conservation programs. On the other hand, one may speculate that even such potential leaders will not appear until a sizable sector of the population is willing to follow them. Whatever the reason, and until the number and the quality of those leaders reach a certain minimum, it is difficult if not impossible to expect changes within a country. As many scientists or scientifically minded people have experienced, it is frustrating to try to convince the political leaders to bring about the necessary changes until these leaders either have sufficient training themselves or are being submitted to pressure from what we can call subjectively the "proper sources" (Beltran, 1962). The same can be said, of course, if these changes are to be forced. It is the top which needs to be strengthened first, later the lower ramifications.

Under present circumstances, education in all its forms is certainly one of the best investments. But education, too, must be carefully channeled. For example, it is a mistake to build a school for forest rangers, as has been done in one Central American country, if no provisions are made to provide for the "cadres" in the form of foresters of high-quality university training.

On the other side, it is just as impractical for a country with limited possibilities to build up several second-class faculties to train the needed leaders. Successful training at a higher level in the field of natural resources is very much dependent on high-quality staff, research, and proper facilities. If these are not provided, university graduates will acquire a false feeling of confidence in their abilities to deal with the numerous and often complicated problems. But the same mistakes are being done over and over.

Technical Aid

It is not the purpose of this short appraisal to criticize technical aid as presently offered by friendly nations or international organizations to so many of the countries of Middle America. Technical aid in the form of missions, advisers, easy loans, or plain grants is a mark of our time. Everyone is aware that political, economic, and social factors are involved to a larger or smaller degree. But in the light of past evidence and working myself in a kind of technical assistance program dedicated mainly to training and research, I cannot help feeling that in order to create an impact, it is not enough to intend to solve some of the immediate problems of the country. It is absolutely indispensable to adopt a long-term policy involving an integral approach to the different problems. To leave one of the facets of these problems out because it is politically or economically risky or dangerous is to ultimately lose the possibility of achieving any result at all. To neglect, for instance, the training of local leaders or to ignore the wild population-growth figures is to doom the whole program to failure.

The implications are obvious. Too often, help is merely prolonging the agony of a country in need of drastic reforms which, on the other hand, can only be achieved through pressure from inside. In fact, external help often creates a false feeling of confidence because it is assumed that a new crisis will be solved by a new apport. And intervals for the appearance of a new and often more violent crisis may become shorter.

Until such aid is used to deal with some of the most fundamental problems such as birth rate, development plans based on sound ecological principles, and education, its long-term effects will be wholly insufficient to help the countries to stand on their own much less to meet the aspirations of the people of Middle America.

BIBLIOGRAPHY

Aguirre, A. 1963. Estudio Silvicultural y Económico del Sistema Taungya en las Condiciones de Turrialba. *Turrialba* 13 (3):168–71.

Bartlett, H. H. 1955. Fire, Primitive Agriculture and Grazing in the Tropics. In: Thomas, William L., Jr., ed., *Man's Role in Changing the Face of the Earth:* 692–720. Univ. Chicago Press.

Beltran, E. 1962. Educación de Dirigentes Políticos e Industriales. In: *Proceedings, Fifth World Forestry Cong., Seattle, Aug. 29–Sept. 10, 1960:* 1248–51.

Budowski, G. 1965. The Classification of Natural Habitats in Need of Preservation in Central America. Presented to the Symp. On Conserva, along the Pacific Coast, Mexico City, Feb., 1964. *Turrialba* 15 (3):238–46.

Cater, John. 1941. The Formation of Teak Plantations in Trinidad with the Assistance of Peasant Contractors. *Carib. Forester* 2 (4):144–53.

CIDA. 1963. *Inventario de la Informacion Básica para la Programación del Desarrollo Agricola en la América Latina: Informe Regional.* Pan-Amer. Union, Wash., D.C.

Cook, O. F. 1909. *Vegetation Affected by Agriculture in Central America.* U.S. Dept. Agric., Bur. Plant Industry, Bul. 145.

Gourou, Pierre. 1953. *The Tropical World: Its Social and Economic Conditions and Its Future Status,* trans. by E. D. Laborde, Longmans, Green, London.

Holdridge, L. R. 1959. Ecological Indications of the Need for a New Approach to Tropical Land Use. In: *Symposia Interamericana* No. 1:1–58. Instituto Interamericano de Ciencias Agicolas, Turrialba, Costa Rica.

Jones, Joseph Marion. 1962. *La Sobrepoblación Significa Pobreza?* Ctr. for Inter. Econ. Growth, Wash., D.C.

Lee, Douglas H. K. 1957. *Climate and Economic Development.* Harper, New York.

Sears, Paul B. 1953. An Ecological View of Land Use in Middle America. *Ceiba* 3:157–65.

Tosi, Joseph A., Jr., and Voertman, Robert F. 1964. Some Environment Factors in the Economic Development of the Tropics. *Econ. Geog.* 40 (3):189–205.

13 The Myth of Fertility Dooms Development Plans

Darryl G. Cole

Darryl G. Cole is administrator of Finca Loma Linda, the focus of the farming efforts described here. Applying ecological principles, Cole and his co-workers in 1971 began to develop a land-use system that established a basis for diversified production under mechanization. This was reinforced in 1973 by a research program that, Cole hopes, will allow him to write a sequel to this selection.

From *The National Observer,* 7 (April 22, 1968), p. 10. Reprinted with permission from the author and *The National Observer,*

Not long ago Adrian Alfaro came to our home to tell us that his wife had been in labor for a week and could not give birth to her baby. Adrian owns a small farm near our coffee plantation at Canas Gordas in southwestern Costa Rica. A young man, he had been married a little more than a year. Now he was haggard and distraught. His wife was dying, he said.

My wife and I went to Adrian's farm and found Mrs. Alfaro resting on a bed of boards in a hut with a dirt floor and walls fashioned of split saplings. A white flour sack had been spread on the floor by an elderly midwife to receive the baby. We carried Mrs. Alfaro on a stretcher to our pickup truck. Adrian, two brothers, and his father held the stretcher in the back of the truck to ease the jolting of travel, and we began the trip to the regional medical station at San Vito de Jaba. It was in the middle of the wet season, a light rain was falling, and Adrian covered Mrs. Alfaro with a large sheet of plastic. The road to San Vito, 13 miles distant, was in bad condition. With heavy chains on its tires, and with a front-mounted winch, our truck would get through the mud. But as the rain fell over the soggy land, and the truck churned on, I felt immeasurably depressed.

Great Expectations

My family and I had come to Canas Gordas 13 years earlier. It was a frontier settlement then, emergent in the highland rain forest. We proposed to establish a diversified farm, employing the latest techniques of modern agriculture. Over the years my parents and my wife and I devoted a considerable amount of money and effort to the project. But our venture has yielded only a small measure of the rewards we expected. The emergency involving Adrian Alfaro and his wife underscored for me how mistaken those early expectations were.

I am frequently dismayed by articles appearing in the Costa Rican press asserting that agriculture in this country must be diversified and that the rich bounty and natural fertility of new land must be harvested. Such assertions seem to be made with all of the best intentions and with an eagerness from their writers suggesting that with this or another government program, with a loan or technical assistance, with the right attitude on the part of farmers, the new lands would yield prodigiously. Well-being and even prosperity would follow. I would like to submit that such hopes have not been realized in the Canas Gordas-San Vito area, that they are not being realized in other areas of Costa Rica, and that, on the basis of our present knowledge of tropical agriculture, they will not be realized in similar new lands in underdeveloped nations.

The Myth Persists

The myth of the fertility of these virgin lands has been too long in dying. It has long been dead for the farmers like Adrian Alfaro, whose subservience to a meager soil leaves them few illusions about the "untapped riches" of virgin lands. But the myth persists, and even thrives, among sectors of government and the public where misinformation has been accepted as fact.

When we came to Canas Gordas we purchased a tract of rain forest in an area widely acclaimed at that time as a future center of farming progress and development. An Italian colony had been started two years earlier in San Vito de Jaba, a project eventually involving the investment of about $2,000,000. We looked at the rain forest on our land, at the moist, dark soil supporting it, and concluded that anything would grow in soil so apparently fertile.

Fertility Restricted to Forest

We felled the forest, cleared the land, and planted the first crops. They were a failure from the beginning. We weren't discouraged; we began experimenting—using fertilizers, lime, manure, insecticides, fungicides, varieties of seed, cover crops, various methods of tillage. We consulted agronomists, farm-research and extension agencies, and farmers' publications, and called upon the experience that had enabled us to farm successfully in the United States. We discovered that other settlers in our area were making similar efforts with equally unsatisfactory results.

Finally, we learned what we might have been told at the beginning, had we been less adamant about our ability to succeed where others had demonstrated no remarkable success. We learned that the fertility of these virgin lands is mostly in the forest and in the thin layer of humus carpeting the forest floor.

The soil nutrients needed to sustain plant life are circulated from the soil through plant tissues. Drawn from the humus and top soil by the roots of the forest vegetation, the nutrients return to the soil in falling leaves and dead branches. At any given time the forest itself is a storehouse of soil fertility.

Cut down, the forest ceases to hold nutrients in suspension; violent rains drive over the land, leaching nutrients below root levels; sunlight invades shade-loving retreats; and, unwittingly, the farmer, by tilling and exposing the soil to the sun and rain, outrages its microbiotic order. Fertility vanishes in a sudden, stunning conflagration.

Diversity Is No Solution

Much can be done, of course, to supplement soil fertility with fertilizers. However, where rain drives month after month through loose, permeable soil, where terrain is hilly and erosion carries soil away, where sunlight burns fiercely into the land, it requires an art beyond the ability of most farmers to grow crops well over sustained periods, even with fertilizers and modern techniques. What is more to the point, such methods applied to soils divested of their original forest cover are seldom profitable.

There is a good deal of discussion today in Costa Rica about agriculture diversification. Hardly a week goes by without a pronouncement by someone on the important subject. It is evident that the traditional mainstays of the Costa Rican economy—coffee, cacao, bananas, sugar, which are embroiled in market surpluses or in rising production costs—cannot be relied upon.

What to do? The solution, most frequently advanced, to diversify, is little better than a restatement of the problem. It too frequently ignores the harshness of the land.

Farmers whose livelihood is dependent upon a monoculture crippled by the woes of international overproduction do not continue to rely upon that monoculture by choice; their dependence on the economic loser is largely the product of conditions that at no time have offered any significant measure of flexibility. Planners and economists, serving up the latest potpourri of diversification, might well consider that, where agriculture has seized

> **In simplest terms, agriculture is an effort by man to move beyond the limits set by nature. . . .**
> **LESTER R. BROWN and GAIL W. FINSTERBUSCH**

upon one or two successful crops over a period of years, diversification has probably been tried and been found to be competitively impractical.

The basis of our planning from the beginning of our work here in Canas Gordas has been to develop a diversified farm. Yet, after 13 years, we have been reduced to dependence upon a monoculture, the production of coffee. It is true that the soil here will produce other crops: pasture grass, vegetables, corn, beans, fruits. But given the conditions of climate, soil, and topography, they cannot be produced and marketed competitively.

Enthusiasm Fades

The enthusiasm and confidence that characterized the beginning steps in development in the Canas Gordas-San Vito area have languished. They have been replaced by a lean skepticism, more closely defined by a need to survive than by a will to succeed. Under the circumstances of these latter years success has become merely the ability to subsist. Yet, what is more disturbing are the repeated assertions made publicly purporting that the limitations of this area either do not exist or can be eliminated with relatively simple measures.

Such arguments abet a form of social irresponsibility. They suggest that solutions to basic problems hinge on but a few steps that, if taken, would lead shortly to sweeping remedies.

Too often the standard of living in more-developed nations is used as an example of what can be accomplished by pursuing this or that technique. This point of view ignores the basic differences between the ingredients of agriculture in the developed nation and those in the nation seeking development. It would be preferable in underdeveloped nations to work within the limits of the environment and recognize that the blessings of fertility and productivity are not everywhere evenly distributed.

Mrs. Alfaro returned to her home not long ago with a baby boy, Adrian Alberto. I would like to believe that when Adrian Alberto goes to school he will follow a road no longer beset with the quagmires that made his arrival in the world the subject of an emergency, that as a man with a family he would enjoy the benefits his father has been denied. I would like to believe this, but 13 years of work in Canas Gordas have left me doubtful.

14 Brazil—The Way to Dusty Death

Peter Bunyard

Peter Bunyard, associate editor of *The Ecologist,* now works full-time on his self-sufficient small farm holding in Cornwall and contributes free-lance to environmental periodicals when time allows.

Reprinted, with minor revisions, from *The Ecologist,* 4 (April 1974), pp. 89–93, by permission of *The Ecologist.*

Brazil has a rapidly expanding population of over 90 million, vast tracts of unexplored territory which make it the fifth largest nation in the world, immense as yet unquantified natural resources, and a programme for development which must be the envy of any nation that has set its sights on the goal of material prosperity. Indeed, under its ruthlessly proficient military dictatorship Brazil is now expanding its economy at a pace that can have few rivals outside Japan, and unlike Japan, Brazil is not likely to be jarred to a halt by a sudden short-fall in basic raw materials. It can only be a matter of years before Brazil has achieved its goal of transforming itself from a backward plantation colony into an industrialised superpower which is the dominant force in South America.

But 2,000 miles of tropical rain forest—by far the largest remaining forest in the world—lie between Brazil's intentions and their fruition. If the conquest of Amazonia is to be successful not only must the forest be cleared but there must be people ready to take over the open spaces. It is hardly surprising therefore to find the Brazilian government absolutely opposed to the United Nations proposed action plan on population.

The forest is already being cut down at a pace that can have few parallels in human history, with the result that Amazonia is near to being encircled and transected by dirt track highways. Already the link between the eastern seaboard and the Pacific Ocean has practically been achieved. The scheme is for farmers, miners and industrialists to move in, working their way outwards from the roads into the jungle, until the entire area is dominated by man. The government is insisting that some virgin forest remain—at least

Brazil has long been known as a land of extreme contrasts and contradictions. Pessimists have called it a land of unlimited impossibilities. . . .

The great majority of Brazilians are convinced that their country is destined to become a world power and that this destiny will be achieved within a short time.

ROLLIE E. POPPINO

one fifth of any area that is being cleared. But over such a vast territory who is going to supervise which trees are left standing and how many?

Two kinds of farmers have already moved in. The big capitalist farmer who has cleared the forest and made for himself a huge cattle ranch; and the poor peasant from the drought-ridden northeast. The government has promised to transplant some 80,000 peasant families into the jungle but only 5,000 have gone, and many of these are in great difficulty through unremitting poverty and through tropical diseases to which they have no resistance. Moreover the peasants have little choice as to where they are put and many have found themselves reaping pitiful harvests from the impoverished soil of Brazil's terra firma.

Foreign investment has been made easy in Brazil, and money is pouring in. The multinationals feel that there are huge bonanzas waiting to be discovered and the entire continent is being explored for minerals. Vast quantities of manganese, tin, bauxite and iron have already been found and the Brazilian government is determined to establish highly competitive manufacturing industries based on these raw materials.

At all costs Brazil wishes to avoid the fate of so many Third World countries which in the past have sold their raw materials cheap to the manufacturing countries and have then bought them back as finished products at highly elevated prices. Over the past decade Brazil's iron ore exports quadrupled to reach 50 million tons a year. But Brazil now wants to get its own steel industry going, and with help from Nippon Steel and possibly the British Steel Corporation another enormous mine is to be opened in the Carajas Mountains near the Amazon Basin. In addition a £1,000 million steelworks is to be constructed on the North Coast; its steel will be strictly for export.

CLEARING THE FORESTS

The jungle for most Brazilians is an obstacle to be overcome; a barrier which keeps them from their heritage of riches and holds back economic progress. Nevertheless the Brazilians have become aware that their intention to chop down the forest has given rise to some outspoken criticism from outside. The response is a mixture of self-justification and indignation. "We have a right to cut the forest down, even if it does mean there'll be less oxygen in the world. Other countries like Britain and the United States have developed themselves and are using up the world's oxygen. Now it's our turn," I was angrily told by a bright young Brazilian woman.

Experts have already poohpoohed the idea that industrialisation has led, or could lead to, a significant fall in global oxygen levels, so at first sight the woman's statement about oxygen would appear to be sheer nonsense. But there is a disturbing element of truth in it. The first question we have to ask is whether the Amazon rain forest produces a net global gain in oxygen which is then absorbed by some other less productive area in the world. According to Howard Odum, a tropical rain forest is capable of relatively high photosynthetic efficiencies, with as much as 10 per cent of the incident light being taken up by the vegetation. The net result of this relatively high rate of photosynthesis must be a high rate of growth, which in fact occurs. But growth is an energy-consuming process and therefore a higher net increase in oxygen production is countered by an equally high increase in oxygen use. The two processes are therefore likely to balance each other out. In the end growth gives way to death and decay and these are also processes requiring oxygen. In addition the tropical ecosystem contains a complete spectrum of animals which eat vegetation and eat each other. On biological grounds it would therefore seem most unlikely that there would be any significant net gain in global oxygen derived from the Amazon rain forest.

So what if one cleared the Amazon of vegetation, leaving large areas as empty of life as a bare mountain? One could argue that so long as there were no fossil fuel consuming industries nor any great animal activity such cleared areas would have no impact at all on the global oxygen levels. And even if there were industries releasing such gases as carbon dioxide into the atmosphere it need not necessarily matter; for there is evidence that the higher the concentrations of carbon dioxide in the air, the greater the rate of photosynthesis. The result then would be a greater production of oxygen from areas which remained covered in vegetation.

But there are some extremely disturbing arguments as to why a wholesale chopping down of the Amazon forests would be totally disastrous. Undoubtedly the world's climate would change and it is possible that instead of being an area of high rainfall the entire Amazon Basin would begin to dry out. One indication that this change is likely is found in Northeast Brazil which was once an area of tropical rain forest but is now a drought-ridden disaster zone.

The most convincing reason of all why it would be foolhardy to destroy the tree cover in the Amazon comes from work by Dr. Mary McNeil. Her speciality is lateritic soils which are largely found throughout the humid tropics including the Amazon basin. These soils are rich in iron, manganese, tin and aluminium and in time, as air gets to these minerals they are oxidized into hard brick-like substances which form an unworkable hardpan over the surface. Although soil laterization is going on all the time—and in the tropics the process proceeds more quickly because of the high temperatures —it is in fact checked by vegetation and particularly by tree cover.

Just what happens when tree cover is removed has been vividly related by Dr. McNeil herself. "At Iata, an equatorial wonderland in the heart of the Amazon basin, the Brazilian government set up an agricultural colony. Earthmoving machinery wrenched a clearing from the forest and crops were planted. From the very beginning there were ominous signs of the presence of laterite. Blocks of ironstone stood out on the surface in some places; in others nodules of the laterite lay just below a thin layer of soil. What had appeared to be rich soil, with a promising cover of humus disintegrated after the first or second planting. Under the equatorial sun the iron-rich soil began to bake into brick. In less than five years the cleared fields became virtually pavements of rock. Today Iata is a drab despairing colony that testifies eloquently to the formidable problem laterite presents throughout the tropics."

An isolated incident of soil laterization may not seem particularly important but it does illustrate what would probably happen to large areas of the Amazon basin should the tree cover be removed. In addition, laterization uses up considerable quantities of oxygen. According to McNeil, if all the protective vegetation were removed from the humid Tropics "the earth's atmosphere would soon be denuded of oxygen." If she is right then the wholesale cutting down of the Amazon forest would be bound to cause a substantial drop in global oxygen levels.

Not only are the Amazon soils lateritic but most of them are very low in fertility. Nevertheless the forest is extremely luxuriant, with an incomparably varied flora, and many people have been misled into believing that the world's food problems could be solved at a stroke by turning the Amazon and other tropical rain forests over to modern methods of "intensive" monoculture cultivation. But in reality the forest's luxuriance is little more than an extraordinary facade which crumbles into nothing once the tree canopy has gone. This canopy provides shade and keeps the soil temperature down so that humus breakdown does not exceed accumulation. The all important soil nutrients are thus held in the soil rather than being washed away in the heavy tropical rains. The trees are also remarkable retrieval systems, taking up nutrients together with large quantities of water—a fact which undoubtedly helps to break the impact of the rains, and diminishes erosion.

SHIFTING CULTIVATION

The dismal failure at Iata is all too typical, and it appears that the intensive cultivation of permanent fields cannot succeed in a tropical rain forest. Yet Brazilian agronomists and indeed experts from FAO still believe it possible to establish such an agricultural system in Amazonia primarily because they want to develop large settlements of population. The terrible irony is that the indigenous tribes, who are the only people to have developed a satisfactory and long lasting system of agriculture in the Amazon, are fast having to abandon their way of life—that is if they survive at all—in the face of the white man's ruthless advance into the jungle. Because the Indians are still virtually in the stone-age many agronomists sneer at their slash-and-burn or shifting cultivation. But not only is shifting cultivation the sole method of agriculture which appears to leave the forest undamaged over any length of time, it can also be as productive, if not more so, than a conventional agricultural system based on permanent fields.

Evidence of the productivity of shifting cultivation is now coming from many different parts of the world. Oscar Lewis, for example, has shown that swidden farmers in Tepotzlan in Mexico, obtain yields of maize which are twice those from continuously cropped lands, and that in spite of the fact that these "extensive" farmers tend to be pushed out onto marginal land. According to Harold Conklin rice production per man-hour from the Hanunoo rice swiddens in the Philippines "compares favourably with labour cost figures for rice production under the best conditions elsewhere in the tropics." In the New Guinea highlands Professor Roy Rappaport has shown that the stone-age Tsembaga get good returns in terms of man-hour production. Thus for every calorie of energy they expend in clearing the forest, planting it and tending their crops, they get 16 calories back in their food. The excess is fed to pigs. Moreover they have developed a system to prevent any site being used for more than a couple of years at a stretch. In the tropical rain forest weeds grow rapidly and the Tsembaga therefore spend a considerable amount of time in their gardens keeping their crops—always a polyculture—free from weeds. But the most formidable weeds of all, the young tree saplings, are left untouched and after two year's growth they have become so big as to make further cultivation of the plot impossible. The plot is then abandoned for a new clearing. The tree saplings are called "duk mi"—the mother of gardens.

It has been claimed that slash-and-burn methods of cultivation can support only a small number of people in any one area. According to E. R. Wolfe "slash-and-burn cultivation usually implies a scattered population, a population unwilling to pay homage to a centre of control." Such a system of cultivation, says Wolfe, "could not provide a stable economic basis for the growth and existence of Maya civilization." It is hard to see how Wolfe could come to that conclusion when the evidence suggests that shifting cultivation was probably the only kind in existence in that part of Meso-America when the Maya civilization came into being. R. L. Carneiro for example has calculated that a swidden farming group of low average efficiency in tropical South America should be able to support nearly five hundred people in a single sedentary settlement, and that the Kuikuru of the Upper Xingu region of Brazil could support a farming population of two thousand people in a single sedentary village. It is also well known that many of the large settlements found in West Africa at the time of the first European contact were maintained by swidden agriculture. In the 1850s, at least nine settlements in what today is Nigeria had populations of between 20 and 70 thousand with craft specialization and social stratification. These settlements were supported by slash-and-burn agriculture.

The Amerindians aside, Brazil has none of the traditions of West Africa and in one generation is determined to shake itself free of anything that seems outmoded and laborious in its attempt to make itself into a powerful respected nation. To the Brazilians the Indians in the Amazon represent more than just backward, archaic man, they represent nature, and nature is something execrably gross and inefficient, to be banished from the earth as soon as man can bring his dazzling knowledge and technological ingenuity to bear upon the problem. For example, the Brazilian government cannot with equanimity contemplate all the energy pouring each day upon the earth and being used so unproductively by nature.

NO LIMITS TO GROWTH

In the words of the Brazilian representative to the UN Population Commission: "The sun throws upon the earth's surface an amount of energy equivalent to about 40,000 times the consumption of energy by man in 1973. This energy is today absorbed, in part, through the little understood, but inefficient method of vegetal photosynthesis, both on land and in water. A fantastic amount of it is wasted, but it is a fair guess that a good deal will eventually be harnessed directly by man-made receptors . . . With unlimited energy and within the evolving cost structures of new mining and recycling of waste resources, including of ocean water, the limits to growth, be it economic or demographic growth, will be shattered out of existence . . ."

Perhaps the Brazilian representative has forgotten that nearly one-third of the sun's energy to the earth dissipates in working the planet; a very necessary function if life is to exist. As for photosynthesis—far from being "poorly understood" it has been analysed in detail. Moreover to think of it as "inefficient" when it is capable of supporting such a veritable galaxy of living organisms and is in fact the only life-support system of man, would be nothing more than foolish if the Brazilian government did not intend to use it as justification for getting rid of the Amazon forest. In that context it is nothing more or less than a recipe for disaster.

There can be no doubt that Brazil has an embarrassing demographic problem. Its population is growing at the rate of 2.9 per cent each year, and despite its incredible economic boom Brazil is still a place of excruciating poverty and of enormous shanty town slums. According to ILO [International Labor Organization], the slum and squatter settlements of all Brazilian cities with more than 100,000 inhabitants are expected to multiply six times in the next 12 years. Admittedly the military dictatorship, which took over 10 years ago, did not create the problems of Brazil and in particular the yawning gulf between the rich and poor. But despite the fantastic fall in inflation from a peak rate of 100 per cent in 1964 to its present level of around 16 per cent and despite the enormous foreign investments the poor are relatively as bad off as they ever were. In 1960 the 10 per cent poorest of the population took 1.17 per cent of the total income. Ten years later the percentage taken by this sector of the population had fallen to 1.11 per cent. The 10 per cent richest in the meantime had increased their share of Brazil's income from 39.66 per cent in 1960 to 47.79 per cent 10 years later. Thus while one man in São Paulo may earn as much as £100,000 per year, another, also in full employment as a labourer may earn no more than £300.

. . . Moreover by putting the emphasis on exports the government has made life very hard for the working classes who are having to pay inflated prices for basic commodities. Thus one government policy caused land to be taken over for the production of soya beans for export. That land had previously been used for the production of ordinary beans which provided basic protein for the working classes. As a result the price of ordinary beans more than doubled—a great burden for people who have to spend more than 40 per cent of their incomes on food.

The present regime in Brazil is aware of these problems, but the belief is that they can be solved only by pushing remorselessly ahead with economic growth. In its headlong pursuit of that growth the Brazilian government has underway a great many gigantic construction schemes and has reached a point where there is a shortage of skilled labour. The answer to this shortage in the government's eyes is simple, a bigger population.

To justify this desire for a greatly expanded working population the Brazilian representative has come up with some interesting calculations which are as meaningful as those relating to the numbers of Englishmen that can stand side by side and back to back on the Isle of Wight.

AN UNDERPOPULATED WORLD

"The Brazilian amazonic area alone, with three and a half million square kilometres and less than one person per square kilometre, could take a lot more people than presently there. If we had the demographic density of the Federal Republic of Germany or of the Netherlands, Brazil would be respectively 2,052 million and 3,082 million people instead of 100 million. If we had the density of the state of São Paulo, our most productive provincial unit in industry, agriculture and services, we would have 630 million people," he said. "The truth is that the problem of geographical space for man's physical support is not at all a problem at this juncture or in the foreseeable future."

The Brazilian representative has overlooked a number of important facts. First and foremost he has forgotten that the population of the Netherlands, or of Britain for that matter, was only able to expand to its present size because it had an outlet in its colonies. Even though the colonies have vanished the same basis of trade exists. Food for example is imported in enormous quantities at very cheap prices, as are other raw materials, such as phosphate for fertiliser. In *World Food Resources* Professor Georg Borgstrom points out that both the Dutch and British depend on large acreages of land abroad in order to feed themselves. According to Borgstrom, the Dutch need more than four times as much land for agricultural purposes as they have at home and the British nearly three times.

These imports of food might not matter if the world were presently able to feed itself satisfactorily. Even in Brazil despite its enormous land area, people are still suffering from malnutrition, and the situation is now appalling for many Asians and Africans.

The Brazilian delegate also seems to have overlooked the difference in structure between his country's population and that of a European country such as Britain. Even though the population density may be high in Europe the rate of growth of the population has never at any time reached the levels found today in many Third World countries, including Brazil. This slower growth rate in Europe has meant that the population was better balanced in the sense that no one age group predominated. In countries such as Brazil the under 15s predominate and they now make up more than 40 per cent of the total population. And what happens when the under 15s grow up and have children of their own? Will the percentage of the under 15s increase still further, especially in countries such as Brazil which appear to have no intention of trying to curb population growth?

In its report *Population and Labour* ILO points out that "between 1970 and 1985—a mere 15 years—the world total of children under the age of 15 is expected to increase by nearly 450 million, 400 million of whom will be in the Third World." By 1980 two-thirds of these children will have reached the age when they should be at school, if that age is assumed to be five years. "If they were to have educational opportunities equal to those of children in more fortunate lands, their governments would, within 10 years, have to build, equip and train teachers for as many schools as now exist in the whole of Europe, the Soviet Union, the United States, Canada, Australia and New Zealand combined."

. . . But there is also another point that the Brazilians should bear in mind. As the population grows so the number of basic services has to increase; indeed new schools, hospitals, farms, workshops all have to be provided to satisfy demand. It has been calculated that a one per cent growth in population will absorb something like 4 per cent of the total national income just in order that these demographic demands be met. With its 2.9 per cent rate of growth Brazil will therefore have to expend nearly 12 per cent of its national income just to stand still.

Although it would be hard to take the Brazilian seriously when he claimed the extraordinarily high disabilities arising in children whose parents had taken chemical contraceptives, his concern at the widespread use of such contraceptives may be justifiable. In fact there is very little evidence that the use of chemical contraceptives is or has ever been a major factor in reducing the fertility of a population. In Britain for example fertility dropped to its lowest point since the industrial revolution began in the 1920s and 1930s when chemical contraceptives did not exist in Europe and there were no such devices as IUDs.

If there are increasing numbers of deficient children being born in Brazil the causes could be much more subtle than teratogenic effects arising from contraceptive use. Indeed stress through overcrowding, poverty, ill-health and malnutrition might be a more important factor. A number of different authorities, Professor McKeown in Britain for example, have shown that the number of malformations such as spina bifida and anencephaly increase during specific periods of social stress. Thus there was a distinct peak in the malformation rate recorded in Birmingham during the years 1940–43 when the bombing was at its worst.

STRESS AND ABNORMALITY

In a study of 102 mentally subnormal children Dr. D. H. Stott found that their mother's pregnancy was much more commonly disturbed by illness or emotional upsets than was the mother's pregnancy of normal children. Many animal behaviourists have made similar findings among animals which are subjected to stress, and Stott suggests that the high rate of malformations and falling fertility among animals as well as humans may well be a natural mechanism for preventing excessive population growth. At what level a human population feels itself to be under stress will obviously vary according to the environmental conditions. A number of Amerindian tribes have practically stopped breeding in response to contact with western civilisation. One Brazilian tribe had only produced two children over a five-year period despite there being a number of eligible parents.

Almost without exception the tribes living in the Amazon have devised cultural controls for limiting their populations to levels well below the carrying capacity of the environment. These methods of population control, which include infanticide, an extended period of lactation with taboos on sexual intercourse and even the occasional murder of related tribesmen from neighbouring villages as practised by the Jivaros of Ecuador, appear to be highly effective and are the hallmark of an incredible adaptation to environmental conditions in the rain-forest. To us they may seem savage and barbaric, but what can be more barbarous than subjecting man in his millions to the unspeakable squalor of a shanty town slum?

Ironically by destroying the way of life of the Indian, the Brazilian government is losing the opportunity to learn how best to use its heritage of tropical rain forest. There is undoubtedly more room for people in the jungle, but rather than force the environment to adapt to them, they should learn to adapt, as the Indians have done, to their environment; for in forcing their environment to adapt to them they will inevitably and irrevocably destroy it.

It is perhaps too much to ask of the Brazilians to abandon their schemes for cutting down the forest and for wresting its riches from it. The big question is whether such actions will put Brazil in the league of powerful nations. Brazil sees the world's markets at its feet and envisages itself to be in an enviable position of both producer of raw materials and manufacturer. But will the rest of the world be able to afford its prices? The world is now in crisis and the long-term future of the industrial society is in jeopardy. What if the energy which the Brazilians claim to be ready for the taking is not forthcoming, what good their steel then or their laterized soils?

In fact the Brazilians can hardly be blamed for embarking on a way of life which has already been pursued with equal avidity and ruthlessness by the industrialised nations of the world. Nor can they be blamed for thinking that the UN Plan of Action on Population is threatening their own right of sovereignty, especially when this plan is so heavily promoted by the industrialised nations. But they should take care to avoid the trap of thinking that industrialisation will solve their problems when instead it is just adding to them and willy nilly to those of the rest of the world.

15 The Galápagos and Man

Charles Gilbert

Charles Gilbert is a free-lance writer and photographer who specializes in marine and coastline subjects. This selection is based on research done during a stay on the Galápagos Islands.

Reprinted from *Oceans,* March 1974, pp. 40–47, by permission of the Oceanic Society.

Not too many years ago the situation looked bleak. The Galápagos Islands seemed destined to share the fate of so many other unique island communities broken under the pressure of man's arrival. Already entire subspecies of gigantic tortoises had been destroyed, and the character of several islands had been irreparably changed by the aggressive browsing of domestic animals gone wild. And the damage was increasing as the islands became ever more accessible.

This was the situation discovered in 1955 by the investigative team sent by the International Union for Conservation of Nature and Natural Resources. With their reporting of the deteriorating condition of this small archipelago, so rare as to be called "The Showcase of Evolution," an international movement to save the Galápagos was launched. Now, almost twenty years later, although several singular species of plants and animals are threatened with extinction, the future of the islands can be secured with the continued concern and support of the international community.

WHY THE GALÁPAGOS DESERVE PROTECTION

The Galápagos Islands are especially deserving of international concern and protection. The simplicity of their biological and physical environments approaches the structured conditions of experimental science, making the Galápagos a natural laboratory of ecology and evolution. In many cases, as with the tortoise and Darwin's finches, a single species evolved into the clearly recognizable subspecies now found in the islands. Other animals and plants, e.g., the marine iguana and flightless cormorant, evolved from their ancestral forms on the American continent into entirely new species, particularly adapted to the Galápagos environment. In 1835 young Charles Darwin visited the archipelago, where his still rudimentary concept of evolution through natural selection began to crystallize. He pondered the deviations in form of the various closely related subspecies. He stated in his journal: "The most curious fact is the perfect gradation in the size of the beaks in the different species of *Geospiza* [Darwin's finches]. . . . Seeing this gradation and diversity in one small, intimately related group of birds, one might really fancy that from an original paucity of birds on this archipelago, one species had been taken and modified for different ends." He noted the same curious variations in the huge Galápagos tortoise. Many years after his historic visit Darwin wrote that "it was such cases as that of the Galápagos . . . which chiefly led me to study the origin of species." The modern visitor to the "Enchanted Isles," as they have long been called by passing sailors, shares in Darwin's wonder as he sees more clearly into the mystery of creation.

The remarkable fearlessness of the native animals makes the Galápagos interesting and valuable to the amateur naturalist as well as the scientist. (Due to the absence of predators fear became an unnecessary response.) Today the animals' behavior is so little affected by the presence of man that study can be conducted at close range, without the usual elaborate camouflage and time spent awaiting the appearance of timid subjects. All that is needed to gain an understanding of animal behavior and interdependencies is patience and a keen power of observation.

Additionally, the Galápagos provide an excellent setting for marine study. The 600 miles separating the archipelago from the South American continent is an effective barrier to shore fish, resulting in their independent evolutionarv development: 23 percent are endemic to the Galápagos. While most marine communities elsewhere have suffered from the effects of pollution, and some are dying from massive dumping of effluents, the Galápagos are nearly unique in the purity of their surrounding waters. The convergence at the islands of the cold Humboldt Current from the south and a warm current from the north further increases possibilities for rewarding oceanographic studies.

Small Isla Española, for example, farthest to the south in the Galápagos archipelago, is the only known breeding ground for the magnificent waved albatross. These large graceful birds are seen along the west coast of South America and far out to sea, but each year they return to this one island for their intricate mating dance and the rearing of their young. Interruption of their breeding pattern would be all too easy, so continual vigilance is necessary if they are to survive.

And of course many species of both plants and animals can be found nowhere else in the world. The Galápagos penguin, the marine iguana, the tree-like *Opuntia* cactus (with a thick trunk similar in appearance to a pine), and the Galápagos snake are but a few which occur only in the islands. Endemism is extremely high, being about 75 percent for animals and 45 percent for plants. Each endemic species, unique in structure and behavior, demonstrates divergence from its continental relatives.

VULNERABILITY OF THE GALÁPAGOS

Unfortunately, island environments are easily disturbed. And the Galápagos are particularly susceptible to disruption by man and his host of intentional and accidental introductions. Introduced plants and animals, adapted to their former more competitive communities, quickly multiply in the absence of their natural checks. As the introduced animals increase, food resources are reduced. The endemic animals, specialized to a particular food in the simplified island environment, cannot compete successfully for the already sparse supply, and their populations decrease in size. Feral carnivores and omnivores feed directly on the endemic animals, who have little protection against these new predators. Since man's arrival in the Galápagos more than 400 years ago, drastic changes have altered the character of the archipelago, and it is only in recent years that the situation has been stabilized and restoration begun.

HISTORY OF HUMAN IMPACT

The first recorded visit to the Galápagos occurred in 1535. The crew of a Spanish galleon, carried off her course from Panama to Peru by strong ocean currents, found the arid barren islands inhospitable. They wanted water, and only after several hot and desperate days of search did they find a freshwater pool. Soon afterwards the ship set out to continue her journey. The report sent to Carlos V of Spain noted the gigantic tortoises, the previously unknown marine iguanas, and the amazing fearlessness of the birds. But these early visitors, concerned more with the treasures of the Andes than with exploration, had little use for so uninviting an archipelago. Few ships returned to the Galápagos until, in the late 1600's, English buccaneers used the islands as a hideout. . . . The buccaneers discovered the palatability of the flesh of the Galápagos tortoise, which they claimed was so delicate and tasty that a man was spoiled for all other meats. Tortoise oil served as a spread equal to butter for their hard biscuits. The bands of buccaneers were small, so even though they had hearty appetities for the tortoise, its population was not severely diminished.

During the height of the whaling industry from 1780 to 1860, however, British and American whaling ships drastically reduced tortoise populations. It was found that the giant tortoises could live in the hold of a ship for up to eighteen months without food or water. Crews, signed on for these long voyages, were eager to provision themselves by spending several days in the Galápagos gathering tortoises, with some ships' crews stacking as many as 900 in their holds. When fresh meat was desired the whalers had only to haul a 200-pound tortoise to the deck. Scurvy no longer plagued even the longest cruises. It is estimated that 200,000 to 300,000 tortoises were removed by the whalers; the present population is approximately 10,000. The tortoises were abundant in 1780, and scarce in all except the most inaccessible places by 1875. The Santa Fe and Floreana subspecies were extinct. It is even contended that the Galápagos tortoise made Pacific whaling feasible, and that it was the decline of the tortoise as much as the whale that ended the industry.

With the whalers came the sealers, and sometimes ships pursued both ventures. As with the tortoise, the fur seal was once plentiful in the islands and now is scarce. The logbook of Captain Benjamin Morrel reports that 13,000 fur seal skins were taken in two months by his crew. The present population is variously estimated at between 500 and 4,000.

When Ecuador officially took possession of the Galápagos in 1831 colonists were sent to Floreana to secure the islands for their distant capital. A small settlement was established at what was later to be called Bahía Post Office. Domestic animals soon escaped the colonists and rapidly covered the entire island. Goats and donkeys were intentionally released so that when their numbers had increased, they could serve as a constant food reserve. However, the staple food of these earliest colonists was the Galápagos tortoise. So many were taken from Floreana by both the whalers and residents, that within three years colonists were sending expeditions to other islands for tortoises. By the time Darwin arrived at Floreana the colony had been abandoned and cattle, dogs, pigs, donkeys, and goats were abundant. A similar pattern developed when Santa Cruz, San Cristóbal, and Isabela were colonized, though the impact of the introduced animals was more localized because of the larger size of these islands.

It had long been the practice of frequent visitors to the Galápagos to set ashore a number of goats, so that on their next trip through the islands fresh meat could easily be obtained. By 1875 large goat populations existed on all the major islands, and had in some cases drastically reduced the native vegetation.

The native Galápagos doves and hawks have also suffered at the hands of visitors and colonists. Both birds are so fearless of man that they can be approached to within a few feet. The hawks were thought to prey on domestic animals and were hunted by colonists. Now perhaps only 200 pairs remain in the entire archipelago. Doves were killed by the thousands to provide food. With such brutal predation the Galápagos doves and hawks relearned the fear response, and today are noticeably more retiring in areas frequented by man than in relatively undisturbed locations.

The California Academy of Sciences conducted the first major scientific investigation of the Galápagos Islands. In 1906, working from the schooner *Academy*, a small team spent more than a year in the archipelago. In reading their report to the Academy on the tortoise, the changing values of science become apparent. Collection was the primary purpose of the expedition. Measurements and anatomical data are presented in the report but almost no information is given on behavior, as indeed little time was spent in observation. The expedition's reptile expert spent most of his time skinning, transporting, and preserving the nearly 300 tortoises obtained for the Academy's collection. Most difficult for the modern reader to understand is the killing of what were thought to be the last remaining members of a unique tortoise subspecies. Prior to man's arrival fourteen subspecies, or races, of tortoises had existed in the Galápagos. One separate subspecies inhabited each of the major islands, except Marchena where none ever existed, and Isabela, where each of its five large volcanoes supported a subspecies. With the wholesale removal of tortoises by whaling ships, the Floreana and Santa Fe subspecies became extinct, and the subspecies of all the islands except Santa Cruz and Isabela had become extremely rare. Fortunately, despite the expedition's thorough search, several members of each subspecies remained

The Galápagos Islands . . . are, in miniature, an alternative to the world we know, just as the nightmare crabs upon [the Atlantic island] of South Trinidad present an unthinkable alternative to ourselves. Each of these island worlds has taken through chance a different turning at some point in the past; each possesses a different novelty. In none has man twice appeared save as a wandering intruder from outside.
LOREN EISELEY

undiscovered, and today they have a chance of survival. It seems likely these scientists saw the continual reduction of the tortoise populations, and had resigned themselves to the extinction of all tortoises during the twentieth century. They therefore wanted a record of the amazing animal for posterity and future scientific study.

The New York Zoological Society in their 1928 expedition also collected tortoises. Besides the more than 180 they killed and skinned, the Society removed what was hoped to be a "breeding stock," that would someday repopulate the Galápagos after the predation by man and his introduced animals could be controlled. It has been found, however, that the tortoises reproduce very poorly in captivity.

During World War II the United States military built and maintained an airstrip on Isla Baltra, primarily for protection of the Panama Canal. Bored servicemen reduced the island's endemic rat and land iguana populations to extinction.

Meanwhile the Ecuadorian population was growing with the impetus of a Galápagos colonization program, sponsored by the Ecuadorian government. Each of four major islands—Santa Cruz, San Cristóbal, Floreana, and Isabela—had a small village on the coast and another in the moist highlands. The highland communities were bringing more land under cultivation each year, by expanding into the areas of native vegetation. In the 1930's cattle were introduced on the slopes of Volcán Cerro Azul on the southwest tip of Isabela, and as the herd increased cattle were exported back to the continent. In 1934 Ecuador designated the Galápagos a wildlife sanctuary, but no provision for enforcement was made, so poaching became more serious as the population grew. The tortoise remained the most prized game of the islands, owing to its fine oil and meat. Until the 1950's hunters killed tortoises for the nearly three gallons of exceedingly pure oil each carcass provides. The tortoise populations were so reduced, however, that the business became economically unfeasible. Today just over 5,000 Ecuadorians reside permanently in the Galápagos.

The Galápagos Islands have been nearly worthless commercially. Except for a small salt industry and insignificant exports of cattle and coffee, little use could be found for them. Even commercial fishing and taking of lobsters were on the decline by the late 1950's. The Galápagos are simply too volcanic and arid to be successfully exploited, and too hostile even to support a human population except in the few milder areas. This harshness is the very thing that has preserved the Galápagos. Because man could make no significant encroachment into this alien world there is a possibility today that the archipelago could be restored to its original state.

CONDITION OF THE GALÁPAGOS TODAY

With the exception of the tortoise, and the few instances of senseless cruelty, man's impact on the Galápagos has been incidental. Since there are no minerals to excavate, no dangerous predators to eliminate, and very little land suitable for cultivation, the destruction has come not from man but principally from his multitude of introduced plants and animals. Today the Galápagos Islands are plagued by feral goats, cattle, horses, donkeys, cats, dogs, rats, and mice. Each of these introduced animals in some way destroys the native flora and fauna. Islands to the north of the main group are free of the introduced animals (with the exception of Pinta where goats were introduced in the late 1950's). Fernandina too has been spared, because of the extreme hostility to life of its barren lava fields and the absence of fresh water. The islands that man has inhabited or frequently visited, however, are populated by numerous feral animals. All of the major introduced animals occur on Isabela and San Cristóbal, and Santa Cruz and Floreana have all but the horse. Santiago has pigs, cats, donkeys, rats, and goats. Baltra has rats and goats. Pinzón only has rats and Española only goats.

Wild dogs and cats are predacious and feed on the reptiles and birds. Adult tortoises have been found with shells gnawed and limbs missing, this apparently the work of dogs and pigs. On southwestern Isabela packs of wild dogs hunt calves of the wild cattle, and have occasionally attacked men. Pigs dig through tortoise nests, eating the eggs and the young. Additionally, in the highland breeding ground on Santa Cruz, pigs have been known to eat the egg, the adults, and the immature young of the threatened dark-rumped petrel. Dogs also attack the mature dark-rumped petrels. The aggressive black rat is particularly threatening to Galápagos fauna. It attacks the eggs and young of many native animals, again including the dark-rumped petrel. Because this petrel nests on the ground in highland burrows, it is critically vulnerable to attack. It is one of the most endangered Galápagos species. Black rats also attack the tortoise and endemic rat young.

Cattle, donkeys, and horses present fewer problems to the ecology of the Galápagos, yet their grazing of the already often sparse vegetation reduces the chances of survival of the native animals, and changes the general character of the plant life. Occasionally they destroy land iguana burrows and smash tortoise nests and eggs. Volcán Alcedo on Isabela, however, has both the greatest tortoise and donkey populations in the Galápagos, so it appears that the donkey presents no immediate danger.

The greatest threat to the Galápagos is the goat. Goats are well suited to the desert-like character of much of the islands, for they need little water and quickly adapt to eating any available vegetation. They are voracious browsers, eating all parts of plants, thus slowing the plant's natural regenerative process. One hundred years ago on Santa Cruz it was nearly impossible to reach the highlands, for the dense vegetation formed an almost impenetrable barrier. Today, in a wide area surrounding the coastal village of Puerto Ayora, one can walk with relative ease where before each step had to be chopped out by machete.

Small Isla Santa Fe has been reduced by goats to an open parkland. Nearly the only vegetation that remains are the high, tree-like *Opuntia* cactus and the palo santo trees. The wild tomatoes and cotton are now nearly extinct, and the land iguana population has decreased because of the reduction in vegetation.

Land iguanas cannot compete with goats. The pads of the *Opuntia* cactus are one of the land iguana's principal foods. Normally the large, succulent pads droop from the main trunk to the ground. For example, on Gardner, an islet just off the coast of Española, no goats exist and the pads reach the ground in abundance. In striking contrast, on Española, where a rather small population of goats occurs, few pads can be found at a height lower than six feet. Goats eat their way up the *Opuntia* until they reach their maximum height of six feet, standing on hind legs with front legs resting on the trunk and necks outstretched. Confined to the ground, land iguanas can only eat what little occasionally drops from above.

Floreana and Santiago islands also have substantially changed in appearance. Much of their low-lying vegetation has disappeared and several endemic species are threatened with extinction by the goats' aggressive feeding habits. Some *Opuntia* trunks on Santiago, up to two feet in diameter, have been completely gnawed through by goats, thus killing the plants. Some native species of *Scalesia* (members of the sunflower family, one species of which grows to a height of thirty feet) are in danger of extinction on Santiago. . . . Land iguanas, once so plentiful that Darwin encountered difficulty in finding enough unoccupied ground to pitch his tent, are extinct on Santiago.

The very rare succulent, *Calandrinia galapagosa,* found only at Sappho Cove, San Cristóbal, has been reduced to a score or so of damaged and sterile individuals. The endemic succulent, *Portulaca howelli,* is found only on islets where goats have never been introduced.

The introduced guava plant has rapidly spread through the Galápagos highlands, decreasing the populations of native plants, particularly the *Miconia.* The 25-foot-high guava trees are distributed along cattle trails, for their seeds are carried by the coats of the wild cattle.

MOVEMENT TO CONSERVATION

In 1957, on the basis of the 1955 report by the International Union for Conservation of Nature and Natural Resources, UNESCO, in cooperation with the Ecuadorian government, sent an investigative team which surveyed the condition of the native flora and fauna, and sought to determine an appropriate site for a permanent scientific research station. At the 1958 International Zoological Congress, appropriately held on the centenary of Darwin's first public statement on the theory of evolution, the Galápagos project was officially recognized. From the Galápagos Committee formed at the Congress, under the acting chairmanship of Sir Julian Huxley, came the Charles Darwin Foundation for the Galápagos Islands. The sole purpose of the Foundation is the preservation of the natural environment and the promotion of scientific research relating to conservation in the islands.

Construction of the Charles Darwin Research Station began in 1960 on Santa Cruz at Bahía Academy near the village of Puerto Ayora. The Station was officially inaugurated in 1964. It offers basic laboratories and equipment, a workshop, dormitories and dining hall, and a modern, scientifically equipped motor cruiser to resident and authorized visiting scientists.

The staff and sometimes visiting scientists conduct educational programs for local residents, schoolteachers, tour guides, and occasionally for interested visiting groups. In addition, the director and staff of the Station work in close cooperation with the National Park Service in making conservation policy decisions for the Galápagos.

In 1959 the government of Ecuador designated all uninhabited areas of the Galápagos a National Parkland, so wildlife would be protected and any further development, if deemed conducive to the preservation of the natural environment, would have to be carefully planned. Hunting and capture of all native animals is expressly prohibited. Collecting of animals or plants is illegal, except for scientific purposes authorized through the Darwin Station. The Ecuadorian National Park Service officially began operations in 1968 with the assignment of several wardens to each of the major islands. Further expansion for agricultural use was prohibited and further immigration to the archipelago restricted.

CONSERVATION PROJECTS

With the establishment of the Darwin Station and the National Park Service, projects were inaugurated to protect endangered species and to restore the Galápagos to their original state. Miguel Castro, an Ecuadorian born in the Galápagos, took upon himself the task of surveying the remaining tortoise populations. A system was devised to number the tortoises by cutting small wedges out of the various shell sections, so their movements, ages, and growth rates could be studied. Using this method Castro marked tortoises in the most difficult locations. The San Cristóbal subspecies of tortoise was thought to be all but extinct by the California Academy of Sciences expedition of 1906, but Castro's diligent search over the dry, volcanic western end of the island resulted in the marking of eighty tortoises.

Even with the end of man's predation upon the tortoise, his introduced animals continue to destroy the tortoise young. Pigs uproot nests and eat the eggs and helpless young, and black rats kill tortoises up to three and four years old. On many islands replenishing of populations by new generations was not occurring. Although no predators have been introduced on Española, so few tortoises remained that males and females no longer encountered each other for mating.

To aid repopulation of tortoises, Miguel Castro, the Darwin Station, and the National Park Service constructed an incubation and rearing house and tortoise corrals near the Station. . . . Today eggs from six endangered races are being incubated at the Station, and more than 250 young tortoises are being raised there. A large tortoise reserve was established near the Santa Cruz highlands for the protection and study of the Santa Cruz subspecies. The tortoise project has been a tremendous success, guaranteeing the survival of one of the most ancient and intriguing animals in the world.

Because introduced animals constitute the greatest danger to the Galápagos Islands, elimination of these feral animals is the primary conservation project of both the National Park Service and the Darwin Station. Small Santa Fe, lying only fifteen miles from the conservation headquarters at Bahía Academy, was the natural place to begin the removal of introduced animals. Given the limited resources available, hunting the goats presented the only feasible method for their elimination. Hunting parties made repeated visits to Santa Fe until nearly all the goats had been killed. Early in 1971 the island began what must be a slow, and in some respects, incomplete recovery. Money has recently been allocated for an elimination project on Isla Pinta, where under the pressure of an estimated 20,000 goats, the vegetation is being devastated.

Santa Fe and Pinta are small islands, and yet the removal of goats is difficult. Santiago, Isabela, Santa Cruz, and San Cristóbal are much larger and consequently the elimination of feral animals is immeasurably more difficult. On these larger islands complete elimination will be impossible. Control of the size of introduced animal populations is all that can be expected. This kind of control requires sufficient funds for a staff that can conduct sustained programs.

Pig control is being practiced now on Santa Cruz and will soon be instituted on Santiago. Methods of control have yet to be devised for the black rat.

CONSERVATION STUDIES

Vital to the success of conservation in the Galápagos Islands is scientific study. Understanding must be gained of individual flora and fauna characteristics and functions, and, just as importantly, understanding of the specific ecologies of Galápagos communities. There are still many gaps in biological knowledge, particularly in entomology and marine biology. Population and behavioral studies still are lacking for some birds and mammals. The breeding ground of the white-vented storm petrel remains undiscovered. Just completed at the end of 1971 was the first thorough study of the famous giant tortoise, and investigation of the endangered Galápagos rat was begun only last December by two Peace Corps volunteers. Much interesting and important study remains to be done. The Darwin Station can accommodate ten scientists, and investigation may also be conducted independent of the Station's facilities with proper authorization. Scientists are encouraged to apply for research permission and assistance through the Charles Darwin Foundation in Brussels.

In 1969 about 200 tourists visited the Galápagos. Well over 6,000 are expected in 1974. This increased volume of visitors potentially threatens the ecological balance, but it also comes as a blessing to conservation projects sorely in need of funds. Besides contributions generated by the visitors' personal acquaintance with the Galápagos, a six-dollar tourist tax is directly channeled into conservation programs. With adequate supervision and size limitations of visiting groups and strict quarantine against introduction of additional plants and animals, many people can come into contact with the rewarding world of the Galápagos without causing their destruction.

As territory belonging to Ecuador the Galápagos are tied to a continent eager to narrow the economic disparity with its North American neighbors. Although Ecuadorians are proud of their Enchanted Islands, few are rich enough to afford the luxury of travel. The majority of the visitors are Europeans and North Americans spending their new leisure time on an interesting biological field trip. To a South American struggling to survive, and a government seeking rapid economic advancement, preservation of a species of bird or lizard can at best be of secondary importance. The economic benefits of a well-conducted tourist program, however, serve the best interests of both the Ecuadorian people and the preservation of the Galápagos. Money generated by tourism can be of great economic importance to Ecuador. Let us be sensitive to the situation and aspirations of our South American neighbors, yet firm in our resolve to protect the Galápagos, knowing the consequences of our own excessive development and exploitation.

EUROPE

16

Comprehensive Planning and Management of the Countryside: A Step Towards the Perpetuation of an Ecological Balance

A. H. Hoffmann

Excerpted and reprinted from United Nations Economic Commission for Europe, *ECE Symposium on Problems Relating to Environment* (New York, 1971) (ST/ECE/ENV/1).

DESTRUCTION (PAST AND RECENT) AND ITS IMPLICATIONS

The history of technology is a story of concentration of widely dispersed energies. At the beginning it was mainly a concentration in space, and later on also in time. Only when man came to practise agriculture was the construction of cities made possible. These cities were provided with food and fuels by the surpluses from the surrounding countryside.

With the advance of agriculture came the deforestation. This made Europe, through the centuries, almost devoid of natural forests which were highly diversified because of climatic, edaphic and biogeographic interactions. Wars and the beginning of the smelting industry were also big reducers of forests, because of their demand for fuel. The forests of England were almost non-existent by 1700, and the whole of Europe was changed to a completely cultivated landscape losing in the process of development an enormous amount of diversity. Contemporaneous landscapes are made up of only relatively few crops—and their weeds, that can now be found over many wide areas.

The whole history of agriculture, and later, forestry, is basically a continuous effort to create simplified ecosystems in which specialized crops are kept free of other species which interfere with the harvest through competition. But often this effort has led to quite disastrous effects. Diversified systems have built-in insurances against major failures, while the simplified systems need constant care. The dry and eroded regions along the Mediterranean are sad reminders of badly managed and over-exploited land, a consequence of which was the shortage of subsistence followed by a growing urban concentration of people. This has occurred throughout Europe since the industrial revolution with our cities growing constantly and with a shortage of housing as a corollary. The ever more intense communication system disrupts the ecosystems and destroys the scenery. Noise and pollution from our transport system again result in a loss of diversity as

> **In the West, our desire to conquer nature often means simply that we diminish the probability of small inconveniences at the cost of increasing the probability of very large disasters.**
> KENNETH BOULDING

species of animals and plants retire to more remote areas. Industries, with little concern for their surroundings, are geared only towards economic targets with fierce competition between the firms to cut down production costs, and scant care about the disposal of their wastes, or the dereliction they are causing.

Dereliction, which is a truly ecological disaster, forms ecosystems which have been set back to beyond their pioneer stage and which may have very little hope of return. Throughout man's history, war has been one of the main culprits in creating dereliction. The post-war efforts at rebuilding which, as a rule, are stimulated by incredible idealism, accelerate the effacement of ruin and give an example of what could be achieved by industry if only the necessary idealism and laws existed and were enforced in this sector. While technically feasible solutions for reclamation have often been dismissed with the argument that they involved excessive costs, the question of at what level costs really become excessive has never been defined appropriately. What about the future, when technology will offer only limited or very costly escape hatches! Therefore, the more thoroughly it is understood that not only is there no free lunch but also no free beauty, the easier it should become to find good solutions.

Hydroelectric plants and the intensive use of ground water have in places lowered the water table by some 80 feet without any serious thought being given to the long-term implications this might have for a whole region. Rivers have dried up as their water has been used for the production of electricity and the microclimate of whole valleys has been changed accordingly.

Tourism which today is a magic word, can and often does contribute as rapidly as other factors to the destruction of the landscape. The auxiliaries of tourism are often more important in their effects on the landscape than the tourists and their actual accommodation. People have to be brought to a particular place which means construction of roads, airstrips, car parks, restaurants, etc., a fact which often has disastrous effects on coastlines. Immediately after the construction of a road which opens up a new stretch of coastline, land prices start booming and it is hardly possible any more to stop development or to acquire land in order to save parts of the unspoiled coastline.

However, the increase in tourism in recent years has had more consequences than merely the easily observable crowded beaches and road congestion. Ecological damage, especially in areas with fragile ecological systems such as coasts, lake shores and so on, has only too easily been brought about. Dunes are especially susceptible since their stabilizing vegetation does not stand up to trampling and is therefore all too quickly damaged. Inland watercourses, an asset to every landscape, are often cluttered with second homes without any proper sewage treatment system or any planned structure or integrating architecture being developed. The same, unfortunately, is also beginning to happen in the countryside which, until recently, preserved its purely rural character.

Pressure of mass tourism on a landscape is also manifested through erosion and litter which accelerates degradation of the whole region. Mountain slopes are often deforested without consideration being given to the ecological, or scenic and aesthetic repercussions resulting from the need to accommodate the ever-growing masses of skiers. That such actions favour avalanches is one among many factors which should receive careful consideration.

The enormous forest fires which each year ravage the southern part of France, parts of Italy, Greece and Spain are mainly started by cigarettes or matches carelessly thrown away by passers-by or tourists. Such disasters are also caused by fires not properly extinguished by campers, by people burning waste, or by farmers burning tough grass on roadsides and slopes. Fires are rapidly propagated in cases where, because of forest economy, monocultures have been planted. Biological productivity and ecological wealth depend on a wide variety of different species. If a natural ecosystem is altered by man, it is generally made less complete and therefore less resistant. These are examples of degradation through lack of planning and through bad management.

A staggering example of short-term planning is that of flood plains. Most rivers have, over thousands of years, formed fairly wide plains where the river floods its banks every two, ten or a hundred years. These plains permit the water to slow down its flow and are generally fertilized at the same time through the mass of silt a river carries from the mountains at times of flood, which accounts for the fact that flood plains are usually extremely fertile. In these plains we have drained marshes at the foothills of the mountains, we have built roads, houses and factories. We have regulated the river by means of dams and in other ways in order to protect the roads and buildings against floods, and at higher altitudes we have constructed regulating reservoirs at enormous cost, also to prevent the rivers from flooding in their lower course. This is a typical example of short-term planning which does not take into account more long-term costs and benefits. Valuable agricultural land is lost through heavy building in the flood plain, the fertile soil once deposited on the plain is now carried to the sea through the river systems or is deposited in the reservoir and is slowly filling it up. The river has to be regulated and, on top of this, new regulating reservoirs have to be built once the old ones are filled up. This example shows that the cost to the community is far greater than it would have been if the roads and other constructions had been laid elsewhere than in the flood plains.

Another example: a poor valley in the south of Switzerland sells the water of its river to an electric company which constructs a dam and shoots the water through pipelines into turbines and from there directly into a lake. Only then do the inhabitants of the valley realize the effects on their own landscape. The old river-bed is dry, waste decays in little pools which are left in the river-bed, tourism begins to decline. The ultimate effect on the whole ecology of the valley can only be guessed at. It is at any rate a great loss of diversity and a long-term economic disadvantage. . . .

COMPREHENSIVE PLANNING OF ALL DEVELOPMENT IN THE COUNTRYSIDE

It has not yet been fully recognized in all instances that the landscape needs more comprehensive planning on a regional basis. Landscape is an organic entity and one part cannot be treated in isolation from another. The purpose of landscape planning is to conserve the essential values of the landscape while adjusting it to the needs of increased population, of industry, and of modern farming methods. Without this deliberate planning, the process of industrialization and growth is likely to produce environmental chaos and impoverish natural assets.

We must find a way to plan for the preservation of the resources of our habitat, the biosphere, which is irreplaceable. To do this, some way of controlling growth must be found, since too much growth will defeat the efforts of any planner. We will have to plan even against what is called progress, in the interest of future survival.

. . . We cannot confine our policy to mere protection because of the great losses of natural variety. We must re-create and maintain the diversity of our landscape and environment.

Planning of the countryside based on ecology presupposes maintenance as well as modifying operations for the whole landscape. Nature does not need to be defended against the species *Homo sapiens* but against the deterioration which modern technology causes, against the levelling down of the variety and the diminishing of the original richness of flora and fauna. Conservation stands for maintenance and an increase in environmental variety.

We have to change from single purpose to multipurpose land use, from segregation to integration. There is an urgent need for research aimed at establishing and enlarging areas of agreement. A demand for research may be said to be just a sophisticated way of avoiding action; however, often enough, it is known to enhance some action. Naturally, this will take time. Meanwhile, new ways must be explored of reaching decisions in which the various voices concerned, the governments and the scientific communities, including ecologists, biologists, health services, agriculturalists, landscape planners, foresters, town planners, and others can be heard.

An integrated and balanced landscape also fosters a two-way link between town and country. Until quite recently the town's main interest in the countryside was the provision of food. Now, townspeople visit the countryside more and more for recreation purposes, something that can hardly be prevented. In landscapes with a more or less balanced ecology which is self-maintaining, the user, whether animal or man, provides most of the maintenance involved. If a method of controlling the holiday-maker or tourist could be found which ensured a similar kind of maintenance to balance the wear and tear on the environment, the tourist could become a part of the ecosystem. It is therefore essential to plan ahead and with some advanced knowledge of the numbers of visitors likely to invade the countryside and to limit these numbers so that the countryside will not be degraded by tourism. Still more important would be injecting the recreation surge from urban agglomerations into the landscape in such a way as to make it compatible with the functioning of the landscape. This would constitute planning on ecological principles.

Many social and technical advances came out of catastrophes, for example the floods in the Netherlands, which probably gave its people an incentive to achieve the greatest total benefit. A plan based on ecological principles for the landscape, however, should be feasible without the spur of a disaster. One main factor which has to be overcome in Europe, in the first instance, is the idea that any kind of planning is an interference with the individual. It is because of this that comprehensive planning for multipurpose land use must involve all the professions concerned with the land: the engineer as well as the agriculturalist, the ecologist and the farmer, the landscape planner, the architect, etc. Once these different professions realize that there can be constructive teamwork and that consultations are indispensable a good step forward has been made towards an integrated and balanced landscape.

A growing problem is that of waste material, the discarded left-overs of our advanced consumer society, which is becoming a real threat to the countryside and strains the facilities of municipal governments. There are three methods of dealing with waste. It can be deposited, burned or composted.

The simple depositing of waste material in the countryside, if not well planned, gives rise to various dangers: groundwater pollution through the washing out of chlorides, sulphates, nitrates and carbonates; smells; flies and rats; and the propagation of illness, quite apart from its unsightliness. Being the cheapest method, it is widely used but is rarely satisfactory.

The burning of waste, especially of the unreturnable bottles and material containing chlorine, releases hydrochloric acid into the air which cannot be overlooked because of its corrosive action on buildings, etc. The mounting acidity of rain comes mainly from the sulphur oxides released through the burning of oil and coal.

The composting method, like incineration, is a volume reduction method. It fosters the conditions necessary for micro-organisms to transform organic matter of plant or animal origin. The aim is to make the waste hygienic. Through the aerobe decaying of the waste material in stacks, which have to be regularly turned over, the organic parts will be changed into humus, and the total waste into useable ground-conditioning material as quickly as possible. The composting method is the most satisfactory even though a certain amount always has to be burned and/or deposited. It therefore always involves a combined system. The Netherlands, where the composting method is widely practised, uses 50 per cent of the available compost for the creation of urban recreation areas. The rest is used for agriculture, which shows that compost can be used more and more for open spaces and recreation areas. This has, however, led to the wrong idea that only the selling of the compost obtained justifies the installation of a composting plant, which does not seem to be the case. In North European countries, agriculture has, because of its climate, little use for compost. Selling of compost is therefore hardly possible. If there are nevertheless many composting plants, the reason lies in the fact that it is a comparatively cheap method of waste disposal. The combined system is about 0.8 per cent cheaper than incineration, which also makes it more interesting from the financial point of view. Even though this method is effective in dealing with waste it is difficult to practise in large towns because of the space it requires and quite often also because of transport difficulties. Considering the constant increase in waste mainly due to the increase of packing materials, the recycling of such material should be advocated.

Considering that recycling and saving of materials have been practised for hundreds of years, it seems almost blasphemy to waste resources the very moment we start to realize that these resources are limited. This is especially true if we keep in mind that such a procedure creates a twofold problem: decline in natural resources and the need to dispose of waste.

The previous statement that there must be planning against growth, also applies to population. It is well known that animal communities, once their population gets excessive, start to be regulated either through famine or disease, or through self-destruction. The best known phenomenon of this sort is the 4-year cycle of the lemmings (*Lemmius lemmus*). History shows that something similar has already happened to man. A recent textbook example is Ireland where two problems are touched upon. When the potato was introduced into Ireland, the country had a diversified farming system which supported a population of about one million in 1670. The potato adapted very well to soil and climate and its abundant yield allowed the farms to be divided into small holdings which could produce enough revenue to support a family. Practically the whole agriculture moved over from a diversified system to monoculture of potatoes. The great potato boom encouraged high population fertility and by the year 1845 the number of Irishmen had increased to over eight million. In 1846, the potato plant was struck by a

The immense duration of certain man-made landscapes contributes a peculiar sense of tranquility to many parts of the Old World; it inspires confidence that mankind can act as steward of the earth for the sake of the future. . . .

. . . In many places . . . the interplay between man and nature results in creative symbiotic relationships that facilitate evolutionary changes. Man continuously tries to derive from nature new satisfactions that go beyond his elementary biological needs—and he thereby gives expression to some of nature's potentialities that would remain unrecognized without his efforts.

RENÉ DUBOS

fungus disease and for six years the crops failed again and again. During these years a million people died of hunger and another million emigrated. The population decreased and adjusted itself at around four million, near which it seems to have stabilized. . . .

REHABILITATION NOW: RESULTS

Naturally we need roads, factories, housing, etc. All these can be achieved and place found for them, provided that appropriate planning and management is introduced allowing for the countryside to be kept as a human environment worth living in. It has been done and it is being done, but by far not enough. It should be a condition for every form of development.

For instance, the mining of brown coal in the Ruhr area was accompanied at the very beginning by plans to reclaim the area of open mining. The results are quite amazing. Numberless acres of land which would have been ruined have been and still are being reclaimed, forming an even more diversified landscape than before. New forests with lakes and nature trails give a good example of a multipurpose land use. The lakes with clearly defined adjacent camp sites give the possibility of recreation, swimming and boating, near the agglomeration of Cologne. Strict regulations of transport and access keep vehicles on the main roads or the specially designed parking lots. The well managed forests produce not only an interesting amount of wood but are also well stocked with game bringing high revenues from hunting. The new areas of agriculture, which have been reinstated through the silt-flooding method, taking the layer of fertile Rhine silt from the areas to be mined and bringing it to the areas to be reinstated, are now more fertile than they were before and produce a higher yield of crops. This is a clear example of a success story of long-term planning.

In the Netherlands, where the saying goes that God created the world but that Holland was created by the Dutch, the face of the country is constantly being changed in a very constructive and inspiring way. In the polders of the Zuidersee, 500,000 acres of new land will be created before the end of the century. First this newly gained land was almost entirely dedicated to agriculture, with only 8 per cent of the total area in use for recreational and forest purposes. Today, however, 20 per cent or more is already being used for non-agricultural purposes. The enormous demand for urbanization, recreation, natural areas, beaches and forests is putting its mark upon the land use plans for these areas.

The Delta region of the Rhine, which was planned as a new strategy against the sea floods and established as a legal entity through the Delta Act of 1958, is now rapidly taking shape. The whole concept of the plan is based on multipurpose land use. Apart from the main purpose of protecting the land from the sea, a new highway system will be constructed on the dams linking the agglomeration of the so-called "Randstadt" to the recreation areas of the Delta. Water control and water supply are also a major factor and around 1980 large freshwater basins will come into being when the sea is cut off by the dams, a vital element for ever-growing population and industry. The newly created recreational areas in the former estuaries and on the coast will undoubtedly be of benefit to economy and society. Planning policy with respect to landscape is directed to the protection of valuable existing elements as well as to a great many creative measures. Sand-banks in the former tidal streams will be transformed into islands, some of them completely protected, whereas others will be partly planted and provided with recreational facilities. The measures taken are no compensation for the loss in natural wealth, but they ensure a new balance in freshly created ecosystems.

Also in the Netherlands, the approach to forestry has been completely changed. The unattractive chessboard pattern of roads is being changed by new roads and trails interspersed with open spaces for picnics and camping. The whole approach of the Netherlands is an approach based on multipurpose land use, which helps to bring diversity to an intensively used man-made landscape.

Sweden is another good example of a very advanced policy of landscape conservation and management. It was able to establish laws which considerably reduced the rights of the individual to use land as he saw fit. For instance, the Beach Act of 1952, for the protection of the sea and lake shores, keeps shorestrips, up to a maximum width of 500 metres, free of construction, in order to make them available for recreation. Furthermore, the Swedish Parliament has voted a new law for the protection of the environment which came into force on 1 July 1969. The law deals with pollution of water and air, damage to the environment due to vibration, noise and influence of light. The preventive element of avoiding disturbance to the environment is especially accentuated in this law. The result of an inquiry by Folksam, a Swedish insurance company, entitled "Fight the destruction of the environment" was striking. One of the questions worth pointing out read: "How do you react to the statement that the Swedish population must in the future renounce the hitherto existing yearly rise of its standard of living, if thereby the increasing stress on the environment can be avoided?" More than 50 per cent of the people interviewed answered that the lowering of the increase in the standard of living was just, reasonable and necessary.

These three examples prove that something is being done, that action has started, that awareness of the value of natural resources exists, and that it is increasing all over Europe.

17 Prescription for the Mediterranean

S. J. Holt

S. J. Holt is director of the International Ocean Institute, Malta.

Reprinted from *Environment,* 16(4) (May 1974), pp. 20–33, by permission of the Scientists' Institute for Public Information.

The *Economist* of London, regarded usually as a "responsible" weekly, published in March 1973 an article on the Mediterranean in which it was said that "its ecological balance was not seriously disturbed until the early 1960s. Now over-population, the tourist boom, industrial development and maritime irresponsibility are combining to turn it into a dead sea." Two years earlier the *New York Times* headed an article "Is the Mediterranean Dying?"; and about the same time Tony Loftas writing in the *New Scientist* on "Mediterranean Pollution—Another Year of Neglect" asserted that "in spite of pollution scares the health of the Mediterranean continues to deteriorate." Reports of incidents leading to pollution of one kind or another get headlines in many parts of the world when they relate to the Mediterranean. This is due partly to the historic associations of this sea and the lands which surround it. But also the area is visited yearly by millions of holiday-makers—and numerous scientists—from outside the region, attracted by the special combination of past glories of civilizations and the present sunshine and warm water.

Neither is the Mediterranean "neglected" by the people who live there—at least many of them are making a noise about its present state, and numerous organizations have been holding international conferences, passing resolutions, and drafting declarations and conventions. The situation is however clouded by a real lack of understanding of just what is "its present state," and complicated by the large number of interested bodies with a variety of ideas about what is to be done about it. Certainly the local people *should* be concerned. In September 1972, Professor W. Brumfitt, a microbiologist, told the "Medicine in the 70s Symposium" that one in ten people living along the

Mediterranean coastline had changes in their blood showing exposure to hepatitis virus. Here, I use the term "pollution" as defined, collectively, by the agencies of the United Nations system, for their purposes: "Introduction by man of substances or energy into the marine environment resulting in such deleterious effects as harm to living resources, hazards to human health, hindrance to marine activities including fishing, impairment of quality for use of sea water and reduction of amenities."

Man can significantly influence the sea in more ways than by "introducing substance or energy." Most commonly he overfishes its living resources; that is to say by fishing intensively he reduces stocks of fish to an excessively low level so that the *harder* he fishes (more and bigger boats, for longer seasons, and more efficient nets) the fewer fish he catches. I mention this here because one of the important, and most difficult, tasks of marine scientists everywhere is to determine to what extent changes in the resources are due to fishing or to other causes—such as pollution.

In the Mediterranean, and in other more-or-less closed seas, man can cause other changes as he still cannot in the ocean. In the Mediterranean he cut the Suez Canal a hundred years ago, and since that time animals and plants native to the Indian Ocean and Red Sea have been finding their way into the eastern Mediterranean; some of these—species of fishes—have become numerous enough to be caught in commercial fisheries (the reverse process—migration of Atlantic/Mediterranean species to the Indo-Pacific—is much less common, but has been reported in the last few years). Much more recently—but having more immediate effects—the high dam at Aswan has been completed. This has substantially changed the southeastern Mediterranean, by stopping both the freshwater inflow there, and the Nile sediments which used to "fertilize" the sea area. So it has been called—not very suitably—an example of "negative pollution"—a *withholding* of materials rather than their *introduction.* Lastly, in addition to these diverse changes caused by man we have continuing natural changes, such as fluctuation in sea level, which we observed from geological, archaeological, and historical studies, and which continue, albeit slowly.

Notwithstanding the rather considerable recent changes in the physics and chemistry of the Mediterranean, major changes in its biological aspect which might be the result of pollution are not so easy to find. There has been no decline in the total fish catch which has stayed near a modest one million tons annually for several years. Even where the substantial hydrological changes caused by the Aswan Dam reach, recent reports indicate no change in the fisheries off Israel, for example, although there seem to have been local changes in the fisheries in and near the delta, as one would expect.

A SMALL OCEAN

The Mediterranean is, unlike the North Sea, and even more the Baltic, a deep sea—it is in reality a small ocean, or rather a small part of the great ocean of Tethys which was cut off and isolated by the closing, in the remote past, of northeastern Africa with Arabia and the southwestern part of the continent of Asia. Its isolation was later ended by a breakthrough of the Atlantic at the Strait of Gibraltar, across which there is still a rather shallow sill. The continental shelf is limited in extent—being broad mainly in the northern Adriatic and off Tunisia. The Mediterranean is practically without tides because its size and shape do not make it a good resonator of either lunar or solar tidal waves. Evaporation in the Mediterranean area is high relative to the nearby Atlantic, and this loss of water is only partially compensated by the Ebro, Po, Rhone, and other large rivers flowing into it. Evaporation also causes the Mediterranean water to be saltier, and therefore heavier, than the Atlantic water near the surface. Evaporation is higher towards the eastern end of the Mediterranean than the western, so the sea level slopes slightly down from west to east. Consequently light water to make up the loss flows in through the Straits of Gibraltar in the upper layer and down the gradient. Currents flowing in the Northern Hemisphere tend to deflect to the right because of the centrifugal force of the earth's spin. The inflowing surface current, which is impeded by the rather shallow narrows between Sicily and Tunisia, therefore follows an anticlockwise path in the western basin. A lesser current continues into the eastern basin and likewise follows an anticlockwise path there.

At a very few locations, particularly in the Gulf of Lion, and at times of winter storms, the surface water which has become saltier and heavier through the summer, becomes colder and thus even heavier and rather rapidly sinks to the bottom. This "reverse fountain" renews the oxygen which had become diminished by the respiration of animals swimming and feeding in the surface layers. Lastly, some of the heavy Mediterranean water flows out into the Atlantic underneath the incoming water. This phenomenon was exploited in the Second World War by submarines which discovered they could drift out into the Atlantic unobserved by submerging and shutting off their engines. The deep Mediterranean water can be traced far out into the Atlantic, almost to its western side.

So if we consider the Mediterranean as a whole we see that there are active processes of oxygen renewal which ensure it will not easily become stagnant. Pollutants which can mix and dissolve in the water can eventually find their way out to the Atlantic, or are precipitated in the deep bottom sediments. On the other hand, spilt oil stays at the surface, and therefore tends to accumulate in the Mediterranean. Floating waste, such as plastics, in the Atlantic can of course also be carried in by the surface current. Thus, when a freighter lost, in the Atlantic, a number of dangerous containers carrying tetra-ethyl lead residue, U.S. embassies warned the authorities of the Mediterranean states that they might appear floating near their coasts.

> **The Mediterranean is not . . . a *single* sea, it is a complex of seas; and these seas are broken up by islands, interrupted by peninsulas, ringed by intricate coastlines. Its life is linked to the land, . . . and its history can no more be separated from that of the lands surrounding it than the clay can be separated from the hands of the potter who shapes it.**
> **FERNAND BRAUDEL**

Not only the oil itself, but pollutants which are more soluble in oil than in seawater, will tend to be accumulated at the surface in the Mediterranean. Oil droplets can be taken into the small organisms which are the food of the fishes, giving the fish catch a bad taste, and with them the more insidious contaminants such as the chlorinated hydrocarbons—pesticides and polychlorinated biphenyls (PCBs).

INCREASED TANKER TRAFFIC

When the Suez Canal was closed it was at first thought that the Mediterranean would at least benefit from a reduction in tanker traffic and hence in spilt oil. But, on the contrary, with the development of oil resources in North Africa, the construction of pipelines, the enormous growth of oil unloading and refining facilities on the northern shore—especially in Italy—and the traffic of supertankers entering loaded by Gibraltar, the situation is now very much worse. Despite some improvement as far as tank-cleaning is concerned, through the application of the international conventions under the Intergovernmental Maritime Consultative Organization, and the "load-ontop" system adopted by oil companies, any tourist will vouch for the fact that in most places beach pollution by oil and by petrochemical residues gets worse each year. In part this is because some tankers do not obey the rules, in part because the international rules still permit tank-cleaning in a limited area of the high seas off the coast of Libya, and because pollution of local origin—engine oil from boats, dumped dirty oil from vehicles and so on—continues to increase, with practically no control.

From west to east, north and south, Mediterranean countries are now exploring for oil on their continental shelves, and soon will be trying deeper than that. Some of them have found oil. Like the seabed off California, the Mediterranean floor is in many parts "active"—volcanoes and earth movements are common. So spills must be expected here, too, no matter how careful and technically adept those in charge of future oil extraction are.

POLLUTION

The Mediterranean is a sea of diverse and seemingly incompatible uses. Swimming and boating there are popular because the water is clear as well as warm. It is clear because the water is not very productive biologically, which is why the fish catches are modest, by world standards. Nevertheless, the catches are valuable—they are consumed directly, mostly fresh; practically none are converted to fishmeal for animal feeds, and they fetch high market prices. The fish resources probably benefited on the whole from the raw sewage pumped into the Mediterranean for many years from the coastal settlements. The populations of Mediterranean countries still increase, more of them are moving to the coastal zone, some to work, some to occupy part-time their scattered seaside houses, and they are joined by "foreigners" from further afield each year. The vastly augmented sewage still mostly goes into the sea untreated, and now spiced with modern additions such as synthetic detergents, with locally most unpleasant results. In fact the detergents in seaspray have been blamed by Professor Lapucci, of Pisa University, for deterioration of the pine woods along the coast of the Tyrrhenian Sea—Neptune's revenge. In some urban areas a beginning is now being made to install sewage-treatment plants, but the capital cost of this throughout the Mediterranean will be enormous. If the sewage is treated and, instead, wastes containing inorganic nutrients are pumped into this sea, some species of fish again may benefit, provided the effluent is reasonably dispersed in time and space. But we can expect the classical water clarity to be diminished and some recreational attraction to be lost—or perhaps changed, with fewer swimmers and divers and more fishing for sport.

However, just as a possible reprieve from urban wastes can be glimpsed, the Mediterranean is being called upon to absorb another order of magnitude of industrial wastes, including warm water—and perhaps low-level radioactive wastes—from power plants located on its shores. Such wastes are entering by the rivers, by coastal outfalls, and by dumping from ships. The latter has been the cause of one of the most recent and serious international incidents, as when it was found, early in 1973, that an Italian company was dumping large quantities of acid and heavy metal wastes from titanium oxide manufacture in deep water near Corsica. So far most of the industrialization is on the north shore, particularly of the western basin. And wastes from petrochemical industry in particular, whether in the atmosphere—as in southeastern Sicily for example—or in the sea, are definitely not compatible with healthy life, be it of fish or man himself. The northern Adriatic is also in a sad state, exacerbated by the regular seasonal discharge of wastes from sugar-refining—another of the "dirty" industries. But North Africa and the Levant are beginning to contribute their share. For example, last year reports were published of very high concentrations of lead in the sea off the Lebanon coast; in Greece, the Gulf of Thermai has been reported, by an expert committee, as receiving "such large quantities of industrial wastes that it is practically doomed to biological death"; and for much of the year the Lagoon of Tunis is now a stinking swamp.

The Mediterranean has, since early times, been famous for its maricultures—the raising of fish and shellfish in ponds, lagoons, and sheltered bays. The shellfish—mussels and clams—are efficient concentrators of just those things in seawater which man puts there and does not want back again—pathogenic organisms and toxic compounds of heavy metals. It was natural, but not necessarily justified, therefore, that when cholera broke out last summer in Naples the mussels would be blamed. At the same time, cholera or no cholera, the authorities in Malta destroyed the new mussel "park" which had adopted a practice not unknown in Italy of fattening the mussels, just before sale, by hanging them near to a main outfall of untreated and nutrient-rich sewage!

WORLDWIDE CONCERN

One could multiply such cases almost indefinitely; it is now time to look briefly at what is being done about the situation as a whole. While a multitude and variety of local cosmetic actions are essential and urgent, the problems of the Mediterranean are overwhelmingly international. There are very many international organizations concerned with the Mediterranean—European ones, North African ones, regional and subregional ones, and worldwide organizations. Furthermore, the headquarters of many of this last category are on or near the Mediterranean—Rome, Geneva, Paris—and its problems come readily to their attention. The General Assembly of the International Union for Conservation of Nature and Natural Resources (IUCN), meeting in Canada in September 1972, pressed the authorities of Mediterranean countries to "demand compliance with the most stringent regulations" concerning the oil industry, and also reaffirmed its support of movements to establish marine parks and reserves in the region. In fact, such movements are expanding and seem to be rather successful, at least in holding back the pollutant tide in very limited areas. In many countries they are justified and even financially supported as areas from which "baseline studies" can be conducted to give the sorely needed scientific evidence of the effects of man here. . . . A regional conference on Mediterranean marine parks [was] convened [in 1973] by the regional government of Campania (Naples and Southern Italy) with the cooperation of the International Ocean Institute of Malta. Many concrete actions may flow from this, but especially training and information about marine ecology to the people of this area. It illustrates, too, another trend: that is, international action at the level of provinces and cities as well as at the level of national government. On the initiative of the mayor of Beirut a "World Intercommunal Conference for the Protection of the Mediterranean Sea Against Pollution" was also held [in 1973]. It adopted a document, now called the Beirut Charter, laying down principles to be followed, cooperatively, by urban authorities, and agreed to have annual conferences elsewhere in the next few years. The Interparliamentary Union has also recently got into the act, although it will, I presume, be able to influence events only indirectly, and in the Mediterranean countries which have effective parliaments—perhaps a minority of them.

Regionally, the Organization for Economic Cooperation and Development (OCED), the North Atlantic Treaty Organization (NATO), the International Commission for the Scientific Exploration of the Mediterranean Sea, and others have recently interested themselves in aspects—scientific, technical, economic, and legal—of the Mediterranean pollution problem. The OECI study is a pilot one under the leadership of Spain. Subregionally there are a number of concrete initiatives. One is the so-called Ramoge project of St. Raphael, Monaco, and Genoa to study and plan antipollution activities along this very important stretch of coast. Another is a proposal by the government of Malta to its neighbors Italy, Libya, and Tunisia, that the four countries jointly operate a pollution monitoring and control ship in the central Mediterranean.

Of the worldwide organizations the Paris-UNESCO-based Intergovernmental Oceanographic Commission, which is also linked with the Food and Agriculture Organization (FAO) of the UN and the World Meteorological Organization (WMO), is concentrating its cooperative scientific investigations of the Mediterranean on research related to pollution. The UN Environment Program, sired by Stockholm and now based in Nairobi, is expected to give financial and other support to promote continuous monitoring of the Mediterranean environment. And the International Atomic Energy Agency, which has maintained for ten years a marine radioactivity laboratory at Monaco, has now been joined by UNESCO in broadening the work of that laboratory to cover also research on heavy metals and chlorinated hydrocarbons. Then, as is well-known, the UN itself is attempting to bring a modicum of global order into things legal and marine, including pollution, through the Conference on the Law of the Sea. . . . The most far-reaching move in this region has, however, been made by the General Fisheries for the Mediterranean (GFCM), an intergovernmental FAO subsidiary. The GFCM published in 1972 an authoritative study of "The State of Marine Pollution in the Mediterranean." To anyone familiar with earlier, similar reviews, such as that produced by ICES for the North Sea, the GFCM study was remarkable for the paucity of concrete information. Nevertheless, it brought together effectively what was known, and on the basis of this the council asked FAO's director-general to consult governments and convene a plenipotentiary conference to consider and adopt international legal instruments for the protection of Mediterranean fisheries from pollution damage. . . .*

INTERNATIONAL OCEAN INSTITUTE

. . . The International Ocean Institute . . . convenes summer schools at which young people from Mediterranean countries become informed about

---

**Editor's note* In February 1976, delegates of 12 countries signed a convention in Barcelona, Spain, "to prevent, abate, and control pollution of the Mediterranean and to protect and enhance the marine environment in that area." The accord was approved by Cyprus, Egypt, France, Greece, Israel, Italy, Lebanon, Malta, Monaco, Morocco, Spain, and Turkey. In providing for joint programs to monitor pollution levels, the agreement put special emphasis on combating dumping from ships and planes as well as harmful discharges from coastlines.

the sea and, naturally, about man's misuse of it. That is not, however, [its] *raison d'être.* This is to bring together, in special "study projects" and in annual "convocations," diplomats, lawyers, technicians, scientists, economists, politicians, and the interested public to discuss the governance of ocean space. This is a unique role. From it has emerged a clear idea that while actions in each sector, such as fisheries, oil pollution, and so on are to be welcomed, they must eventually be seen as part of a whole international regime linking, regulating, promoting, and, as far as possible, harmonizing all human activities and interests in this vast area of potential conflict. From such a view crystallized the concept of the "common heritage of mankind" as eventually endorsed last year by the UN General Assembly with respect to the peaceful uses of the seabed and ocean floor. Many people would now like to see that concept extended to all ocean space and its resources—a much more difficult pill to swallow by many governments and especially by the maritime powers. In trying to see more clearly what form the governing and consultative organs pertaining to a comprehensive ocean regime might take, the Council of the International Ocean Institute two years ago selected the Mediterranean for a pilot study. In many ways it is a model for the world ocean. Its bordering states are some "developing," others "developed." It has minerals, fish, tourism, and maritime transport; the interests and potential conflicts among these are manifest by states outside the area as well as within it. So it is possibly a model for a regional regime within the global one. The main proposal which emerged from the pilot study was for an Interim Council of Mediterranean countries, with many-layered representation from national and municipal government, industry, science, and law, to serve initially as a consultative body with respect to all marine problems. This proposition has been put to the Mediterranean governments; it remains to be seen what they will do with it. We are cautiously optimistic, especially as we see looser bodies of citizens—the "consumers" of the sort of world we are all making—and especially students, bringing pressures to bear on authorities, to act before this very special environment really is grossly and irretrievably damaged.

18 First Aid for a Half-Dead Sea

Trudy West

Trudy West is a free-lance journalist and author as well as an architectural historian. One of her latest books, *The Timber-Frame House in England,* is recommended reading for environmental studies in English universities.

Reprinted from *The Ecologist,* 5 (March-April 1975), p. 128, by permission of *The Ecologist.*

Brochures issued by holiday resorts on the Federal German Baltic Sea coast still describe a perfect holiday scene. They promise "a bay glittering in the sunshine, sparkling, sea-green and crystal-clear water." In reality the tourists' holiday paradise has serious flaws. Every day two hundred rivers bring filth in their wake; waste water from sixty major cities along the 20,000 km. coastline flows virtually unpurified into the "Adriatic of the North" and the 70,000 ships sailing in these waters annually—including over 3,000 tankers—burden the sea with their refuse. The result—the FAO, the World Food and Agriculture Organisation, has declared the Baltic to be one of the most heavily polluted seas in the world.

The "Convention for the Protection of the Baltic Sea's Environment," signed recently by the seven neighbouring countries and which will come into force in a few months, is not a moment too soon. Denmark, Sweden, Finland, the Soviet Union, Poland, the German Democratic Republic and the Federal Republic of Germany held laborious negotiations in order to act in unison as countries neighbouring the Baltic Sea regardless of differing political, military and economic interests.

None of the countries that began negotiations a year ago had a clean record. Everyone had contributed to a greater or lesser degree to the Baltic almost being "out of breath" due to lack of oxygen. It was threatened with a similar fate to that of other suffocated inland seas; it was rotting from its bed upwards. The upper layer of the water rests like a lid on the one below so that an unsatisfactory exchange of water is achieved. It has so far been possible to limit but not to stop the salination and fertilisation of the upper layer of water and decay in the deeper regions caused by sulphuric hydrogen. Data have been collected by the "International Council for the Exploration of the Sea" by scientists at the Marine Research Institute from Kiel to Leningrad.

According to cautious estimates, ten per cent of the 4,000,000 sq. km. bed of the Baltic is already devastated and the situation is deteriorating daily. Poisons such as DDT contained in agricultural waste are destroying marine fauna. Organic substances present in waste water lead to the formation of blue algae. Broad slicks of oil lie on the beaches. First-aid measures were urgently needed.

The agreement which has now been reached in Helsinki is the first convention between East and West which, so to speak, overrides systems and straddles frontiers in the cause of environmental conservation. It came about because the Baltic jibbed at its role as industrial sewer and struck back. Fish

It is a curious situation that the sea, from which life first arose, should now be threatened by the activities of one form of that life. But the sea, though changed in a sinister way, will continue to exist; the threat is rather to life itself.

RACHEL CARSON

in the Baltic today contain ten times as much DDT as those from the North Sea. Large areas of the inland sea have been closed off because the marine animals were so full of quicksilver that the catch had to be destroyed.

Now the neighbouring countries intend to combat pollution jointly on the land, from the air and ships. It is the only agreement of its kind so far in the world.* A High Seas Convention devised in London in October 1973 concerned ships only. In addition, it will probably not come into force until three years' time. For this reason, the Baltic Sea countries blazed the trail with comprehensive and forward-looking controls.

Everything prohibited or permitted is contained in thirty articles. The regulations mean new standards of cleanliness for industry, coastal towns and merchant shipping. In short, the Baltic Sea will no longer be a rubbish dump.

A warning is given in a 15-point list of especially harmful substances

---

**Editor's note* A similar agreement was reached in 1976 regarding the Mediterranean Sea. See editor's note, p. 260.

such as heavy metals, carbolic acid, cyanide, various insecticides, radioactive materials and plastics. Cadmium and quicksilver are well to the top of the list and DDT and PCB are to be prohibited in principle. The agreement obliges the partners to undertake all suitable measures necessary to control and reduce pollution of the marine environment caused by the country concerned. Of particular importance are regulations concerning industrial waste water and that of major communities. Stricter guidelines have also been created for cooling water from atomic reactors. Pollution caused by gases from the atmosphere also come under restrictive regulations.

The rule that oil tankers and other ships of 400 tons and above are no longer permitted to discharge oil and oil mixtures into the Baltic is particularly severe. Other rubbish and water for washing the tanks is also not to be dumped into the sea in future. When an oil slick is sighted behind a ship, an investigation is to take place immediately. In future, no industrial waste is to be submerged in the Baltic.

A Secretariat with its headquarters in Helsinki is foreseen as the administrative organ for realising the agreement. It is subordinate to a "Marine Environmental Conservation Commission for the Baltic Sea" which meets at least once a year. Members are constantly to supervise the realisation of the agreement, recommend suitable measures and define criteria for control.

19

Environmental Disruption and Its Mechanism in East-Central Europe

Leslie Dienes

Leslie Dienes, associate professor in the Department of Geography at the University of Kansas, is a research specialist on the Soviet Union and Eastern Europe, particularly Hungary. He has contributed numerous articles on East European industrial location problems, energy policy issues, and regional development to professional journals and is the author of *Locational Factors and Developments in the Soviet Chemical Industry* (1969).

From *The Professional Geographer*, November 1974, pp. 375–81 (most footnotes have been omitted). Reproduced by permission from *The Professional Geographer* of the Association of American Geographers, Volume 26, 1974, L. Dienes.

A highly controversial notion of comparative social studies is the assumption of a broadly universal highway to modernization. The application of this "convergence theory" to the evolving pattern of *social* order under different political systems may be strictly limited. There is little doubt, however, that the costs and side effects of modernization, stemming from an explosive technology, have been strikingly similar under widely different political ideologies and organizations. Environmental disruption is just such a toll to modernization in general and urban-industrialization growth in particular. The magnitude of this problem, the mechanism of disruption, and the societal responses to such deterioration in the People's Democracies of East-Central Europe are still little known and this paper is intended as an exploratory effort.

Air

Smog was first reported in Budapest only in 1958. Today, smog sits over the Hungarian capital 20–25 days of each year, inflicting an estimated 800 million forints ($27 million) damage, not counting cost to human health. Compared to control points outside the city limits, yearly solar radiation is reduced by 7–8 percent. From November to March, radiation is reduced by over 15 percent. The yearly *average* SO_2 content in the atmosphere in 1969 reached 0.41 milligram per cubic meter as opposed to the international norm of 0.15–0.20 milligram. In some provincial cities the quality of the atmosphere is even worse: every square kilometer of Tatabanya, Hungary, receives nearly 400 tons of dust and ashes, together with large amounts of SO_2 (Figure 1).

Air pollution is very serious in Poland and Czechoslovakia, especially in such bastions of heavy industry as the Silesian and the North-Bohemian coal fields. Average daily deposition of particulate matter over the Silesian industrial conurbation was claimed to equal that over London and Berlin combined, with solar radiation reduced to well below 90 percent. Over ten major cities of the District, the measured amount of deposition in the mid 1960's ranged from 500 to more than 1200 tons per km^2 per year. The domestic use of sulfurous coal greatly increases the emission of harmful matter and in winter the SO_2 content of the air frequently exceeds the designated safety limit of 0.25 mgr per cubic meter several fold. On a yearly basis, heat emission from man's activities can approach half of solar radiation.

In Czechoslovakia, the emission of SO_2 and H_2S during the 1965–1970 period increased nearly as fast as total industrial production and the discharge is expected to be double the 1965 amount by the end of the current decade. In the worst affected region of the state, the North-Bohemian coal field, environmental damage caused by atmospheric pollutants from power stations alone exceeded 1.5 billion Czech crowns ($120 million) during the 1960s and life expectancy has been reduced 3–4 years below the Czech average.

Water

Misuse and mismanagement of water resources are widespread and international cooperation for the control of effluents is as yet nonexistent. Hungarian industry treats only half of its contaminated industrial water and applies secondary treatment to only one sixth. Budapest discharges into the Danube two thirds of its 1.2 million m^3 of sewage without any purification, and treats much of the rest only partially. The problem is aggravated by the fact that 95 percent of Hungary's surface flow originates outside the country's boundaries and some rivers arrive in an already polluted state. The small streams in the Northern Industrial Region are the most seriously affected since, at the border, they already carry the unassimilated wastes of the large East Slovak Iron Works (at Kosice) and various chemical plants.

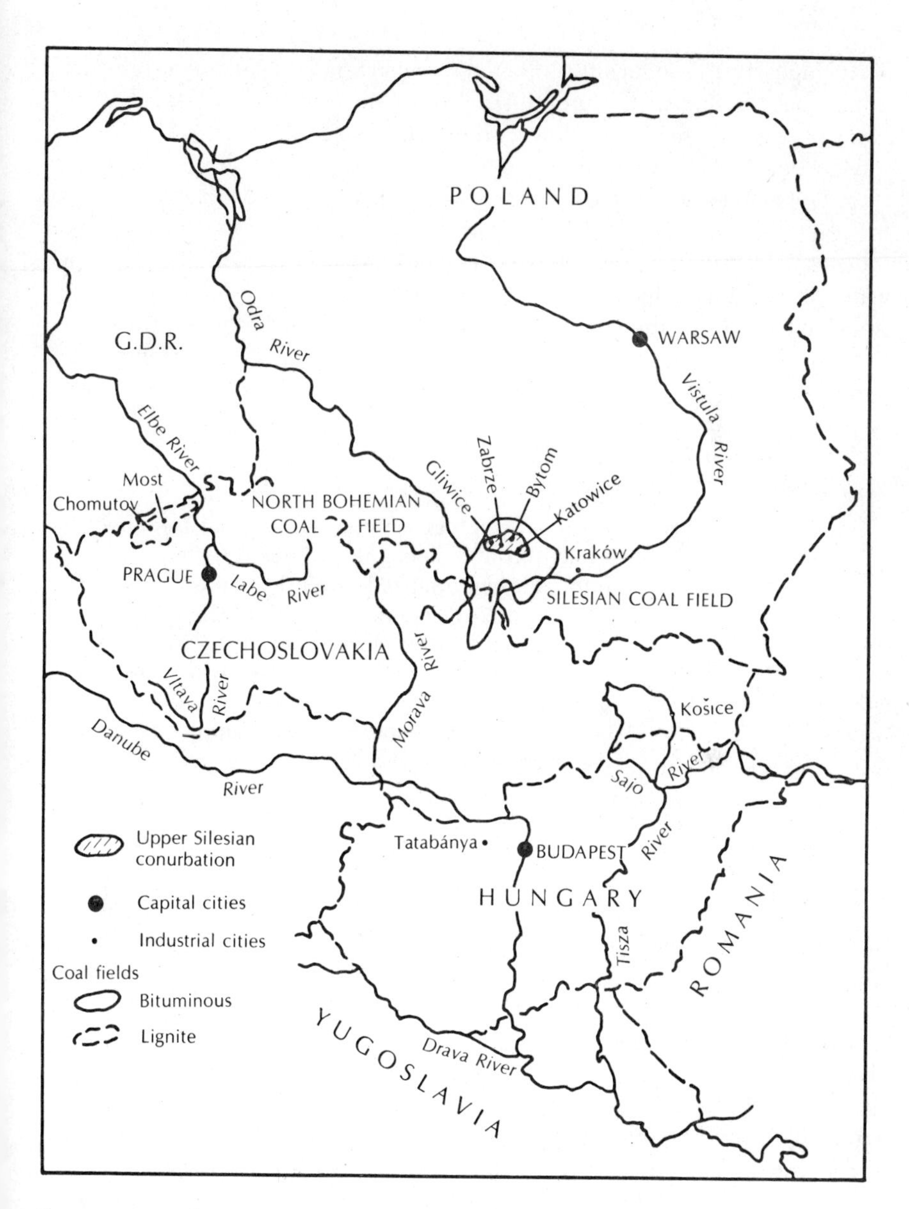

Figure 1 *Polluted areas in East Central Europe*

That Yugoslavia is so far little afflicted by upstream Hungarian waste-water is no doubt owing to the unindustrialized nature of the southern half of Hungary and the far larger flow of the three rivers which cross the Yugoslavian border.

In Upper Silesia, the great concentration of mining and industry results in serious interference with both surface and subsurface hydrology. The Industrial District, situated on a watershed, has been chronically short of water. Its streams deliver to the Vistula and Odra some 15 billion cu. ft. of water, one-tenth of the combined mean flow of rivers supplying the Ruhr District, and even on a per capita basis, less water is available than in the latter region. The radial pattern of Silesian streams rules out the approach so successful in the Ruhr, where the smallest of the three principal rivers, with roughly parallel courses, has been converted into the *cloaca maxima* of

For nearly eighteen hundred miles the Danube winds through highland, plain, and delta. Although immortalized in music as the "Beautiful Blue Danube," its color scheme is rather black and red, for it rises in the Black Forest to empty into the Black Sea, while too often its current has been red with human blood and the flare of burning villages.

FERDINAND C. LANE

the area, greatly reducing the burden on the other streams. In Silesia the large quantities of mine water and urban-industrial effluent discharged equal 80 percent of mean river flow into the Odra and 110 percent of the very much larger flow into the Vistula system. The effluent of the Industrial District pollutes the Vistula with ammonia, nitrates, sulphates, chlorides, oils, phenols, etc., as far as Cracow some 120 km. downstream. . . .

Czechoslovakia's record of wastewater treatment is no better. Annual discharge into the Labe, Odra, Morava, and Danube alone equals half the mean flow of the Vltava at Prague. Less than 20 percent of industrial sewage is treated in any fashion. The situation is most critical in the North-Bohemian and Ostrava industrial districts, regions with scarce water resources, where—as in Polish Silesia—the scanty surface flow is unable to handle the vast amount of urban-industrial effluent.

The Influence of the Economic and Spatial Structures

Whatever may be the precise link of environmental degradation with our anthropocentric belief structure, a view shared by Communist ideologists, the economic causes of such abuse are now well enough understood. The total cost of the production process, which includes such diseconomies as environmental and social dislocation, surpasses the sum of costs entering into the monetary expense calculus. The consequences of harmful spillovers are thus shifted to the community and are not borne by the producers. In market economies, the neglect of such "externalities" in determining exchange value leads to the underpricing of output and to a consumption level and consumption mix which are far out of line with true social costs. Despite the vaunted "comprehensiveness" of socialist planning, to date harmful spillovers rarely have been included in the cost calculation of firms in Eastern Europe either. In the North-Bohemian coal field, for example, power stations produce at a self cost of 0.18 crowns per kwh (the cost of all inputs an enterprise must pay for) while causing 0.27 crown environmental damage per kwh through air pollution alone. The true cost of power to the national economy, therefore, falls between 0.40 and 0.50 crowns per kwh, yet the present selling price is set much lower than that and, at such rates, electric stations cannot afford effective pollution measures.

Control of the economic structure is, of course, the hallmark of central planning. The leaders of these countries aim at not simply maximizing total GNP, but GNP of a predetermined structure, and it could be argued that, even without internalizing social costs, a socialist economy could effect an output mix which would minimize environmental disruption. That socialist countries have long de-emphasized disposable products, fancy packaging and synthetics, is well known. Thus, as Goldman pointed out, there is less to discard, and low labor costs encourage the collection of waste, a lively junk business and clean streets.[1] Similarly, the shortage of private cars has so far spared these states from the worst environmental effects of the automobile.

Yet, the output mix of these countries is not one which would minimize environmental disruption. Since the beginning of the socialist era, the East European governments have emphasized heavy industry and neglected services and light manufacturing—sectors palpably "easier" on the environment. Moreover, the autarchic policies of the past, and for the COMECON as a whole even of the present, meant a reluctance to depend on outside sources for fuel and primary raw materials and resulted in the intensive exploitation of low quality, domestic resources. In Northern Bohemia, for example, surface mining for lignites has been forced to ever greater depths, with the newest works aiming at 200 meters, and the ratio of overburden to coal has now reached 5:1, the highest in Europe. Even in such mineral-poor states as Hungary, the mining sector in the mid 1960's employed 10.5 percent of total industrial labor force. Mining, metallurgy and power stations combined still employ some 16 percent of industrial employment, a far higher share than in most West European countries. Not counting prospecting for oil and gas, the fuel and energy industries received nearly one third of all industrial investment between 1965 and 1970, and in the 1950's well over two fifths.

Presently, the East European countries are increasing the share of the service sector, consumer durables, oil refining and chemicals in their GNP at the expense of coal and primary metals, but petroleum and chemicals are already bringing as many environmental problems as the old-line heavy industries. Excepting Bulgaria and Romania, East Europe is also rapidly becoming motorized. In Hungary, for example, the growth of car ownership has outstripped expectations, with 37 percent of all automobiles concentrated in the capital. Within 3–4 years, there will be one car for every tenth person in Budapest and by the mid-eighties one for every fourth or fifth even by conservative estimates. With the common use of two-cycle engines, the automobile already affects the quality of the air in some East European cities and will increasingly do so in the future.

In theory, socialist planning should also prevent the excessive regional concentration of production and wealth, and a more even distribution of population, cities and industry should lessen environmental damage. One sees but little evidence of such decentralization in the small states of East-Central Europe. The desire for rapid industrial growth tended to give priority to manufacturing expansion in large cities and established industrial regions. Between 1950 and 1967, Budapest and its immediate hinterland received over a third of all Hungarian state investment, though much less of *industrial* capital outlays alone. Postwar development in Polish heavy industry is, in Hamilton's words, a classic example of "spatial inelasticity," involving "the increasing centrifugal scatter of industrial growth in new plants located outside but near existing agglomerations."[2]

As in Western countries, suburbanization and industrial expansion in and around metropolitan centers are devouring agricultural land, creating sprawls and making proper land management difficult. The very slow pace of housing construction and administrative restrictions on metropolitan residence resulted in unskilled migrants from the provinces inundating the surrounding villages where restrictions on settlements did not apply or were very laxly enforced. During the 1960's for example, the 44 villages surrounding Budapest had to accommodate some 60,000 migrants. Virtually no zoning regulations have yet been enforced and land development has been chaotic, made still more so by the increasing proliferation of weekend cottages and shacks.

State Ownership of Resources and Environmental Disruption

Soviet writers insist that "in a society with public ownership of the means of production—there is no contradiction between the interest of society and the interest of individuals—and environmental disruption will invariably be accidental."[3] This is the official position in the East European People's Democracies also but scholars in these countries tend to be less dogmatic when discussing the dissonance between the technological and natural environments. One Hungarian writer feels that *by the turn of the century* a much greater reconciliation between the two can be effected. He admits the existence of a strong pro-production bias in investment allocation even today which results in a reluctance to divert funds to pollution control and impotency in the enforcement of regulations. He even concedes that at the present stage of socialism, state ownership often encourages embezzlement and an indifference to public property.

As Goldman remarked, state ownership of resources and productive facilities means that "there is usually an identity of interest between the factory manager and local government officials."[4] This is no doubt one of the reasons why the recent decree to relocate highly polluting enterprises from Budapest have met with only modest success. This general identity of interest is also reflected in the usually low and laxly enforced fines on polluters and, until recently, in the absence of emission norms in some countries. In Czechoslovakia, the payment of fines has proved much cheaper for enterprises than the cost of installing anti-pollution equipment. In Hungary, air pollution norms for the most dangerous 8 or 9 compounds have only been drawn up in 1973 and these vary immensely (e.g., from 1.5 mg/m^3 to 240 mg/m^3 for benzine) according to geographic areas. Industries are also categorized according to their impact on the atmosphere and must now pay a yearly environmental protection charge which will form a national clean air fund. The charge, however, amounts to a mere 1.2/1000 of gross revenue for the most polluting industries and the clean air fund will at best come to $5 million annually at the official exchange rate.

Growth versus the Environment: The Last European View

The East European nations are generally unimpressed by arguments concerning the impending environmental limits on economic development. They reject the conclusion of the controversial MIT publication, *The Limits of Growth* and regard such forecasts as misconceived and erroneous. Nor is that only the official position. To the average citizen, pronouncements by Americans on the evils of the automobile appear gross hyprocrisy, at least when they are made in support of restricting car ownership in Eastern Europe. I can also attest that pictures of super-highways and double-decked cloverleaves evoke admiration, not dismay.

Those concerned with man's impact on the environment admit to limits on demographic growth (hardly a problem in East Europe today) and concede the existence of ecological constraints to the type of "extensive" economic growth which in the past tended to disregard ecological questions altogether. However, they perceive equally real economic constraints to environmental protection. In the words of a Hungarian sociologist, the violation of these economic constraints for the sake of the environment is not an alternative that can be entertained: it is "national suicide." East European scholars, however, tend to be optimistic with respect to the relationship between nature and technology. They maintain that the relationship between economic development and environmental quality is nonlinear and is not a simple inverse one. One writer even claims that in each historical era it was mankind's reaching and pressing against the "limits of growth" which called forth the creative powers of humanity and made the expansion of global *lebensraum* both a possibility and an existential necessity. In this optimistic view, it should be possible to advance towards an ecological civilization in which, after the attainment of modest comfort, human wants are increasingly directed towards cultural, health, and social services and grace and quality of life style.

As in other fields, East European scholars are looking to Western examples for introducing environmental costs and benefits into industrial and technological growth models. But the reliability of technology assessment itself depends on modern technology, such as high-speed computers, of which these countries are painfully short. For the near future I am also skeptical concerning the rejection of the ideal of mass consumption. The narrow range of consumer goods and sophisticated synthetics in Eastern Europe is viewed not merely as an inconvenience but also as a sign of backwardness, a sore point for these proud nations. A move towards a nonexpansionist, concentric view of the world may be in evidence today in the most developed countries that are becoming satiated with gadgets. "Zero Growth" has no appeal yet in Eastern Europe. Still one hopes that even as these nations have telescoped the abuse of the environment into a shorter time-span, they will resolve the worst aspects of that abuse with less wavering and delay than countries more economically advanced.

NOTES

1 M. I. Goldman, "Environmental Disruption in the Soviet Union," in Thomas R. Detwyler, *Man's Impact on Environment* (New York: McGraw-Hill, 1971), p. 72.

2 F. E. I. Hamilton, "The Location of Industry in East-Central and Southeast Europe," in George W. Hoffman, ed., *Eastern Europe: Essays in Geographical Problems* (London: Methuen and Co., 1971), p. 198.

3 I. Petryanov, "Public Greed? No! Public Weal!" *Soviet Life,* No. 12, 1971, pp. 42–43.

4 Goldman, op. cit., p. 68.

20

Environmental Problems in the U.S.S.R: The Divergence of Theory and Practice

John M. Kramer

John M. Kramer is assistant professor in the Department of Economics and Political Science at Mary Washington College.

Reprinted from *The Journal of Politics,* 36 (November 1974), pp. 886-99, by permission of the Southern Political Science Association.

The effective management of the economy is incompatible with the capitalist system. This is manifested most clearly by the vast amount of environmental degradation in most capitalist countries. It is clear that within the framework of a capitalist economy there is no point in even raising the question of the management of the environment on a nationwide scale. But such a formulation is logical and necessary in conditions of a planned socialist economy.

Voprosy ekonomiki, no. 10 (October 1972), 74.

Soviet commentators often argue that the capitalist system is incapable of halting such adverse consequences of industrialization as environmental degradation. They view capitalist systems as having a "spontaneous" mode of development wherein no one is concerned with the interests of society as a whole. The capitalist entrepreneur supposedly is interested solely in maximizing profits and indifferent to the damage that his activity may inflict on the environment.

The Soviets do not deny that the Russian environment has also experienced some deterioration as a consequence of industrialization. However, they do suggest that the socialist system is far superior to its capitalist counterpart in controlling and minimizing the adverse consequences of industrialization—even if it cannot completely eliminate them.

The socialist system has supposedly attained this superiority because it is a monolithic entity that pursues the "true" interests of society, and thereby can formulate public policy that ensures environmental quality. Since the means of production in a socialist system are state owned, socialism is not hindered by that divergence of interests between society and individual producers that supposedly makes the effective management of the environment under capitalsim almost impossible. Rather, all economic and political actors in the USSR are said to subordinate any parochial interests that they might have to pursue the broader interests of society as embodied in national economic plans.

The following analysis examines the validity of the Soviet argument by focusing on various aspects of environmental disruption in the USSR. Our examination suggests that the reality of the Soviet system differs substantially from the theory of a centrally planned monolithic state pursuing the "true" interests of society. In particular, government bureaucracies in the Soviet Union appear to be the functional equivalent of the capitalist entrepreneur who greedily pursues his private gains to society's detriment. The Soviets have in fact tacitly recognized that a divergence exists between theory and practice, and have labeled the phenomenon "departmentalism."[1] In Soviet usage, the term refers to the tendency of bureaucracies to formulate and pursue policies from their own narrow functional perspective and ignore or devote insufficient attention to the interests of the system as a whole.

ENVIRONMENTAL POLLUTION IN THE USSR

The Soviet Union is in the midst of a severe water crisis.[2] Certain areas of the country are experiencing substantial water shortages, including the Urals, the southern Ukraine, the north Caucasus, and the central Black Earth region.[3] These shortages have caused serious dislocations in the economy and have frequently prevented the further expansion of agricultural and industrial capacities.[4] Water pollution has played an important role in creating such shortages because many bodies of water are so choked with sewage as to be unusable even for industrial purposes.

The USSR Academy of Sciences estimates that almost 100 million cubic meters of completely unpurified sewage daily enters Russia's waterways.[5] This figure represents an increase in the amount of unpurified sewage of almost 90 percent as compared to 1959.[6] Actually the total amount of water pollution is considerably higher than the Academy of Sciences' estimate, which considered only the amount of *completely* unpurified sewage. In fact, even water that has undergone purification frequently contains a significant amount of wastes.[7]

Many of Russia's most famous waterways have been especially hard hit by water pollution. The Caspian-Volga Basin, for example, receives more than ten billion cubic meters of unpurified sewage annually.[8] One source estimated that if all of the unpurified wastes that daily entered the Volga were put in railroad tank cars, the resulting train would stretch from the White to the Black seas![9]

Such major arteries as the Northern Dvina, the Dnepr, and the Oka are increasingly clogged with wastes. A Soviet commentator facetiously noted that there is a wonderful restaurant on the banks of the Oka. There, he reported, "you will be served a royal dish: carp with a 'rose' aroma, perhaps, or a pike with a 'magnolia' scent. You can have still tastier dishes! Perch cooked in benzene, bream in kerosene, or turbot in first-class lubricating oil."[10]

Beautiful bodies of water such as Lake Ladoga and Lake Baikal have also experienced considerable pollution. Pulp and paper combines have especially polluted Lake Ladoga. The color of the lake's water is often a rust hue "as far as the horizon . . . and even farther with a good wind." The lake's bottom is covered with a layer of fiber more than 2.5 meters deep.[11] Lake Baikal, the home of many unique flora and fauna, is also heavily polluted by the Baikalsk Cellulose Combine. Between 1967 and 1969, there were almost 2,000 recorded instances of the plant's sewage exceeding the maximum permissible norm. As a result of the pollution, scientists note that in the water near the Baikalsk plant the amount of flora and fauna has decreased by one-third to one-half.[12]

Naturally, the heavily industrialized areas of the Soviet Union have suffered the greatest water pollution. Yet even the remote Central Asian republic of Kazakhstan has a significant pollution problem. A number of the republic's rivers are said to be completely devoid of living matter. In fact, many of Kazakhstan's water bodies are so polluted that they cannot even be used for irrigation.[13]

The authorities have focused a good deal of their attention on Russia's water crisis and have devoted less attention to the nation's air-pollution problem, in part because it is smaller in scope. The absence of a large number of automobiles in the Soviet Union helps to prevent atmospheric pollution from assuming the dimensions of Russia's water pollution.

Yet, in some areas of the country, air pollution has reached alarming proportions. Steel and chemical plants are among the heaviest polluters; their discharges are said to "rise luxuriantly from the smokestacks, painting the sky a rusty color."[14] The residents of Volkhava are periodically subjected to "gas attacks" from an aluminum plant that "continually belches clouds of sulfur and fluoride gases."[15] At the center of many of Russia's largest cities the content of silicon dioxide is two to two and one-half times greater than the maximum permissible concentration. Air pollution is continuing to grow not only in large urban centers but even more so in small cities where efforts to control it are frequently nonexistent.[16]

The Party Central Committee and the Council of Ministers have responded to this growing pollution problem by promulgating several legislative acts and decrees. Thus the regime has formulated all-union legislation on water use as well as decrees to improve the environmental regime of Lake Baikal, the Caspian Sea, and the Volga-Ural Basin. While the regime has not yet promulgated all-union legislation on air pollution, several legislative acts regulate atmospheric emissions by industrial enterprises.[17]

Industrial ministries and enterprise managers have frequently frustrated the realization of these central directives. Ironically, the imperatives of Soviet development plans work against enterprises and ministries observing pollution regulations. In the Soviet Union quantitative fulfillment of plan targets determines an enterprise's success, and bonuses and premiums are based upon quantitative criteria. Ministries and industrial managers are, therefore, reluctant to engage in any activity which diverts them from achieving plan targets. Unfortunately for the environment, the imperatives of the plan often conflict with the needs of a sound antipollution policy. One harried industrialist graphically illustrated the tension between the plan and the environment when he exclaimed, "You think we do not see? But what is to be done? . . . What about the plan? Are you going to order the plants to stop? That is the dialectic. One has to choose between civilization and one's love of nature."[18]

Because of the need for plan fulfillment, ministries have done little to purify industrial wastes. First, they have allocated few funds for the research and design of advanced purification equipment. Even the purification system of the Baikalsk Cellulose Combine, supposedly the most advanced in the country, has serious technological deficiencies.[19] Inspectors making a random check of purification equipment found that almost all of the installations had "structural imperfections."[20] The creation of advanced air-purification equipment has especially suffered from the neglect of industrial ministries. Thus, the All-Union Gas Purification and Dust Removal Association, the only all-union body conducting research into the causes and effects of air pollution, has been severely hampered by being attached first to the USSR Ministry of the Chemical Industry and currently to the USSR Ministry of the Petroleum Refining and Petrochemical Industry. The following quotation indicates the manifold difficulties that this organization has experienced:

> The association's research facilities are developing extremely slowly, as for its design facilities they have been in a frozen state for many long years. At present the association is able to fulfill only an insignificant part of the most necessary design and research projects.[21]

Second, industrial ministries devote few funds to actually building purification installations, preferring instead to increase production capacities. The result is that in 1967 between 60 and 75 percent of all industrial sewage was not treated at all,[22] while in 1968, 60 percent of the enterprises that polluted the air had no purification installations whatsoever.[23] Further, in their eagerness to have plants put into operation as quickly as possible, ministries engage in the common practice of planning the construction of decontamination installations as a second or even a third stage, after the main shops have already worked full-blast for years. Nobody appears to be concerned that such practices violate the laws on conservation.[24]

Third, enterprise managers do not always utilize funds for purification installations, even when ministries have allocated such funds.[25] Managers know that so long as they fulfill plan targets their grateful ministries will not complain about unutilized funds for purification equipment. In fact, many protective ministries even give enterprise managers bonuses by way of "compensation" if the latter are fined by the courts for pollution violations.[26]

A January, 1973, joint decree of the Party Central Committee and the Council of Ministers attempted to remedy many of the problems described above.[27] The decree called on ministries and departments to substantially increase their efforts to reduce the discharge of industrial wastes. The decree also established a special administration for the design and production of gas purification equipment. Inspectorates will be subordinated to this chief administration to ensure that enterprises properly operate this equipment. This decree gives hope that the Soviet Union will eventually produce a sufficient amount of quality purification equipment to significantly reduce pollution emissions. At present, however, both in quality and quantity, such facilities appear inadequate to meet the task assigned to them.

Industrialists have aggravated the pollution problem and frustrated attempts to solve it in other ways. Thus, ministries frequently construct new enterprises in densely populated areas, despite regulations prohibiting such practices.[28] The ministries do so, of course, because it is cheaper to build in a large urban center where such things as a transportation network already exist. The result of such practices, however, is that the "people in these districts cannot open their windows or relax in a square somewhere, for everything around them is covered in soot." On occasion the long-term costs of locating plants in residential areas may exceed any short-term economies derived from such a decision. For example, industrial enterprises so fouled the air of Gubakha that it became impossible to live in the city, and so at considerable cost, Gubakha's entire residential area had to be moved.[29]

The Soviets have made several efforts to relocate heavily polluting plants to less populated areas.[30] Yet many ministries and enterprise directors vigorously resist such efforts. The most obvious reason for their resistance is the expense and inconvenience involved in moving the plant. Further, while the plant must bear the cost of relocation, it does not directly bear the cost of its pollution: the population bears this cost in the form of a threat to public health. Finally, such a move may bring unwanted headaches to the enterprise. If, for example, the plant is relocated to a remote area, it may have difficulty attracting a sufficient labor force. With the existence of such obstacles, the campaign to relocate polluting enterprises will most likely continue to encounter stiff resistance. While this resistance is understandable from the viewpoint of the ministries and enterprises involved, such a "departmental" approach can only aggravate the environmental conditions of Russia's urban centers.

Finally, industrialists have hindered attempts by government inspection agencies to verify the implementation of antipollution regulations. The two primary agencies for dealing with pollution and polluters are the USSR Ministry of Public Health's Sanitary-Epidemiological Service and the USSR Council of Ministers Chief Administration for the Hydrometeorological Service. The Sanitary-Epidemiological Service has been especially ineffective in carrying out its mandate as detailed in the *Principles of Public Health* (1969). According to the *Principles* (Article 21), the Sanitary-Epidemiological Service can halt a plant's operation if its wastes are a danger to public health. The Soviet press contains numerous reports of the difficulties this agency experiences in achieving its assigned tasks. *Izvestia,* for example, reported the case of a cement plant that was spewing great amounts of dust over the surrounding residential area. The local sanitary station informed the plant director that he would have to cease operations unless the situation was remedied, whereupon the plant director "ordered the guards to throw the sanitary inspectors out and keep them off the premises." Later that night the inspectors slipped inside the plant and sealed the furnaces.

> When informed of this the Director, right before their eyes, broke the seal with his own hands, and had the inspectors pushed out through the main gate. When reminded that there was a nature protection law that had to be obeyed, the Director declared, "There is only one law for me—the production program![31]

In reality, the zeal shown by the Sanitary service in this example is not typical. Since the Service realizes that it does not have the bureaucratic strength needed to enforce its directives, it has usually adopted a live-and-let-live policy toward any industrial polluters.

Not with the mind is Russia comprehended,
The common yardstick will deceive
In gauging her: so singular her nature—
In Russia you must just believe.
FYODOR TYUTCHEV

The level of air, and especially water, pollution in the USSR has assumed serious dimensions. Such pollution has involved significant costs for society in the form of economic losses, threats to public health, and a deterioration in the quality of the physical environment. The nation's highest decision-making bodies have issued a number of authoritative decrees that attempt to halt the environmental disruption. However, ministries and enterprise managers have frequently defied the directives of central decision-makers. Industrialists have usually been far more concerned with fulfilling plan targets and pursuing private gain than in worrying about the potential damage that their activity man inflict on society.

LAND RESOURCES

An examination of land use in the Soviet Union graphically illustrates the indifference frequently exhibited by government bureaucracies for any goal that extends beyond their own narrow functional task.

The Soviets have become increasingly concerned with the efficient management of their land resources. Although the Soviet Union has almost five million more square miles of land than the United States, much of the land is in remote areas and unsuitable for agricultural or industrial development. The Soviets have therefore been compelled to establish land-use priorities for this resource. *The Principles of Land Legislation* (1968) has given agriculture the highest priority in the use of land.[32] . . . The alarming decline in the arable land fund has made immeasurably more difficult and costly the feeding of the Russian people—a task that the regime, even under the best of circumstances, has never performed well. Despite the critical need for the careful utilization of Russia's land resources, the industrial sector has often engaged in land-use practices that conflict with national priorities.

Hydroelectric projects perhaps best illustrate the indifference of industrial bureaucracies for the needs of agriculture. So far these projects have flooded an area equal to the combined territory of Armenia and Moldavia.[33] Frequently the state invests substantial capital to improve the quality of the land for agricultural use. Yet these expenditures are often wasted because hydroelectric projects then flood much of the land.

Of course, hydroelectric projects must flood some land, but critics contend that planners make no attempt to reduce the flooding area. Planners act in this manner because land has no economic evaluation in the Soviet Union, and therefore no economic incentive exists to conserve the land.[34] When planners are compelled to reduce the amount of land to be flooded, they frequently can do so without impairing the efficiency of the hydroelectric project. The problem, then, is usually not that planners are unable to reduce the amount of land to be flooded. Rather, it is that hydroelectric project designers often take a "departmental" approach to their task, considering only the needs of their branch while ignoring the needs of other sectors.

The misuse of land by industrialists also contributes to Russia's pollution problem. Since purifying industrial wastes is expensive, producers often choose to simply dump the wastes on nearby land that they do not pay for. Not only do the rains frequently wash these wastes into streams, but the land itself becomes unusable. The Soviets call these sewage dumps "dead zones," and they are often of enormous proportions.[35] Mining enterprises are especially guilty of this practice. For example, the runoff from coal mines has polluted substantial amounts of land. Further, mine trash dumps cover thousands of acres of formerly valuable agricultural land.[36]

Mining operations also despoil the land through open-cut and strip mining. Such methods create what one observer described as a "lunar landscape."[37] Throughout the USSR, mining enterprises have "disturbed" (to use the official term) hundreds of thousands of acres in this manner. In addition, open-cut and strip mining contribute to air pollution, because the earth that they displace is formed into loose piles that the wind often sweeps away. Although the *Principles of Land Legislation* enjoins mining enterprises to restore land that they utilize to its original state, few appear to heed this injunction.[38] A member of the USSR Academy of Sciences argues that it makes good economic sense to recultivate abandoned mining land because profits from such land soon pay for the expenditures needed to put the land in proper condition.[39] However, this argument misses the point. While it may well be that it is profitable from the standpoint of the national economy to restore mined-out land, it certainly is not profitable for the mining enterprises that must perform the recultivation work. This is so because recultivation work is funded by the mining enterprises themselves. Therefore, the mining industry pays for recultivation work, but the agricultural sector derives the benefits of such work.

Finally, the ecological problems experienced by Russia's Black Sea coast area are a most bizarre and revealing illustration of "departmentalism's" impact on the Soviet environment.

The Black Sea is of special significance for most Russians because it is the warmest resort area in the USSR. Russians come there from all parts of the Soviet Union to spend their vacations. The area has recently experienced a building boom to accommodate this influx of visitors.

It is precisely this extensive construction that has been the major source of the problems afflicting the Black Sea coastline. To provide the materials for construction, builders used the pebbles and sand found in abundance along the shore. Since these were "free," the builders naturally preferred them to materials that would have to be shipped from other areas. One source estimates that builders removed 300 million cubic meters of sand and gravel from the beaches along the Black Sea during the postwar years.[40]

Unfortunately for the beaches, the pebbles serve as a buffer against the violent storms that often occur in the area. With the protective pebbles removed, the beaches have experienced erosion of enormous proportions. By 1960, the beach area along the Black Sea coast was reduced by almost 50 percent. In several areas, resort hotels, hospitals, and, perhaps most ironically, the health spa of the Ministry of Defense collapsed as the shoreline receded. In addition, officials fear that the main railroad line will also wash away. Finally, the erosion has led to numerous landslides on the nearby hilly terrain. The landslides have covered roads, destroyed homes, and even threatened the streets of Yalta.

The Soviets have reacted to the problem by passing a law banning the removal of the pebbles from the beach.[41] Yet the builders, far from ceasing their activity, have actually increased their removal of the pebbles. Presumably, local officials have remained indifferent to the removal, if they have not actively colluded with the builders in this process. In the absence of strong administrative controls, the builders themselves continue because they have no economic incentives to stop their operations. While the builders incur no financial losses, the state itself has expended huge sums in an unsuccessful effort to halt the erosion. The Georgian Republic Council of

There will come a time, I know, when . . . those will be accounted the best who will the more widely embrace the world with their hearts. . . . Then will life be great, and the people will be great who live that life.
MAXIM GORKY

Ministers has formulated a ten-year program to save the Black Sea shoreline that will cost more than $130 million; the Ministry of Transportation has already spent approximately $45 million to strengthen the coastline.[42] The monetary value of the efforts to save the beach far exceeds the value of the beach pebbles as building materials. The lack of an economic evaluation reflecting the true value of the pebbles for nonbuilding purposes has thus resulted in a classic example of "departmentalism" and disruption of the environment.

CONCLUSION

An examination of the Soviet Union's ecological problems must evoke among many Western readers a feeling of déjà vu. Soviet industrial ministries and enterprise managers behave in a manner that is disturbingly similar to those who despoil the environment in capitalist countries. Industrialists, regardless of political system, appear to be concerned primarily with private gain, and peripherally, if at all, with any ecological damage that they might cause.

Yet according to socialist theory, Soviet industrialists should be infused with a sense of purpose that extends beyond private gain to encompass the true interests of society. Why, then, has there been such a divergence between theory and practice in the Soviet system?

We can attribute this divergence largely to the invalidity of certain assumptions upon which socialist theory is based. First, the theory assumes that there is such a thing as the "true" interests of society and that subsystem actors will subordinate their parochial interests to pursue these. In reality, the Soviet system is dominated by numerous industrial bureaucracies, each of which has a different set of priorities and goals determined primarily by its own functional mandate. While protecting the environment may benefit that nebulous entity known as "society," it is not within the functional scope of most industrialists.

Second, the argument assumes that achievement of the various components of the public interest are not incompatible with one another, but rather that maximizing the attainment of one component furthers the attainment of all. In fact, one cannot make such an assumption. In theory, many would agree that rapid economic growth and protection of the environment are both in the public interest. In fact, however, the pursuit of economic growth, especially in the short run, may adversely affect the quality of the environment. Central decision-makers have in effect frequently confronted Soviet industrialists with incompatible tasks: to fulfill high plan-targets and simultaneously to protect the environment. Most industrialists have chosen primarily to stress plan fulfillment, probably sensing (correctly) that while central decision-makers may have an abstract commitment to environmental quality they are far more concerned with maintaining rapid economic growth.

Finally, the ability of the planners to make cost/benefit calculations that reflect the "true" interests of society assumes that there are universally acknowledged standards against which public policy can be measured. However, such universally acknowledged standards almost never exist in a public policy debate. Rather, participants in the debate base their arguments on different, but often equally defensible, conceptions of the public interest. In the absence of agreed upon policy standards, policy output largely becomes a function of the amount of political power exerted by the various participants in the debate. Unfortunately for those interested in protecting the Russian environment, the political power of Soviet industrialists often appears decisive in determining the priorities emphasized by the Soviet system.

NOTES

1 For a typical criticism of the phenomenon of "departmentalism," see B. Bogdanov, "Conservation and Economics," *Ekonomika selskogo khozyaystva,* no. 2 (February 1970) in *Current Digest of the Soviet Press* (hereafter referred to as *CDSP*), 22 (June 9, 1970), 9.

2 *Pravda,* May 13, 1971, 3, provides a general discussion of this problem.

3 M. Loiter, "Economic Measures for the Rational Utilization of Water Resources," *Voprosy ekonomiki,* no. 12 (December 1967), 76.

4 *Selskaya zhizn,* May 19, 1970, 3.

5 *Izvestia,* June 27, 1970, 4.

6 *Ibid.,* April 14, 1959, 2.

7 Thus, in 1968 approximately one-third of all waste water underwent treatment that removed only 40 percent of the impurities, while only 10 percent of the waste water received biochemical treatment that removed 80–95 percent of the impurities. *Nauchno-tekhnicheskiye obshchestva SSSR* (February 1968), in *Joint Publications Research Service* (hereafter referred to as *JPRS*), no. 45,666 (June 1968), 12.

8 *Izvestia,* July 9, 1968, 2.

9 *Ibid.,* June 10, 1967, 3.

10 *Ibid.,* April 15, 1965, 2.

11 *Pravda,* Nov. 15, 1966, 2.

12 *Ibid.,* Feb. 16, 1969, 1. [*Editor's note* See the next selection for a more optimistic view.]

13 *Kazakhstanskaya pravda,* Nov. 18, 1971, 3.

14 *Literaturnaya gazeta,* June 14, 1972, 11.

15 *Pravda,* July 4, 1966, 2.

16 N. Koronkevich, "The Water Problem Can Be Solved," *Seriya nayka o zemle (Berech i Umnozhat Prirodnie Bogatstva),* no. 8–9 (1970), in *JPRS,* no. 51,609 (October 1970), 11, 18.

17 See, for example, Article 21 of the *Principles of Public Health, Izvestia,* Dec. 29, 1969, 3–4.

18 *Pravda,* June 26, 1970, 3.

19 *Komsomolskaya pravda,* Aug. 11, 1970, 4.

20 *Rabochaya gazeta,* June 27, 1969, 4.

21 *Pravda,* March 24, 1969, 3.

22 *Ekonomicheskaya gazeta,* no. 4 (1967), 37.

23 V. Shkatov, "Prices on Natural Resources and Improvement of Planned Price Formation," *Voprosy ekonomiki,* no. 9 (September 1968), 67.

24 For example, Article 4 of the Russian Republic conservation law stipulates that "it is

forbidden to put into operation enterprises, shops, and installations that discharge sewage without carrying out measures that will ensure the purifying of it." *Pravda,* Oct. 28, 1960, 1.

25 *Izvestia,* March 2, 1973, 3.

26 *Pravda,* Aug. 23, 1970, 3.

27 *Izvestia,* Jan. 10, 1973, 1.

28 *Ibid.,* Sept. 18, 1966, 2. The general plan for Moscow prohibits the construction of new facilities or the expansion of existing ones within the city except those needed for direct services to the population.

29 B. Svitlichny, "The City Awaits a Reply," *Oktyabr,* no. 10 (October 1966), in *CDSP,* 18 (Dec. 21, 1966), 15.

30 For several examples see *Izvestia,* Sept. 20, 1972, 2–3.

31 *Ibid.,* Aug. 6, 1971, 3.

32 *Pravda,* Dec. 14, 1968, 2. See especially Article 10.

33 *Selskaya zhizn,* July 14, 1970, 3.

34 I have examined elsewhere the adverse impact that the lack of an economic evaluation has on the conservation of Russia's natural resources. John M. Kramer, "Prices and the Conservation of Natural Resources in the Soviet Union," *Soviet Studies,* 24 (January 1973), 364–373.

35 *Pravda,* Feb. 17, 1967, 2. The "deadzone" of just one enterprise covered almost 25,000 acres, much of it formerly irrigated land.

36 *Trud,* June 28, 1970, 2. In just one area of the Kuzbass region, mining enterprises have polluted almost 75,000 acres.

37 *Literaturnaya gazeta,* May 31, 1972, 10.

38 *Ibid.,* Aug. 12, 1972, 2.

39 *Trud,* June 28, 1970, 2.

40 Marshall Goldman, *The Spoils of Progress* (Cambridge, Mass.: The M.I.T. Press, 1972), 156.

41 *Pravda,* Feb. 26, 1969, 3.

42 Iya Meskhi, "The Sea Threatens," *Ogonyok,* no. 25 (June 1972), 26–27.

ASIA

21 Baikal Survives as a Prize for the Whole Planet

Rudolph Chelminski

Rudolph Chelminski is a free-lance writer living in Paris. From 1962 until its demise in 1972 he worked as a journalist for *Life,* serving as the magazine's bureau chief in Moscow in 1968–69.

In the middle Fifties, Soviet technicians began searching for a site for a large and modern cellulose plant, theretofore lacking in the USSR. After considering several locations—a high-grade pulp plant needs plenty of water, wood and electricity—they finally selected a forest tract on a body of water in the depths of eastern Siberia, not far from the city of Irkutsk. From the engineering point of view, it was ideal. The body of water was Lake Baikal.

From the moment the choice became official, trouble started, as unexpected as it was vexing: Soviet officials are not accustomed to opposition. The outcry which arose locally soon spread to Moscow, found an echo in the press and became known to the world at large. Still, the plant was considered too important to be delayed by grumblings. Construction began in 1959. Uncharacteristically, though, the controversy continued after this fait accompli, fed by scientists, journalists and artists. The lake, they charged, was being needlessly sacrificed at the altar of industry. Even more uncharacteristically, the planners listened, looked at their blueprints again and added more filtration equipment. When the plant was finished it had what the Soviets now call the most extensive and costly waste filtration apparatus ever installed in a cellulose plant. It stands today as a monument to one of the rare occasions in Soviet history when spontaneous public pressure, however limited and cautious, actually changed the course of events ordained by the central government.

That usually compliant Soviets would stand up and fight for a lake might be surprising if it were any but this one. Lake Baikal is extraordinary. Little-known outside the Soviet Union, mostly because of its isolation and relative

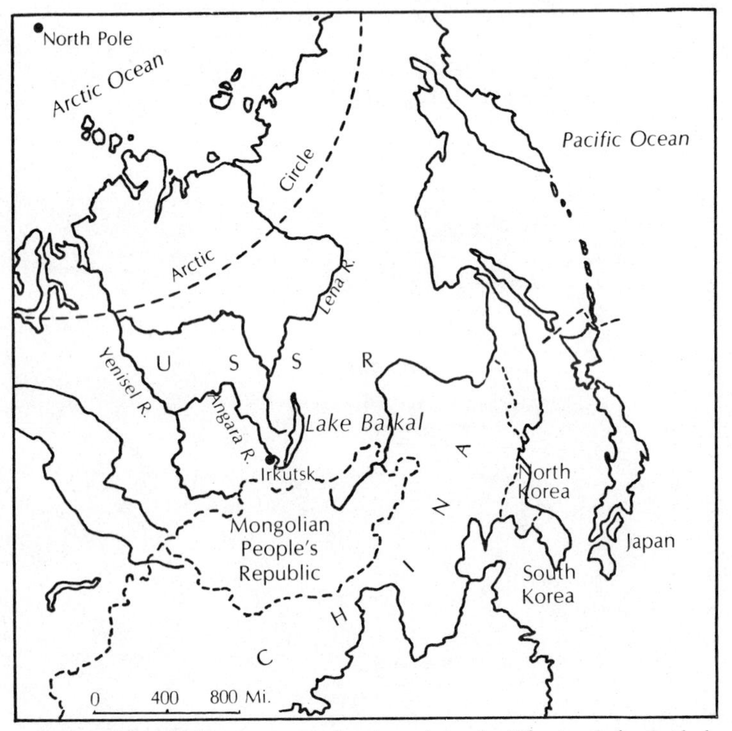

Located along a belt of strong seismic activity in Siberia, Lake Baikal could be the forerunner of a new ocean that will split Asia into two continents.

inaccessibility, Baikal ranks as one of the great natural wonders of the Earth's physiognomy. To begin with, it is the world's biggest lake, not by surface area but by volume. It is, in fact, by far the largest body of fresh water in the world: The 5,518 cubic miles of water trapped in the Baikal trench represents about one-fifth of our planet's liquid fresh water and a much larger percentage of the Soviet Union's. Baikal's water volume is nearly as great as the Baltic Sea's and is greater than that of all the Great Lakes combined. Its colossal depth (5,314 feet to the lake bottom itself, and then thousands of more feet of sediment) makes for endless superlatives, which Siberians never tire of enumerating: There is five times more water in Baikal than in all of Europe's and Asia's large and middle-sized lakes combined; if all the 336 rivers which feed the lake were blocked and the Angara, its sole outflow, continued its normal flow, 500 years would pass before Baikal went dry.

Because of its remoteness, Baikal did not come under even the most cursory scientific study until the middle of the 18th century, and no systematic, modern research was carried out until the latter half of the 19th. The Soviets have naturally continued the work of their czarist predecessors, but they do not hesitate to admit that there is still much to be learned, particularly about the deepest abyssal sections, where no light ever penetrates, where the temperature hovers near 38 degrees the year around (warm compared to the lakeside air temperature, which averages 30 degrees), and where colorless, flabby-bodied mutant sculpins (a kind of bullhead) join tiny, pink-eyed freshwater shrimp to roam over a mournful kingdom of viscous blue mud.

From the very first, though, it was evident that Baikal was a treasure. For one thing, it is one of the world's oldest lakes. Dr. G. I. Galazy, the good-natured giant who directs the Baikal Limnological Institute, dates its present shoreline and bottom configuration to about 25 million years. (Most lakes, including the Great Lakes, have lives measured in mere tens of thousands of years.) The forces that created the lake and the mountains that ring it began 80 million years ago, he says, in the late Mesozoic era, the age of dinosaurs and flying reptiles, when the climate of the area was subtropical. (At about the same time Baikal's mirror-image "sister," Lake Tanganyika, was taking shape in what is now the continent of Africa. It is the second deepest lake in the world.)

Baikal is the center of a tectonic belt running in a line southwest to northeast, and the 30 major earthquakes recorded over the last couple of centuries attest to a permanent instability. In 1861, some 77 square miles of a plain called Gypsy Steppe subsided cataclysmically into the lake, killing 1,300 people and forming a deep new pit now known as Proval Bay.

"Some scientists think that Baikal is the beginning of the formation of a new ocean," Galazy told me, "which is already denominated as the Siberian Sea. The same kind of splitting-apart forces which caused the continental drift in the Atlantic are active here."

Like similarly ancient lakes (Tanganyika and Ohrid are other examples), Baikal holds numerous animal and plant species found nowhere else. Of the lake's 1,700 indigenous species, 1,200 are endemic. Understandably enough, visiting tourists and other nonspecialists are most fascinated by the famous Baikal freshwater seals. Some 60,000 of the silvery-gray *Phoca sibirica,* called *nerpy* in Russian, inhabit the rocky crags at the north of the lake in summer, migrating south with the advancing ice in autumn.

The *nerpy,* Galazy says, came to Baikal 12,000 years ago, following the courses of the Yenisei and the Angara, when the Arctic Ocean reached farther inland than it does now. Direct descendants of the Arctic seals, the *nerpy* adapted so well and so quickly to the lake that they are now considered a new species.

"Their bone structure and biology and speed of growth are different from the Arctic seals," Galazy said. "The color of the fur is different. Whereas the Baikal seal is silvery-gray, the Arctic is yellow with black marks. The *nerpy* have completely changed from what they had been. They are a good example of the creation of a new species in a very short time."

The seals' diet consists almost exclusively of another endemic beast, the weird fish *Comephorus baicalensis,* which Siberians call *golomyanka.* The *golomyanka* is a favorite of Galazy because its highly specialized characteristics prove that it went through considerable evolutionary change to reach its present form.

There is so much fat in a *golomyanka's* body that it is translucent; when taken from the water and held in the sun, it literally melts away to skin and bones in one's hand. It has neither scales nor swim bladder. *Golomyanka* swims from the surface to great depths and then up again without harm. Unlike other fish, the female gives birth to larvae rather than eggs; normally the mother's belly bursts asunder at birth time. If childbirth is painful for most creatures, it is usually fatal for *Comephorus baicalensis.*

Along with the seals came another migrant to Baikal, *Coregonus autumnalis migratorius,* locally known as *omul,* a salmonlike whitefish which left its ancestors behind in the Arctic to become, along with lake sturgeon, Baikal's most prized delicacy. In all, the lake harbors 50 species of fish, the most numerous being different sorts of sculpin, but also including grayling, whitefish and carp.

Baikal's water is oxygen-rich and astonishingly clear—"practically the cleanest on Earth," Galazy says with pleasure, pointing out that a standard color disc can be seen as much as 44 yards below the surface. Light, and in consequence the algae and plankton it nourishes, penetrates 800 feet or

> **So much about Lake Baikal is different from anything else man has encountered on earth that the lake is sometimes thought to have supernatural powers by those whose lives are touched by it. The area is venerated both by the local inhabitants and by Russian conservationists. But what is sacrosanct to the believer may be seductive to the nonbeliever. The lake is also seen as a voluptuous virgin whose resources are ripe for ravaging. All those fish, all those trees, all that pure water would whet the passion and set the juices running of any red-blooded promoter or industrialist.**
>
> MARSHALL I. GOLDMAN

more. One species of microplankton, *Epischura baicalęnsis,* filters the water down to a depth of 150 feet.

Because of its inordinate size, Baikal is known to natives as "the sea," and the name is more apt than mere folklore. That Baikal has tides is in itself oceanic behavior, and fishermen caught in its storms have described waves more than 15 feet high. Fogbanks, created when a warm front moves over the lake's frigid waters, can last for weeks, making navigation virtually impossible. Like an ocean, Baikal influences the continental climate around it, resulting in a small zone far more temperate than the usually exaggerated Siberian climate. In January, for instance, the air temperature on the shores of the lake is an average of ten degrees warmer than the city of Ulan Ude, 45 miles to the southeast. The lakeside is about seven degrees cooler in summer than such nearby cities as Irkutsk.

Aside from a fishing industry, Baikal remained undisturbed after the arrival in the 17th century of the first Russian colonists in the area. What man has come to call "progress" reached the lake in the waning years of the 19th century, when the steel ribbons of the Trans-Siberian Railway came up against its eastern and western banks, near the southern tip where the land is relatively flat. Only 28 miles separated the two lines, but their junction was to prove the most difficult part of the whole undertaking. To pass around the northern end was patently impossible, because the Stanovoyi Mountains presented a barrier of solid rock. The south was hardly much easier; though the mountains were certainly lower, the loop route was a chaos of crags and ravines that was sure to cost the czar a fortune in bridges, tunnels and escarpments. The decision-makers back in St. Petersburg opted for a ferry. The great, barnlike four-stacker, naturally christened *Baikal,* was launched in 1899 and entered service in 1900. It was the biggest and most frightening man-made thing Siberia had ever seen, but it was virtually helpless against the lake's four-feet thick ice. The behemoth had to be laid up for the winter months; construction began on the southern loop.

Unhappily for the luckless czar, it was not completed in time for the Russo-Japanese War, when Lake Baikal became the scene of one of the strangest engineering triumphs in railroading history. War matériel had to be brought to the beleaguered east at all costs, and the engineers gambled. In February, the coldest month of 1904, they laid track across the ice. When the first test locomotive plunged through the ice and fell hissing and bubbling to the bottom, the engineers improvised a system of supplementary ties—basically, tall cedars stripped of their limbs—placed every 30 feet. This wider distribution of weight enabled trains to pass safely. . . .

When the southern loop was completed in late 1904, Lake Baikal returned to relative anonymity and safety from mankind's depredations. Over the years the local fishing cooperatives grew more efficient (to the point that the dwindling schools of *omul* and sturgeon are now protected) and a few small and nonpolluting industries were built.

By 1960, though, the boom in developing Siberia had signaled the end of Baikal's serenity. Because the big new Irkutsk hydroelectric facility could now provide the electricity—the millions of acres of taiga the wood and the lake itself the water—it was inevitable that Gosplan, the Soviet Union's central economic planning commission, would covet the lake for industry. Work began on the Baikal cellulose plant in 1959; it was finished by 1966. Throughout construction and even after completion, the controversy continued.

There were several points to the controversy, according to Leonid Shinkarev, an *Izvestia* correspondent in Irkutsk. "When construction began, the scientists and engineers had not sufficiently studied the lake. Also, no one had any previous experience in the design of such plants. The controversy was about unknown quantities.

"At first it was clear-cut—the technicians were for the plant and everyone else was against. But then scientists began examining the project more deeply and came out on both sides of the question. The government considered the various opinions and took some new decisions, about how to build the plant while still preserving the lake. Thanks to the controversy, it allotted more capital for this purpose.

"The debate played a positive role. It caught the attention of the public, of the scientists and of the government planning organizations. Because of the cellulose plant, everyone has begun to pay attention to conservation. Baikal has been a catalyst."

The filter system works very well, apparently—about as well as present technology can devise. The standard public relations stunt is to hand visitors a cup of the water from the plant's outspout. Yes, it is drinkable.

Even so, the filter system can never be perfect. Galazy pointed out that although the water now discharged into Baikal is relatively innocent, it is different, and has already changed the environment in an area of 15 square miles or so. Any changes in water chemistry are bound to affect the microplankton.

Barring the happy prospect of a technological breakthrough in filtration, it appears that Baikal will continue to be affected by the cellulose factory, but in a scope that is, everything considered, limited. What next? Can desirable industrial property like this remain safe in the last quarter of the 20th century?

Back in Moscow, I tried unsuccessfully for nearly a week to meet with any Gosplan official who could speak with authority about Baikal's future, but it became evident that I was asking for too much too fast. I eventually left Moscow, having deposited my questions with APN, the state press agency that handles visiting foreign journalists. A month or so later, I received a written statement. After reviewing much of what I had already learned about filtration, the statement, at its very end, finally addressed itself to the crucial point:

"It would hardly be right to stop utilizing timber resources so abundant in the trans-Baikal area. Only 60 to 65 percent of the calculated timber resources are utilized now. But timbering will be conducted with due account given to lake protection.

"There will be no industrial construction on the shores of Baikal, even the enterprises which are not considered harmful to the environment will not be built there. An industry of quite a different nature will flourish on the lake shores, the industry of recreation. Spas, holiday homes and tourist bases will be built."

Baikal's best vocation is simply to remain as beautiful as it is, to be a base for scientific study and to be visited by tourists from around the world. And in case all this strikes Gosplan as less useful than heavy industry, let them consider Galazy's reasoning.

"Think of it," he said, enthusiastically gesturing at a detail map of the great lake. "Twenty percent of the world's fresh water! And the cleanest and the purest you can find. At a time when Americans and Japanese are seriously talking about towing icebergs to water-poor areas, why not use Baikal to furnish drinking water? The time will come when a water pipeline from Baikal will be more profitable to the state than all the oil pipelines in Siberia."

22 Japan's Ecological Crisis

Charles A. Fisher and John Sargent

Professor C. A. Fisher is head of the Department of Geography in the School of Oriental and African Studies, University of London. John Sargent, of the same department, is reader-elect in geography with reference to Asia, in the University of London.

Reprinted from *The Geographical Journal*, 141, pt. 2 (July 1975), pp. 165–76, by permission of the authors and the Royal Geographical Society.

I

. . . The relationship of human communities to their natural environment is a fundamentally different one from that of plants and animals. For while human beings, whose purely physical needs are basically akin to those of other living creatures, cannot for this reason simply opt out of the ecological process of adjusting their behaviour to the physical and biogeographical area in which they live, they can, as self-conscious and at least partly or intermittently rational beings, deliberately adjust their environment to satisfy their own needs and desires, both material and spiritual. Moreover, they can, and to some extent do, learn from the experience of successive generations of achievements and setbacks in the pursuit of such endeavours.

It is out of this process that human societies advance, for, as the early twentieth century Chinese philosopher Hu Shih has observed, "the civilization of a race is simply the sum total of its achievements in adjusting itself to its environment." However, since environments differ very widely in both the opportunities and the problems which they present, the civilizations of some communities advance farther and faster than those of others. And unfortunately there is a danger that those which advance most rapidly and spectacularly may conclude that, thanks to their own technological virtuosity, they have succeeded in opting out of all environmental constraints. But by comparison with human activities, the major natural processes work much more slowly, and for this reason human societies, particularly the most apparently successful, have often failed to appreciate the significance of the kind of situation which modern ecologists have conceptualized as the climax, and hence have not reacted constructively when eventually they were confronted by it.

In this connection it is relevant to quote briefly from the presidential addresses of two of the most eminent scholars in their respective fields, one a historical geographer and the other an agricultural scientist. First then, Professor Carl Sauer (1941), towards the end of his 1940 presidential address to the Association of American Geographers, poses the question:

> Is there in human societies something like an ecological climax, a realization of all the potentialities inherent in that group and its state? What of limits of population growth, of production attained, of accumulation of wealth, even of increment of ideas beyond which the matured culture does not go? . . . The rise and fall of civilizations, which have interested most historically-minded students of man, cannot fail to engage the historical geographer. A part of the answer is found in the relation of the capacity of the culture and the quality of the habitat.

Secondly, Sir Joseph Hutchinson in his presidential address to the British Association in 1966, commented:

> The ecological concept of the climax involves a stable population, within which individuals grow and mature and die, while the community maintains its vigour and stability. In economic thinking the slow start is the characteristic situation of the under-developed countries. The point of inflection of the curve is the point of take-off, to which so much attention has been devoted . . . it is my contention that the lack of attention by economists to the climax situation is one important reason for our lack of success in the management of our older industries . . . we have no conception of the economic principles of the climax by which to guide our policy.

Indeed, we may go further, and say that this ignoring by economists—and others—of the problem of the climax situation is a major reason for the failure not only to manage our industrial economies, but also to cope with the social crisis which now threatens all the most advanced industrialized countries. And in this connection it would seem that Japan may represent the nearest approach yet seen to the human-ecological climax situation.

Nevertheless, more than a century before either Sauer's or Hutchinson's pronouncements were made, there had been at least one widely publicized approach to the concept of the human-ecological climax, namely that outlined by Malthus in his essay on population in 1830. Although he did not use that term, Malthus described the kind of climax situation which was reached when, in any given habitat, population reached the limit which its agricultural and related resources could adequately provide for with the available techniques of the day. But if population rose above that level, counteracting forces—famine, pestilence, natural calamities such as floods, often resulting from over-taxing of the environment, and human violence bred of the distress caused by any or all of these miseries—came into play. These brought the population back to the climax level, or in other words to the limit which it could support.

In biological terms such a sequence is essentially similar to the natural regulating measures by which plant and animal communities are kept within the supportable limits of their habitats. But unlike plants and animals, self-conscious human beings are repelled by the apparent harshness of these natural means of thinning out population, and have tried various ways of escaping from them, of which the most acceptable have been to acquire more land, either near at hand or overseas, or to make some revolutionary improvement in methods of cultivating the land they already possess, or else to turn to overseas trade in order to obtain extra food and other necessities from areas with available surpluses.

This brings us to Japan's reaction to what seems to have been a Malthusian-type climax situation a century and a half before Malthus's time. The first Western Europeans to visit Japan in the sixteenth century were astonished at its very much higher density of population than that with which they were familiar at home. This high density had arisen because Japan, 1000 years earlier, had adopted the highly intensive type of wet rice cultivation first developed in the adjacent mainland of monsoon Asia, but owing to its own exceptionally low proportion of flat land suitable for such cultivation—some 15–20 per cent of its total land surface—it had by the early seventeenth century brought virtually the whole of this into use* and indeed had greatly improved it, and was cultivating it very intensively. Much the greater part of such land consisted of coastal bay-head deltas and adjacent areas which had been recently reclaimed from the sea.

It was at this juncture, when land shortage was already incipient, that

*Except in the northern island of Hokkaido which, in latitudes similar to those of Newfoundland, was considered to be climatically unsuitable for rice cultivation. However, rice was successfully introduced into Hokkaido in the late nineteenth century.

for wholly extraneous reasons, mainly the fear of Spanish intrusion and aggression via the Philippines, Japan decided in 1637 to shut itself off from the outside world, and so virtually denied itself all opportunity of emigration, territorial expansion, or participation in overseas trade. However, by widespread but very laborious recourse to hillside terracing and the related cultivation of new Latin American dry crops, notably sweet potatoes and maize, Japan managed to feed a population which rose from some 23 million in 1650 to nearly 30 million in 1720. But by then virtually all the slack had been taken up and, within the context of the agricultural techniques of the time, the ceiling had been reached.

Thus it was that, during the century and a half which followed, the Japanese peasants evolved their own response to an otherwise intolerable situation and, instead of accepting the indiscriminate harshness of natural checks on population growth, made their own selective thinning out by means of abortion and infanticide. These practices were known as *mabiki,* the same term as that used to refer to the thinning out of the young plants in the paddy field. This was a spontaneous practice at the local domestic level, and not a matter of organized policy, but until the early nineteenth century it virtually succeeded in stabilizing total population at around 30 million. But such an arbitrary stabilization of numbers and, in effect also of wealth, created other pressures, and these became intensified when, in spite of such demographic restraint, severe food shortages and related troubles became endemic in the late eighteenth and early nineteenth centuries.

As so often happens, if one accepts Hollywood's view of the world, the answer here, as elsewhere, to the maiden's prayer appeared in the shape of the United States Navy. For Commodore Perry had been sent, in 1853, to compel Japan to abandon its seclusion and open its doors to trade and other relationships with the outside world. And while the more traditionally-minded Japanese were greatly alarmed by this, others more forward-looking saw in such a course a possible solution to Japan's problem, though in fact it took 15 years before the Japanese finally decided on what their response should be, and within that interlude (from 1853 to 1868) American interest in Japan had been temporarily eclipsed by the problems of its own Civil War. But the Japanese meanwhile had been learning all they could about the newly industrialized West, of which Britain was then the pacemaker. They were quick to sense the parallel presented by this Western island kingdom, off the opposite shores of the same Eurasian landmass as Japan, which had managed to support a rapidly growing population at a rising standard of living by large scale manufacturing and overseas trade. This course, therefore, the Japanese decided to follow, and in most respects they had succeeded beyond their wildest dreams by the end of the first half-century under this new policy.

Nevertheless, in this process of hectic industrialization, which was later intensified in the period of reconstruction after the Second World War, the Japanese again ran into difficulties arising from their acute shortage of land for industrial as well as agricultural purposes. Since, unlike Britain, whose nineteenth-century industries grew up mainly on the coalfields mostly away from the main agriculture areas, Japan, with only meagre coal resources, concentrated its industries in the existing large towns, which were originally the market and regional-administrative centres of the main agricultural lowlands. This greatly increased the pressure on these same narrowly limited coastal plains, particularly those lying along the Pacific and Inland Sea coasts between Tokyo and Kita Kyushu, a problem which became further intensified during the 1950s by the way in which post-war Japan, like London after the fire of 1666, tended to rebuild in essentially the same places as before.

Growth for the sake of growth is the ideology of the cancer cell.
BERTRAND RUSSELL

In purely economic terms this practice had obvious advantages, for here, in this already highly urbanized zone, were the greatest concentrations both of labour and of purchasing power, and also it was only in such coastal locations that further new areas of flat land, large enough for modern industrial complexes, could be obtained by reclamation of the adjacent foreshore. Moreover, all these great conurbations could be economically linked by sea transport along the coast, and enjoyed similar locational advantages in respect of the import of raw materials and fuels and the export of locally manufactured goods. In commercial terms, therefore, all this paid handsomely, and indeed the late Prime Minister Shigeru Yoshida openly stated that his country's small area, which had once been regarded as a major liability, was in reality "a blessing in disguise" (Yoshida, 1967, p. 103).

Nevertheless a price of another kind has had to be paid for this blessing. In particular, the exceptional congestion, arising from the juxtaposition of tightly concentrated industrial plants with no less crowded residential areas, has led to some of the most acute environmental pollution anywhere in the world. Moreover, this has gone hand in hand with the widespread loss of amenity for, ironically, the area which has been the principal focus of such environmental despoliation comprises one of the most exquisitely beautiful examples of coastal scenery in the world.

However, during the last decade, a remarkable reaction has occurred. The severity of environmental problems in Japan has been increasingly recognized by the Japanese themselves, and in recent years there has emerged a widespread consensus of opinion to the effect that, from now on, Japan must adjust to a less hectic rate of economic expansion which will allow for increased spending not only on the development of a welfare state but also on measures aimed at coping with the environmental costs of industrial growth. . . .

II

As has already been implied in Part I, what has made Japan's environmental problem so seemingly intractable has been the exceptionally rapid development of its now extremely large industrial economy—currently the third largest in the world—within the confines of a country which suffers from an unusually acute shortage of land. Indeed over 80 per cent of the value of Japan's industrial production and most of the Japanese population are now concentrated within a narrow coastal zone stretching from the Kanto plain to northern Kyushu. This zone, termed the Pacific belt by Japanese geographers, contains the three conspicuous bay-head industrial regions of Keihin, Hanshin, and Chukyo, which centre respectively on the cities of Tokyo, Osaka, and Nagoya, and although southwestwards from Hanshin the proportion of flat land becomes even smaller, this area includes two further large industrial nuclei in Hiroshima and Kita Kyushu. Taken together, Keihin, Hanshin and Chukyo account for 60 per cent of the value of Japan's industrial output and 47 per cent of its population. Within each major industrial region rapid industrial expansion has been accompanied by a fast rate of population growth. Thus the population of Keihin, a region which occupies approximately 13,400 km^2, rose from 13 million in 1950 to 23 million in 1972.

Although . . . new industries . . . have been able to enjoy economies of agglomeration, these advantages have begun within recent years to be outweighed by diseconomies brought about by such factors as growing traffic congestion and rapidly rising land prices. Moreover, the location of

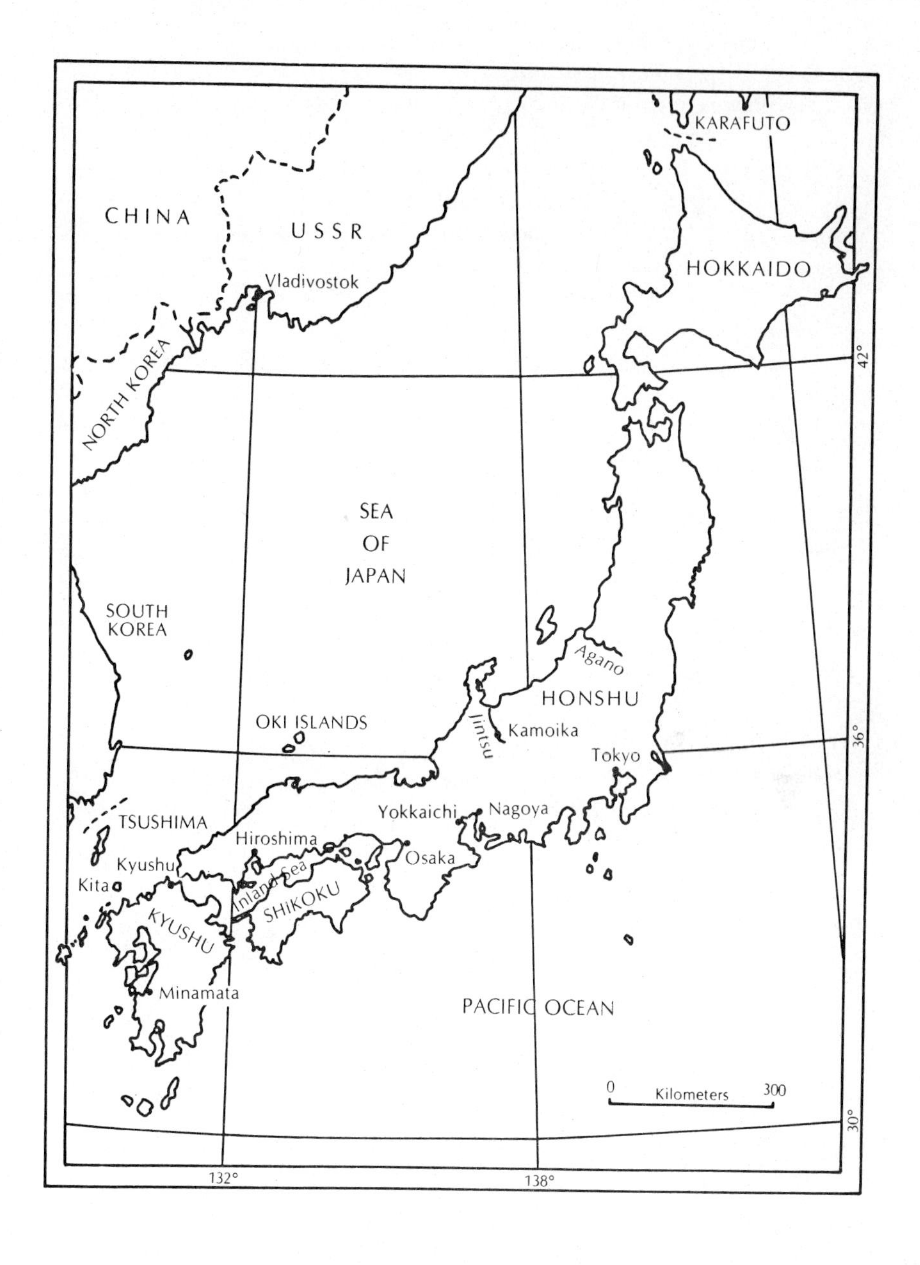
KARAFUTO
CHINA
USSR
HOKKAIDO
Vladivostok
NORTH KOREA
42°
SEA
OF
JAPAN
SOUTH
KOREA
Agano
HONSHU
Jintsu
Kamoika
OKI ISLANDS
36°
Tokyo
Yokkaichi
Nagoya
TSUSHIMA
Hiroshima
Osaka
Kyushu
Kita
Inland Sea
SHIKOKU
KYUSHU
Minamata
PACIFIC OCEAN
0
Kilometers
300
30°
132°
138°

the bulk of Japan's industrial output along a single axis has enabled the state to achieve considerable savings in the provision of economic infrastructure. In respect of transport facilities, for example, one motorway and one high-speed railway line suffice to link together most of Japan's major centres of industrial production. In the United Kingdom, by contrast, a far more complex network of roads and railways is necessary to serve the needs of industry.

The post-war tendency of manufacturing industry to locate within the major industrial regions is not only a function of economies of agglomeration, but also, to some extent, a result of local government policy. Especially between 1955 and 1968, many prefectures and cities within the Pacific belt competed in offering generous inducements to large manufacturing companies, which were regarded as major sources of local government revenue. In many cases, local authority budgets were badly distorted by heavy expenditure on the provision of economic infrastructure for the benefit of industry. Throughout this period, local authority spending on education, housing, and welfare had to take second place to investment in expensive land reclamation schemes, and in industrial roads and railway lines.

By the end of the 1960s, the mounting geographical concentration of industry in the Pacific belt was widely identified throughout Japan as the major cause of serious environmental problems, ranging from atmospheric pollution to land subsidence. Since 1968, growing concern over the deterioration of the urban living environment has forced many local authorities to make a complete *volte-face* in their policy towards industry, and most of the prefectures in the major industrial regions have by now adopted various restrictions aimed in particular at limiting the proliferation of pollution-producing factories. In many cases these developments also reflect political changes in the conurbations where, in recent years, the opposition parties have won control of several local governments.

Meanwhile the central government has attempted to undertake regional development planning aimed at restricting the growth of the major industrial regions, and at encouraging factory location in the peripheries of the country (Sargent, 1973). At both local government and national government levels there is widespread agreement that the geographical concentration of industry has caused unacceptable environmental problems . . .

Problems of the Urban Living Environment

Before examining Japan's urban problems in detail it is first necessary to look critically at a popularly held assumption concerning life in Japanese cities. It is sometimes argued that Japanese cities exhibit a dangerously high degree of overcrowding which, if allowed to persist, must ultimately result in some form of social breakdown. It is of course true that Japanese cities are densely populated and that they suffer from a shortage of adequate housing. Nevertheless, it must not be forgotten that for many centuries the Japanese have lived in settlements which from the Western point of view must seem overcongested. Thus the typical nucleated village of premodern Japan, with its many farmhouses clustered together to form a tightly-packed settlement, exhibited an extremely high density of population. To a large extent, Japanese society, with its strong emphasis on the group as the basic unit of social organization, is well equipped to cope with the pressures created by life in densely-peopled settlements. This proposition seems to be borne out by the post-war experience of Tokyo which, perhaps alone among the capital cities of the industrialized world, has experienced a gradual fall in its crime rate since 1945. Indeed, by contrast with many Western cities, the twin problems of crime and poverty are of negligible importance, not only in Tokyo but throughout the large cities of Japan.

Nevertheless, it cannot be denied that Japanese cities suffer from serious problems, some of which seem almost intractable. Land shortage may be regarded as the root cause of many of these problems. Since 1955, industrial growth has caused a marked acceleration in the competition for land within the major conurbations. In response to this competition, land has been reclaimed from the sea, but the sites so provided have been monopolized by heavy industries such as iron and steel, petrol refining, petro-chemicals, and shipbuilding, and for the foreseeable future, further land reclamation cannot be regarded as a feasible solution to land shortage. In the last decade, then, the demand for urban land has greatly outstripped supply, and in consequence, land prices have risen steeply. Thus, whereas between 1955 and 1973 wholesale prices increased by 1.3 times, and consumer prices by 2.1 times, the price of land within the built-up area of Japanese cities rose by an average of 22.9 times, or at a rate of 19 per cent per annum. Since 1969 this already spectacular rise in the price of urban land has been further accelerated by heavy investment in land by large corporations. In 1973 alone, the price of land in the built-up area of Japanese cities rose by an average of 31 per cent, and in the built-up area of Tokyo by 34 per cent.

In turn, rising land prices have brought about an increase in the intensity of urban land use. Thus within the large cities of Japan open space is at a premium. By contrast with Western cities, public parks and recreational areas are few and far between, while Japanese cities also lag far behind their Western counterparts in the provision of pavements and pedestrian precincts. High intensity of land use is also reflected in the unusually low proportion of the city area taken up by roads.

Rising land prices have likewise exacerbated the difficulties of commuting within the city regions of Japan. Thus the rapid rise in the price of land within striking distance of the centres of Tokyo, Osaka, and Nagoya has resulted in a widening of commuting zones around these cities. Within the Tokyo region, land selling at ¥100,000 per m^2 in 1972 could be found, on average, only at a minimum distance of 40 to 50 km from the city centre, a distance involving between 1½ and 2 hours of commuting (Tanaka, 1972). Meanwhile the notorious overcrowding of Japanese commuter trains is also partly attributable to high land prices, for to relieve pressure during the rush hours would necessitate the construction of extra railway tracks, which would involve the purchase of expensive urban land. Thus the construction costs of the Ryogoku to Tokyo stretch of the new Sobu line amounted to ¥6.2 billion per km, much of which reflects high land prices. Similarly, high land prices hinder the realization of road widening and construction schemes needed to relieve the serious traffic congestion from which most large Japanese cities suffer.

A concomitant of the widening of commuter zones around the large cities is the acute housing shortage within the cities themselves. Here again high land prices made the provision of cheap municipal housing prohibitively expensive. At the same time, rising land values have brought about a sustained increase in rents of existing properties. Between 1960 and 1965 rents in Tokyo rose by 5.4 times, in contrast to a rise of 1.7 times in the commodity price index.

Other urban problems stem from the close juxtaposition of zones of heavy industry and residential areas. Thus Tokyo is not only a city of nearly nine million people, but also the nucleus of a congested industrial region which contains a high proportion of Japan's output of electricity, steel, refined petrol, and petro-chemicals. The dependence of such industries upor crude oil, which now provides 70 per cent of Japan's energy supply, help. to explain the high concentration of sulphur dioxide and other noxious pollutants in the atmosphere over the capital region. Meanwhile rising industrial demands for electricity, coupled with the restrictions imposed by local authorities on the construction of new oil-burning power stations, have brought about increasingly frequent power shortages, especially in Tokyo and Osaka.

Industrial demands for water have greatly exceeded the rise in water supply, and in Tokyo and Osaka the summer months are almost inevitably accompanied by water shortages and water rationing. Furthermore, growing demands have led to the over-extraction of water from underground aquifers, which in turn has caused widespread land subsidence in the major conurbations. Extensive zones of Tokyo and Osaka now lie below sea level, and hence are vulnerable to flooding at high tides and during typhoons.

Apart from their vulnerability to typhoon damage, the large conurbations of Japan are also under permanent threat from earthquakes. Although it is frequently claimed that the tall buildings recently constructed in Tokyo, Osaka, and Nagoya have been designed to withstand the most severe earthquake shocks, the structures have yet to be put to the test. In any case, in the event of a major earthquake, fires are likely to be a major hazard, especially in the densely populated districts close to refineries and petrol storage yards. In 1967 the Tokyo fire department estimated that an earthquake of 1923 dimensions could cause up to 500,000 casualties within Tokyo.

From the above summary it seems obvious that there can be no simple or rapid solution to the manifold problems of Japan's conurbations. Nevertheless, there are hopeful signs that the Japanese government is now preparing to tackle the fundamental problem of high land prices. In 1974, for example, the National Land Agency was established, together with a Land Bureau empowered to control land prices. During 1974 land prices within a 50 km zone centred on Tokyo fell by an average of 11.4 per cent, although it is possible that this decline was more the result of economic recession than a reflection of the work of the National Land Agency.

Secondly, there are indications that the rate of population growth and industrial expansion within the conurbations of Japan is beginning to slow down. During 1973, the highest rates of population growth were registered not in the conurbations but in provincial cities. Rural depopulation, itself a result of large-scale migration to the cities, is now less widespread. Thus, whereas in 1969 28 prefectures lost population, in 1972 and 1973 the totals were 14 and 9 respectively.

As regards industrial growth, diseconomies of agglomeration, including the costs of traffic congestion, high labour costs, and high land prices, are driving companies to seek locations outside the conurbations, and even in some cases overseas. This trend explains the gradual decline in the relative importance of the three major industrial regions since the mid-1960s. . . .

Pollution

Of all the environmental problems of Japan, that of pollution has received the most publicity, both in Japan and abroad. In particular, attention has been focused on four major incidents, three of which have arisen from industrial pollution in predominantly rural districts.

From the early 1950s onwards, a strange illness became apparent in fishing villages around the shores of Minamata Bay in western Kyushu. The chief symptoms of the disease, soon to be known throughout Japan as "Minamata disease," were paralysis and derangement, and by 1971 121 casualties had been identified, 46 of whom had died. Scientific investigations suggested that the cause of Minamata disease was the contamination of fish by organic mercury discharged by an industrial effluent from the local factory of the Chisso Corporation. In 1969 an organization representing the victims took the matter to court, and in 1973 the Chisso Corporation was found guilty of negligence and forced to pay damages amounting to ¥900 million.

A second, smaller outbreak of Minamata disease was identified in villages near the estuary of the Agano river in Niigata prefecture, this time as a result of mercury discharge from the Kanose factory of Showa Denko Limited. In 1967 a group of victims began a legal battle which culminated in a court decision of 1971, whereby Showa Denko was required to pay damages of ¥270 million to the plaintiffs.

Meanwhile a second type of illness attributable to industrial effluent had long been apparent among inhabitants of Toyama prefecture. This ailment, termed "Itai-itai disease," was associated with the discharge of cadmium from the Kamioka mine of Mitsui Mining and Smelting. In 1968, an organization representing 515 victims sued the company, and in June 1971 the court decided in favour of the plaintiffs. Although the company appealed against this decision, the Nagoya High Court upheld the contention of the plaintiffs and the Mitsui company was ordered to pay damages amounting to ¥1360 million.

The fourth major incident, by contrast, concerned the effect of atmospheric pollution in the large industrial city of Yokkaichi . . . , which contains one of the largest concentrations of petrol refining and petro-chemical manufacture in Japan. Here the inhalation of sulphur dioxide fumes has given rise to a peculiarly severe form of asthma, known as "Yokkaichi Zensoku." In Yokkaichi an organization representing victims of the disease appealed in the local courts, and in 1972 six local companies were ordered to pay damages of ¥200 million to the plaintiffs.

Meanwhile in the conurbations the concentration of car exhaust fumes has caused severe outbreaks of photo-chemical smog, especially during the summer months. To combat this problem, the government has recently introduced rigorous controls on the emission of exhaust fumes, with encouraging results. Thus, between 1968 and 1971, the measured level of carbon monoxide in Tokyo fell from 5.0 parts per million to 2.4 ppm.

Water pollution also has assumed disturbing proportions, and many rivers have become lifeless through the uncontrolled discharge of industrial effluent. The pollution of sea water, particularly severe around the shores of the Inland Sea, has seriously threatened the well-being of the Japanese inshore fishing industry. In particular, the seas around Japan, which is the world's biggest importer of crude oil (250 million tons imported in 1972), are especially vulnerable to oil pollution on a massive scale.

The examples of pollution outlined above may give the impression of impending doom. Yet in the field of pollution control, the Japanese government has scored several successes, and the outlook in this respect is quite promising. Thus the government has imposed limits (set in terms of ppm) on the discharge of most poisonous gases and effluents. In some instances these limits are more rigorous than in other industrial countries. Thus, the limit for PCB (polychloride biphenyl) is set at 3.0 ppm in Japan compared with 5.0 ppm in the United States. Moreover, in 1971, the government established the Environment Agency which, together with the Health and Welfare Ministry and the Ministry of International Trade and Industry, carries out constant monitoring of pollution in Japan.

Despoliation of Scenery

In common with other advanced nations, Japan must also cope with the vexed problem of how best to preserve areas of outstanding natural beauty while meeting the needs of a mature industrial economy for land and other resources, and simultaneously satisfying the demands of an affluent population for access to the countryside.

In Japan extensive land reclamation schemes, often in coastal areas long renowned for their fine scenery, and the location of factories and housing estates without regard for their impact upon the landscape, have already caused irreparable damage. Yet at the same time, many areas of outstanding natural beauty have been designated as national parks, and as such benefit from protective legislation (Simmons, 1973). Here, the main problem is often the intense pressure of visitors during the tourist season. Thus in 1973 the total number of individual visits made to national and quasi-national parks reached the extraordinary total of 617 million. Between 1970 and 1973, the number of visitors to Tateyama, possibly the most frequented peak in the Japan Alps, rose from 770,000 to 1,060,000.

The example of national parks in Hokkaido illustrates the conflict between economic demands and the need to preserve areas of natural wilderness. Here, unspoiled forest country is threatened not only by the increasing demand for timber but also by the need for more grassland, in association with the rising Japanese demand for meat and dairy products (Simmons, 1973).

When one is suddenly confronted by . . . such comments as that of the *Japan Times* (27 May 1968) to the effect that the new urban policy "is also aimed at revamping the Japanese island chain to become a highly efficient,

Heaven and earth and I are of the same root,
The ten-thousand things and I are of one substance.

SŌJŌ

balanced, and extensive urban sphere," one's instinctive reaction is to ask what has become of the intense and deeply ingrained Japanese sensitivity to the exquisite natural beauty of their homeland, though against the background of nineteenth-century desecration of so much of our own finest scenery it ill behoves British observers to criticize the Japanese.

In fact, however, the Japanese have by no means lost their traditional appreciation of natural beauty and indeed have shown a sturdy initiative in their attempts to preserve it. Thus, following growing evidence of widespread local concern, of the kind which in April 1969 led Utsonomiya District Court to rule that "preserving a stand of 350- to 400-year-old cedars is more important than widening a road to smooth traffic in Nikko" (*Japan Times,* 10 April 1969), the Japanese government proceeded to establish the Environment Agency, which has repeatedly intervened to prevent schemes which have threatened areas of outstanding scenery, as for example in 1971 by stopping the construction of a proposed road through the southern Japanese Alps on the grounds that it threatened the survival of a 500-year-old forest. Moreover, since the enactment of the Nature Conservancy Law of 1972, all schemes for the construction of new industrial areas must include assessment of the implications of factory development for the natural landscape.

While obviously such measures as these cannot undo all the damage which has already been done, there can be no doubt that the Japanese are now acutely aware of the need for preserving the best of what remains of their natural heritage. Meanwhile the spectacular decline in their net population increase, from 2.2 per cent in 1948 to an annual average of 1.1 per cent during 1963–71, at least makes the situation significantly less daunting than it was. If, therefore, the remarkably persistent Japanese sense of community, which more than any other single factor made possible the phenomenal transformation of their country during the past hundred years, can now be harnessed to restoring its image to that of Yasunari Kawabata's "Japan, the beautiful," there is good reason to believe that this can be accomplished.

REFERENCES

Hutchinson, Sir Joseph, FRS. 1966. Land and human populations. *The Listener,* **23,** 1904:310.

Oriental Economist. 1973. *Japan economic handbook* 1973. Tokyo: Oriental Economist.

Simmons, I. G. 1973. The protection of ecosystems and landscapes in Hokkaido, Japan. *Biological Conservation* **5,** 4:281–89.

Sargent, J. 1973. Remodelling the Japanese archipelago: the Tanaka plan. *Geogrl. J.,* **139,** 3:426–35.

Sauer, Carl Ortwin, 1941. Foreword to historical geography. *Ann. Ass. Am. Geogr.,* **31,** 1:1–24.

Tanaka, Kakuei. 1972. *Nippon Retto Kaizo-ron.* Tokyo: Nikkan Kogyo Shimbun-sha.

——. 1973. *Building a new Japan* (English translation of Tanaka 1972, above). Tokyo: Simul Press.

Yoshida, Shigeru. 1967. *Japan's decisive century.* New York, Washington and London: Praeger.

23 China's Environomics: Backing into Ecological Leadership

Leo A. Orleans

Leo A. Orleans is the China research specialist at the Library of Congress. He is the author of *Every Fifth Child: The Population of China* and numerous articles dealing with China's population, science, public health, and education. He visited China in the summer of 1973 as a member of the American medical delegation.

This selection is condensed, with the author's permission, from a paper submitted in June 1975 to the Joint Economic Committee of the U.S. Congress. The paper has been published in a government publication, *China: A Reassessment of the Economy;* a slightly condensed version has appeared in three parts in *Environmental Policy and Law,* 1 (February 1976), 2 (April 1976), and 2 (June 1976).

In the People's Republic of China, the world's largest population and impressive industrial development combine to create conditions that could, potentially, cause serious environmental degradation. Rightly or wrongly China has a reputation for having a real concern for the well-being of the individual, and, in this regard, for following a course of economic development that has not ignored its consequences on environment. This paper will attempt to examine China's policies relating to the environment, to consider how successful she has been in implementing them, to look at the economics of the problem, and finally, to comment on whether China will succeed in avoiding the environmental problems currently afflicting most Western industrialized nations.

HEALTH AND SANITATION

Because of the size of China's population and the incredibly poor conditions under which the people lived prior to 1949, the new regime's primary concern was not with fancy ecosystems—a distant concept of trivial priority—but rather with the simple and immediate requirements of improving the human environment at its most basic level. It is here perhaps more than in any other field that China has achieved the type of success to be envied not only by all the developing countries but also by many of the industrially advanced nations.

The Mass Campaigns

All visitors to China are duly impressed by the cleanliness of the country. What they are seeing are the results of a long and difficult struggle which began in the early 1950's and is still going on.[1] Environmental sanitation was recognized as the single most important health problem in China, and prevention, rather than cure, as the practical approach to solving it. Chairman Mao called the people to "get mobilized, pay attention to hygiene, reduce disease, and improve health conditions"—and the first of many mass campaigns was on its way. Responding to both moral incentives and political pressures, hundreds of millions of people set out to clean up the country and learn the basic facts about sanitation, from washing hands, to the sanitary preparation of food, to the maintenance of clean homes and individual hygiene. Although a more conventional battle against disease was also waged by "epidemic prevention stations" which were established throughout the country to conduct massive immunization programs and by medical researchers in hospitals and at the institutes of the Chinese Academy of Medical Sciences, practical considerations led the leadership to oppose "the erroneous viewpoint that large sums of money, more chemicals and apparatus and numerous experts are needed to improve sanitary conditions"[2] and to place major emphasis on the organization of intensive mass participation in sanitation.

One aspect of the early campaigns which received the most waggish publicity outside China was the goal to exterminate all flies, mosquitoes, rats, and sparrows. Few observers believed that the four pests or "four evils" could be eliminated in the first place, and some of the fantastic statistics reporting success in terms of numbers or weight of the "kill" usually elicited only smiles. China did, however, mobilize every household for this task and over the years managed essentially to achieve the impossible and rid the country of these pests. As a matter of fact, the successful reduction of the sparrow population resulted in an unforeseen disruption of the natural equilibrium of the environment. The annual grain available for human consumption was initially increased as anticipated, but the campaign did not take into consideration the service the sparrows perform by eating injurious insects. The insects multiplied and losses of grain increased. Belatedly, after most had been killed, sparrows were proclaimed "rehabilitated" and deleted from the "four evils" list; their place was taken by bedbugs or other vermin —depending on local conditions.

Parasitic diseases, which affected the health and lives of literally hundreds of millions of people in China, were also attacked through education and mass mobilization. Probably the best example of the efforts may be seen in the fight against schistosomiasis which, in the past, affected entire communities along the Yangtze River. People were urged to "bury, burn, boil, and chemically dispose" of the snails that carry the disease—and they did. Where but in China could the authorities of just one province (Anhwei) mobilize 1.5 million persons to spend 20 million man-days in such a task over a period of just several months?[3] And Anhwei is not an unusual province, nor was that year, 1956, an unusual year.

Control and Disposal of Night Soil

The ignorance and low level of personal hygiene and the widespread use of animal and human wastes as fertilizer, combined to make fecal-borne diseases primary causes of death in rural China. The collection of both rural and urban fecal material and its spread over most of China's agricultural lands resulted in such pollution of the fields that it was impossible for people to avoid infection.

After 1949 the control and disposal of night soil became another major responsibility of health and sanitation personnel. An extensive education program was initiated to teach people the danger of contact with raw excrement. Literally millions of newly covered latrines were built and covered cesspools were dug for the collection of human and animal waste on the outskirts of villages. An effort was made to distribute chemicals for the treatment of excrement, peasants were taught not to use it as fertilizer until it had been stored from two to eight weeks, depending on the season, and were urged to "observe each other in order to achieve this purpose." One interesting calculation reported in the *People's Daily* stated that ". . . if the manure and urine of the entire populace of China are fully utilized, these ingredients will correspond to some 10 million tons of ammonium sulfate, or about the annual output of scores of chemical fertilizer plants."[4]

Despite the increasing production and importation of chemical fertilizers, their availability is still limited and China's farmers are as dependent as ever on animal dung, household manure, green manure, and mud from rivers and ponds to fertilize their fields. Every spring, for example, a mass movement to accumulate manure is initiated in rural China. In some localities well over half of the total labor force is rounded up to move manure accumulated during the winter months to the fields. During the height of the mass effort to "accumulate and deliver manure in a big way," large numbers of urban workers and employees are called to participate in the campaign—what better way to shed possible remnants of elitist thoughts.[5] The difference, however, is that now most of the spreading is done under strict directives and supervision of public health personnel and great precautions are taken to insure that the manure accumulated over the months is properly cured, chemically treated, and environmentally safe.

Urban Sewage Disposal

Information on urban sewage disposal facilities in China is scanty indeed. Prior to 1949 only the largest cities in China had sewerage systems and even they served only a small part of the population that lived in the central, more modern sections of the city. In most of China's urban areas, untreated sewage was dumped directly into the rivers or was transported via "honey buckets" to fertilize adjacent fields. In northern cities only solid fecal material was collected, carried to the edge of the city and dried into thin cakes containing considerable amounts of nitrogen.[6]

Although large-scale construction of sewerage systems began in many cities during the 1950's, for the most part it was associated with new workers' housing developments on the periphery of the cities.

The first attempt to irrigate extensively with urban sewage was made in 1956 around Chuchow in Hunan Province.[7] Apparently it was a successful experiment to assist with agricultural production and at the same time to decrease pollution of rivers: A couple of years later, during the Great Leap, there was a rapid increase in the use of urban sewage for irrigation and fertilization of fields adjacent to large cities—an ideal example of transforming "what is harmful and useless into something that is harmless and useful."

It was soon discovered, however, that this process was not harmless. The widespread use of raw urban sewage on the fields increased the number of flies and mosquitoes and resulted in a serious setback in the battle to control fecal-borne diseases. Furthermore, tests proved that sewage irrigation tended to contaminate water wells and drastically increase the bacterial count of well-water. New directives had to be issued warning against over-irrigation with sewage, against the use of untreated sewage on fields just prior to harvesting and introducing new technical controls and sanitary standards for wells.[8] The basic problems were obviously solved; by 1966, 43 Chinese cities were developing agricultural irrigation by utilizing treated urban sewage and claiming substantial increases in farm yields.[9] One of these cities—Changchun, in Kirin Province—now reports that 90 percent of its waste water is carried by pipe and irrigation channels into the fields. The city leadership estimates that the nitrate and phosphate contained in this waste saves the communes 3500 tons off chemical fertilizer a year.[10]

China has made significant advances in the disposal of urban sewage but, as in all other accomplishments, this one too must be placed in perspective. The most impressive accomplishment is simply that urban waste produced by some 150 million people is disposed of by essentially sanitary methods, both modern and primitive.

At present it would appear that all large cities have a sewerage system to serve the center of the city and the new apartments, offices, and factories built on the outskirts. The old districts, however, populated by most of the workers, still rely on "honey buckets" or trucks to carry the waste out of the city—and it is not unusual to see this process in operation just a few blocks from Tien An-men Square in the center of Peking. China's middle-sized and small towns, of course, also employ the more traditional methods. Furthermore, there have been interesting reports based on refugee interviews in Hong Kong about a black market in human excrement—arrangements made by representatives of rural brigades to buy human waste directly from individual urban households. If true, it is certainly a comment on the shortage of fertilizer even in close-in communes and on the continuing private initiative and enterprise of Chinese peasants.

A few words about garbage. Since China is not a consumer-oriented society, neither the volume nor the content of the garbage in Chinese cities compares to that in the West. There are few plastic or metal containers and throw-away bottles are unknown. Everything that is fixable is fixed and anything that is reusable is salvaged. For the most part, then, garbage is organic in nature.

Peking, of course, has a larger volume of garbage than most Chinese cities. The method of handling its waste, however, is probably typical of that used in other large cities. Inhabitants in streets and lanes dump their garbage at appointed spots every evening where it is picked up daily. One method is to collect it by truck and take it to the communes where the peasants clear it of hard materials, add to it farmyard manure and then seal the dumps with clay to ensure fermentation within 50 to 60 days. The 22 such sites on the outskirts of Peking convert the 2,700 tons of garbage picked up daily into 2 million tons of compost annually, which is distributed to 36 communes and 80 state farms.[11] Another report says nothing of sealing the dumps with clay; rather, the garbage is simply packed firmly in large heaps where fermentation takes place. There are 48 garbage disposal yards of this type located on the outskirts of Peking.[12]

THE ENVIRONMENT AND CHINA'S LAND

It has been said that "to govern the water is to govern the country," and the battle with water has been going on throughout most of China's long history. Her excellent historical records show, for example, that in the past 2,000 years, the Yellow River Valley has suffered 1,500 floods and 1070 droughts. Since 1949 the government and people have expended enormous amounts of money and labor on innumerable and varying-sized projects aimed at controlling China's water resources in order to prevent major disasters resulting from floods and droughts, to stop erosion and alkalinization of the soil, and to reclaim and expand the land that, over hundreds of years, had been lost to agriculture.

The problems are huge and progress is often slow because of poor planning and management (the Chinese themselves admit that "some of our water projects went off on the wrong track") but, although the effort is likely to be a never-ending one, impressive progress is being made, and gradually China's environment is changing.

Irrigation and Water Conservation

Although several major projects were initiated during the early 1950's, most of the early emphasis was on temporary measures such as the strengthening of existing dikes and on other "key point" projects seeking to prevent major disasters. Most of this activity centered in the flood-prone areas of central and northern China and met with only limited success and many problems. China did not yet have enough specialists to study the local topographical and climatic conditions and to plan and coordinate the work. Too many irrigation canals and not enough drainage canals resulted in seepage, waterlogging, and accentuated alkalinization. Many of the water-control projects were abandoned before completion.

More care was given to the larger water conservation works, many of them started during the First Five-Year Plan (1953–57). Special organizations were named to draw up plans and lead the work for the control of major river systems—the Yangtze, the Huai, the Haiho, and the Yellow ("China's Sorrow") Rivers: It was also during the early 1950's that work was started on the enormous People's Victory Canal which, over the years, has grown to include a 7,500-kilometer network of irrigation and drainage canals, and some 2,600 pump wells, irrigating some 600,000 mou of farmland.[13]

In general, China's approach to water conservation (as in most everything) is to "walk on two legs." The first "leg" concentrates on small projects that are the responsibility of local administrative units, such as communes and production brigades, using local resources for the task. The other "leg" is undertaken by the central government and includes "large-size backbone projects," such as large reservoirs, dams, flood-diversion projects and other major undertakings which require considerable capital and resources. Theoretically, both types of projects are undertaken with a broad plan in mind, somehow linking the various projects into a unified water control scheme—an ever-present predicament for planners and builders.

During the slack agricultural season every commune in the country becomes involved in rural capital construction; this usually has to do with water conservation and land reclamation or improvement. Over the period of a year, as many as 200 million people with picks and shovels are likely to spend literally billions of man-days moving immense mounds of earth in baskets and handcarts. They dig canals, aqueducts, and new riverbeds; they spend years trying to cure alkali lands; they level ground and terrace hillsides; they deep-plow. They do all this mostly in the winter and early spring, which poses still additional problems in the north and northeast where much of the ground is "so frozen it has to be cracked with sledge hammers."

The Chinese point to some important advantages to be gained by concentrating on "small water conservation projects built on a large scale"—that is, projects initiated locally under the principle of self-reliance. The projects get quick results at minimal expense (the masses bring their own tools and grain rations), they are easy to carry out and popularize, they can be accomplished at high elevations to bring water to odd pieces of land and they are easier to manage and safeguard.[14]

Not all the water conservation efforts rely on local resources and initiatives; the other leg, which relies on large government-funded projects, has also been very active. Most of the comprehensive efforts have been concentrated on China's great rivers, with vast national schemes designed to transform whole regions by controlling floods, irrigating extensive areas, creating navigable waterways, and generating large amounts of hydroelectric power. The achievements have been impressive,[15] but since China's needs still appear almost limitless, tremendous amounts of capital will yet have to be spent on major water control projects.

Land Reclamation

Intimately related to water conservation and irrigation are China's efforts to increase agricultural acreage, which is still estimated at not much more than 11 percent of the country's total area. In the early 1950's, considerable optimism was expressed in many Chinese publications about the possibility of expanding arable land. Experience quickly proved, however, that large reclamation projects in the more remote, sparsely settled provinces had to overcome too many natural obstacles and were economically impractical. Primary emphasis, therefore, is placed on increasing the productivity of existing agricultural lands, while most of the reclamation is limited to extending these lands gradually through terracing up the slopes of mountains and into the more marginal, but still adjacent, regions. It is worth noting that when the Chinese refer to reclaimable wastelands they include not only agricultural land but any piece of land which, through the application of labor and perhaps limited capital, can in some way be made productive. As in the case of water conservation, most of these activities take place during the agricultural off-season and with a large labor force that does not have to be moved to distant areas.

Typical are the numerous projects to reclaim alkali and marshy lands—efforts that require tremendous amounts of human labor and perseverance. There are also many projects undertaken to control the spread of the deserts, such as was recently reported in Tunhuang county in Kansu Province. Here they have rescued agricultural land from the encroaching desert by controlling the shifting sands first by planting bushes, then trees, and by digging canals to bring water from nearby mountains.[16]

Some of the most difficult land reclamation projects seem to be undertaken by the various units of the People's Liberation Army (PLA), which are often located in remote areas. These units are able to devote more time and effort to activities considered to be economically impractical in order to prove that, through hard work, it is possible to become self-sufficient in agricultural needs. One Production and Construction Corps which was stationed in distant Sinkiang Province near the Soviet border has, since

1949, reportedly transformed 10 million mou of "primitive desert into fertile lands surrounded by dense forests."[17]

The model for all of China when it comes to land reclamation (as well as perseverance and hard work) is the Tachai production brigade located on poor, badly eroded slopes in Shansi Province. Over a period of many years and after numerous hardships and disasters, the brigade has leveled hilltops, built stone walls, terraced fields, carried compost, and deep-plowed, irrigated, and fertilized the land until it has become a rich brigade producing surplus grain, growing fruit and timber, and raising draft animals and pigs. "Learn from Tachai" is a slogan one hears throughout China. For every

Those who would take over the earth
And shape it to their will
Never, I notice, succeed.
The earth is like a vessel so sacred
That at the mere approach of the profane
It is marred
And when they reach out their fingers it is gone. . . .
At no time in the world will a man who is sane
Over-reach himself,
Over-spend himself,
Over-rate himself.

LAO TZU

reported success story such as Tachai there are undoubtedly several failures, but most such projects probably fall into the "less than success and more than failure" category.

Afforestation

Over thousands of years, the growing population of China, the expanding area cultivated by farmers and the ever-increasing demand for lumber for fuel and construction have caused widespread deforestation of the country with predictably adverse effects on the environment. It is estimated that less than 10 percent of the Chinese land mass is covered by forest; the percentage of forest cover in some of the eastern, densely-populated provinces such as Hopei has become insignificant. Deforestation, in turn, has resulted in flooding and soil erosion with serious effects on agricultural production. Thus, the need for afforestation was well recognized by the regime as an important economic need and correlate to soil and water conservation programs.

By the end of the First Five-Year Plan, the afforestation program was already in full swing, and it was reported that between 1953 and 1957, 169,350,000 mou of trees were planted.[18] In many parts of China, and especially in the western provinces, wide tree-sheltered belts were planted. Forestry experts, with the help of tens of thousands of people, have created shelter belts between western Inner Mongolia and the Ninghsia Hui Autonomous Region,[19] and on the sandy wastes created by the flooding of the Yellow River.[20]

Afforestation work continues and Chairman Mao's call to "make the motherland green" recurs every year. The mass planting campaigns are supplemented by more professional activities of the numerous forestry bureaux and state forestry farms. National Forestry Conferences annually disseminate the latest information on aerial sowing of tree seeds, grafting, and a variety of new techniques developed for the scientific management of forests.

The Chinese admit that much of the early, labor-intensive efforts to plant trees was wasted and that the survival of planted but uncared for seedlings was low. It seems, however, that the management of afforestation is improving. Mass planting continues, but much more attention is now being given to the care of trees after they are planted. Whether or not China meets her rising demands for forestry products is immaterial within the context of this paper, but there is no doubt that the billions of trees planted in the past 25 years have served to reduce damage by drought and flood and have helped to conserve soil and water.

Pollution by Pesticides

In her perpetual battle to increase food production, China has greatly increased her production and use of insecticides and other chemical substances in agriculture which are hazardous to man and animal life. The production of all pesticides increased rapidly during the late 1950's[21] and probably through the 1960's. China now recognizes this source of pollution as one of its three primary categories of pollution. What is being done about it?

Considering the overwhelming priority China must give its agriculture, she is not likely to restrict the use of DDT or other chemicals to control her pest and disease problems unless she can find an appropriate substitute. In the meantime there is an attempt to introduce certain precautionary measures. It has been suggested that violent poisons such as organic phosphorous compounds be forbidden or limited, that pesticides not be used during a certain period before harvest and that close watch be maintained over quantities, concentrations, and frequency and method of application of various pesticides.[22] How widely these suggestions are disseminated and how closely they are complied with is not known.

At the same time, China's concern over this type of pollution is perhaps most clearly revealed by the research her scientists are conducting to find alternatives to chemicals now used in agriculture. "Energetic research on a pesticide to replace DDT is needed," they assert, and "the search for low-toxicity and low-residual toxicity agricultural pesticides is a subject at present being actively pursued."[23]

Chinese scientists seem to have made significant progress in the use of biological pest control methods. In south China's Leicho Peninsula a production brigade is reported to have achieved good results by breeding parasitic bees to destroy insects harmful to rice. These bees are put into paddy-fields to lay their eggs at a time when the insects are proliferating. Ten days after they are hatched the young bees feed on the insects' eggs, destroying them.[24]

Nature Conservation

China has many unique areas that have fauna and flora worthy of protection, but one finds little information on nature conservation in Chinese publications. The People's Republic claims that after 1949 the country adopted a wildlife policy of "preservation, breeding in captivity and hunting in a planned way" and the few references available suggest that she is making some effort to protect rare animals. Not surprisingly, the preservation of the giant panda gets the most publicity and a number of natural reserves have been created where pandas are protected and where scientists can observe and research the animals. Other animals which were close to extinction and are now likely to be protected include the golden monkey in southwest China, some small herds of wild horses still remaining in Inner Mongolia and Sinkiang, the great paddlefish of the Yangtze River and the giant salamander, the goat antelope along the Tibetan border and certain species of pheasants and other rare birds. More and more natural preserves are apparently set aside to protect the habitat of rare animals and plants.

INDUSTRIAL POLLUTION

The growth of China's industrial activities after 1949 exacerbated local pollution problems and initiated some concern over effects on health. During the First Five-Year Plan, the various health departments became increasingly involved with controlling industrial wastes, issuing directives relating to health standards of industrial plants, identifying technical processes considered to be dangerous and urging that the location of new plants take into account environmental considerations.[25] Although with the installation of some purification and recovery facilities the serious aspects were usually dealt with, in general the health departments, so successful in other sanitation programs, were waging a losing battle against industrial pollution. Industries were growing rapidly and long-term or future health problems could not compete with more immediate economic priorities.

It was not until the mid-1960's—when the problems of industrial pollution were integrated with the intensification of Mao's often-quoted exhortation to "struggle against waste," and when an improvement in the

. . . a sound man is good at salvage,
at seeing that nothing is lost . . .
LAO TZU

environment became the responsibility not just of the public health departments but also of the various economic ministries and production units—that the battle against the haphazard disposal of inorganic industrial waste began to show some results. Considerations of health and of air, water and soil quality (which gradually became dominant in many of China's official statements) were considered to be almost a bonus, the primary impetus for the program being overwhelmingly economic.

The thrust of the current campaign may be seen from the slogans which constantly remind the workers of their responsibilities. They are urged to wage a battle against the "three wastes": waste liquid, waste gas and waste slag. They are told that with some effort on the part of the population it is possible to "change wastes into treasures and turn harmful into beneficial." Factories are called on "to take one trade in the main and run various undertakings," which means that in addition to producing their primary product they should collect all waste materials from the production process and, through local innovation and experimentation, try to make full use of them. The production units are further encouraged to do this at a minimal cost, by using local resources and indigenous methods. Let us take a look at just how China is "turning the harmful into the beneficial" in her drive to make comprehensive use of waste materials.

Air

It would be easy to assume that in a country where only 15 percent of the population live in urban areas and where vehicular traffic is insignificant, air pollution would not be a serious problem. In fact, it isn't, for most people—but those who live in the large industrial centers have had to breathe air that is as bad as is likely to be found anywhere in the world. The normal pollution of industry is greatly intensified by the widespread use of coal for both power and heat, and China's northern cities in particular are notorious for the constant heavy pall of pollution, which becomes incredibly heavy in winter.

In the 1950's, factories were urged to take steps to control unnecessary air pollution, suggestions were made to the people to take preventive health measures, ministries were encouraged, when possible, to build new factories "on the opposite side of the city from which the wind usually blows." Although some local achievements were reported, economic priorities predominated and, not unlike their Western capitalist counterparts, Chinese managers of factories were reluctant to spend limited capital in ways that would not increase productivity.

During the 1960's, there was relatively little talk of pollution in China, and after the Cultural Revolution the authorities admitted that "efforts to control industrial pollution were not very successful due to interference and sabotage by swindlers like Liu Shao-chi."[26] It was in 1970 and 1971, then, that the real push for improved air quality began to gather momentum—undoubtedly reducing but by no means eliminating air pollution. Thus, there is both good news and bad news to consider.

A Shanghai newspaper reports that 1,500 chimneys in the municipality no longer spew black smoke. They represent a fraction of an unknown total number of chimneys in this large city and the article therefore goes on to challenge other units to do likewise.[27] But this is not an easy challenge to meet. From the numerous reports one surmises that China does not yet produce (at least not in any quantity) any gas-cleansing equipment, and that each enterprise or unit must handle the problem through self-reliance.

In general, the enterprises that derive economic benefit from the elimination and recycling of gases and smoke probably are more successful in implementing the policy against air pollution. For example, the Chinese claim that "over 100 chemical materials are now recovered from coke oven gas alone."[28] The Liaoyuan Chemical Plant in Shanghai now manages to recover 250 tons of polyvinyl chloride resin a year from carcinogenic vinyl chloride gas that used to be discharged into the air.[29] The "reliance on the masses" is always a dominant theme and whenever possible there is also a jab at foreign methods which require a lot of money.

Finally, a personal note relating to air pollution in China. In the summer of 1973, as a member of the American Medical delegation, I visited the Institute of Hygiene in Peking. This institute functions under the Chinese Academy of Medical Sciences and is responsible for research and the setting of standards relating to industrial hygiene, working conditions and environmental health. The leading cadres told us in some detail of their work at the institute and of the efforts to improve air quality throughout China. A few days later, while visiting China's largest iron and steel works (in Anshan, Liaoning Province), the smoke was so heavy that visibility was limited to no more than a couple hundred yards. I attempted to inquire about the air pollution and what was being done about it. No one seemed to know what I was talking about and, of course, no one had heard of the Institute of Hygiene nor of any prescribed standards. Perhaps this was an exception; perhaps I spoke to the wrong people; perhaps things are better now. But that was my experience in the summer of 1973.

Water

Because the pollution of rivers and streams has a much more direct and immediate effect than does air pollution on public health, on the local economy, and on the environment, it received the highest priority. Urban sewage seems to be well-managed: through expanded waterworks and controlled waterwells, safe drinking water is available to the great majority of the Chinese people.[30] Urban sewage is still mentioned, however, among the three categories of water pollutants currently listed by the Chinese: "industrial waste and sewage from cities; agricultural pesticides washed into the river by rain; harmful materials from mines and rocks which are dissolved by water and washed into the river."[31] Of these, industrial waste currently is cited as the most harmful.

Since early concern about water pollution was entirely health-related, the Ministry of Health and the local public health personnel had basic responsibility in this field. They exercised considerable clout in attacking the basic problems of health and sanitation, but when they had to tangle with industries which polluted rivers and streams they had a much more difficult time. Although some initial progress was made during the 1960's, statements since the Cultural Revolution make it clear that there is considerable dissatisfaction with what has been accomplished so far.

Since 1970, however, the success stories have far outnumbered the complaints and their number has been growing rapidly. One success story that received considerable publicity in China is the Nun-chiang, a river in Heilungkiang on which Tsitsihar—an industrial city with more than 1 million inhabitants—is situated. With the construction of numerous industries in the city, some 250,000 tons of water was being dumped daily into the once clean river, killing off most of the fish and reducing the catch in 1969 to about one-fifth of what it had been in 1960. After identifying the problem and discussing possible solutions with the masses, a program was approved requiring individual enterprises to recover the harmful substances in their industry's waste water, to make comprehensive use of these substances and then to divert waste water to a reservoir for eventual use in irrigating fields. Between June and November 1970, more than 5000 workers, peasants, Liberation Army soldiers, Red Guards and city inhabitants took part daily in the project. Among the various materials recovered from the waste waters were cadmium, oils, acids, alkali, paper pulp, and silver; the oxygen content of the water was raised and fish returned to the Nun-chiang; and fields irrigated by waste water produced greater yields than ever before.[32] Since Tsitsihar is the only large city along the banks of the Nun-shiang, the fairy tale success story described above is very believable. The problems are naturally much more difficult in the more industrialized regions of China. Despite reported successes, Shanghai's Whangpoo River, for example, is not likely to achieve any easy triumph over her capitalist counterparts.

It is quite evident that during the past few years the regime has begun pressuring industrial enterprises which are polluting waters. As in the case of air pollution, the greatest progress so far has been made by those enterprises which are able to transform liquid industrial waste into "treasures"—those that can get some capital return on their investment to process waste waters. A chemical plant in Pang-fou (Anhwei Province) used to dump the waste liquid from its saccharine production into the Huai River, but is now treating it and recovering large volumes of methyl alcohol, fat, and copper sulfate.[33] Detailed instructions are disseminated for the recovery and recycling of mercury-containing liquid wastes harmful to the environment.[34] These and scores of other examples testify to the widespread activities aimed at controlling water pollution; they also illustrate how little was done in the past and how much more there is to do.[35]

Solid Waste

Whereas concern for the first two of the "three wastes"—waste gas and waste water—combine environmental and economic considerations, the third—slag and other waste materials—is almost entirely an economic matter of "turning wastes into treasures"; the emphasis in the first two cases is on prevention, and in the third, on utilization. The recurring mass cleanup campaigns, initiated soon after the Communist regime came into power, also collected waste that was sometimes recycled. The difference between those campaigns and the current drive is not only intensity, but the fact that, once again, primary responsibility is being placed on the enterprises themselves, with most of the collection and much of the recycling being done by those who originate the waste.

An important aspect of the present effort not to waste waste is the involvement of the commercial departments throughout China which have "actively gone to factories and countryside to tap latent potentials in purchasing cast-off materials."[36] The key word here is "purchasing," which provides incentive for capital-short industrial enterprises and even residential neighborhoods to cooperate in this endeavor. Some urban neighborhoods have established "purchase posts" and there are reports of so-called "economy boxes" put up in villages to collect reusable waste, simplifying the work of purchasing agents who must cover large rural areas. Apparently many commercial bureaus now have a "junk purchasing department" which not only collects and purchases various types of industrial waste, but whose members actually go to the factories to explain to the leadership and the workers how waste should be recovered and managed and how it might be utilized. As one example, it was reported that an oil refinery under the direction of a waste material company in Shanghai succeeded in extracting enough fat from waste to produce 1.7 million cakes of soap.[37]

The importance of the commercial departments in the recovery and use of waste may be seen from their accomplishments in 1972. They recovered more than 5 million tons of waste and used materials worth 830 million yuan, including "2,650,000 tons of scrap iron, more than 20,000 tons of copper, 760,000 tons of raw materials for making paper, 180,000 tons of bones of various kinds, and 120,000 tons of used rubber."[38]

The impressive figures quoted above represent, however, only a part of the total effort. Factories, enterprises, government organs, schools, and the masses in general are also involved, either as part of their daily routine or through participation in special collection drives. All large cities report such drives—some organized by an individual enterprise for its own use and others organized by revolutionary committees and involving larger segments of the population.

There are many examples of enterprises responding to the call to "take one trade in the main and run various undertakings." It is reported that in the city of Cheng-chou (Honan Province) more than 140 large-sized enterprises, to make full use of their waste liquid, gas and residue, are operating more than 200 small industrial plants which produce more than 600 different products, including chemical and industrial raw materials.[39]

POLICIES AND POLICY DETERMINANTS

For almost 20 years China did not have a policy aimed at protecting the environment—she was much too preoccupied first with survival and later with the basic goals of economic development. Nonetheless, in retrospect, many of the policies pursued by the People's Republic had a very fundamental impact on the country's human and natural environment. The two most basic of these policy goals focused on the improvement of the health of the people and the improvement of the land in order to increase food production.

Actually, it was not until after the Cultural Revolution (circa 1970) that China initiated serious pollution control measures that were consciously an outgrowth of concern for the environment. Considering the improvement in the health and welfare of the Chinese population, even before the Cultural Revolution, this may appear to be a picayune distinction. It is, however, an important conceptual shift and one that the Chinese themselves recognize. Undoubtedly directives were issued and channelled down through the appropriate institutions, but to the outsider the only indication that a new policy was being implemented comes from articles in major Party publications such as the *People's Daily* and the *Red Flag*. In the early 1970's, such articles talked about "embarking on work to prevent and eliminate environmental pollution" and about the "developing mass movement," also admitting that such a movement was not possible before the Cultural Revolution, when Liu Shao-chi's counterrevolutionary, revisionist line was dominant.

By 1970 China's basic priorities were more or less under control and she could concentrate on some important refinements. At the same time, because of industrial growth in the cities and the construction of small- and medium-sized industries in rural areas where haphazard waste disposal could and did adversely affect agricultural production, there were increasing pressures to check pollution. International concern was peaking in anticipation of the Stockholm Conference on the Human Environment: Peking was not about to pass up the opportunity to revile the "capitalists" and "revisionists" for failing to deal with severe environmental problems, and, at the same time, make points with the developing countries. This meant that domestic measures had to become more rigorous and more visible. Finally, it is doubtful if the antipollution drive would have been mounted with the same intensity if its mainspring were not essentially economic—if it were not so closely integrated with the mass movement to eliminate waste and to recycle everything that was recyclable.

This is not to say that concern for the "quality of life" was not also a motivating factor. A better environment improves people's health which makes them more productive which improves their standard of living. It is therefore quite reasonable that in pursuing a better environment, Peking's priorities would be focused primarily on productivity and resource conservation, rather than on pollution per se. That is why policy statements printed in national publications and reprinted for international consumption will usually mention that the elimination of three wastes (air, water and solid pollutants) will "help raise the level of the people's health," while at the working level and in reports by specific enterprises, the health aspect is subordinated to "promoting the development of the national economy."

The "turning waste into treasure" approach is also the most effective way of motivating the managers of production. China's industrial production still depends on profit and loss accounting, and a manager of an enterprise tends to resist new demands that disturb his cost-benefit balance.

An article in the June 16, 1973 issue of the *People's Daily*[40] discussed the interrelationship between environmental protection and economic development and again emphasized the damage which industrial waste might cause to production and the economy. It pointed out that since "pollution of the environment is rapid whereas its elimination takes a longer time," preventive measures are essential. At the same time the article had this warning to anyone who might favor a slower development of industry to minimize pollution problems: "[W]e can only solve the problem of environmental protection by developing the economy, and not seek a good environment by slowing down economic development, or by other negative methods."

This is another factor to be considered in discussing the determinants of pollution control—that of population. Although there is a natural and obvious correlation between numbers of people and pollution in its broadest sense, the Chinese refuse to acknowledge this relationship except in an indirect way. China's basic philosophy with regard to population was repeated at the Stockholm Conference, namely: "Since of all things in the world people are the most precious, it is wholly groundless to think that population growth in itself will bring about pollution and damage of the environment and give rise to poverty and backwardness."[41] To reinforce this point Tang Ke, the chief spokesman at Stockholm, went on to note that the population of the People's Republic of China increased by some 200 million between 1949 and 1970 but "because we had driven out the imperialist plunderers and overthrown the system of exploitation" the living standard of the people and the living environment have improved. In other words, it is not the number of people, but the political system under which they live that is significant.

There is no doubt about the improved living standard of the Chinese population, but behind all the bravado there is a very clear official understanding that a slower population growth of say 10 million rather than 15 million per year would not only ease economic burdens, but would also be less damaging to the nation's environment. Certainly China's all-out effort to reduce population growth[42] would suggest a more conventional view of the cause-and-effect relationships among population size, economic and social development and the consequent problems of pollution.

The only departure from the "people are precious" line is with regard to urban concentrations—when the density of population becomes a factor. City people are still "precious," but they create more pollution problems than their country cousins. Controlling urban growth has been a fairly consistent policy started soon after establishment of the People's Republic, and Peking has been quite successful in limiting city population.[43] Economic, social, and political considerations predominated in establishing controls over "blind infiltration" of peasants into the cities, but even in the 1950's some mention was made of related environmental problems, including industrial pollution. But since China could not very well afford to shut down operating plants, the reports of factories being moved from some of the larger cities probably referred only to obsolete enterprises that could no longer produce with even limited efficiency. This was also in line with the general policy of concentrating new development in industrial zones on the outskirts of cities, where new housing could be built and where pollution problems could potentially be handled with greater ease.

By the 1960's, at least four other considerations provided impetus for industrial (and, therefore, population) dispersal. The first stemmed from increasing tensions between China and the Soviet Union and China's desire to decrease industrial vulnerability in case of a war. The second was to devise a way of economically absorbing the millions of youth who enter the working ages every year. The third was to encourage local self-sufficiency in agriculture. And the fourth was to develop the backward areas and to raise the living standard of people who lived there. These were the primary reasons for China's program of building small industries in rural areas. Initially, this policy was completely independent of environmental considerations, but since the Cultural Revolution the call for the "rational distribution of industries" and for limiting urban growth has been blended with the need for pollution control. No one can argue with the contention that it is easier to handle pollution in smaller towns with a smaller population, but at the same time it does substantiate the obvious: The greater the density of "precious people," the greater the damage to the environment—even in China.

Although the factors discussed above continue to underlie present environmental policies, and pollution control is still to a large extent a stepchild of economic development, some changes may be in prospect. It may be a sign of things to come that in mid-1974 Peking created an Office of Environmental Protection and placed it directly under the State Council.[44]

As of this writing nothing is known about this organization, but what other purpose could it have except to plan and unify an environmental protection program? If it is patterned after China's similar national organizations, it will have parallel offices in all the provinces and municipalities of the country. There have already been references to regional organizations, such as the Tsinghai Provincial Environmental Protection Commission and the Kirin Municipal Environmental Protection Office, which by now, along with perhaps 40 or 50 other such branch offices, would be standardized and subordinated to the National Office of Environmental Protection. Uncoordinated conferences concerned with various environmental problems, formerly held by municipal revolutionary committees, by industrial ministries, by health and other organizations, in the future will, presumably, be coordinated by the new Office in Peking and its branches.

Although the creation of a new government agency presupposes a growing attention to environmental problems, the question still to be answered is: Will industries have to continue to be "self-reliant" in combating pollution, or will additional incentives be provided through some form of state subsidies? As in other areas, China until now has been "walking on two legs"—attempting to use both modern and traditional methods to fight pollution. From a technological viewpoint, China has the know-how for most of the modern techniques of water treatment and is capable of producing mechanical collectors, wet scrubbers, and electrostatic precipitators to control air pollution. She has been doing so on a very selective basis. Following the pattern in other technical fields, a great deal of encouragement in pollution control has been given to worker innovation and to the adaption of indigenous Chinese methods, rather than "seeking after big and foreign things." This, of course, is a frugal approach but some significant innovations have been reported. At the same time, the regime now calls on scientific research organs, health organizations, institutions of higher learning, and other units to cooperate in eliminating the "three wastes" and "to combine the revolutionary spirit with a scientific approach."[45] It seems possible that the emphasis on science and technology might result in establishment of a new national research institute specifically for environmental protection.

CONCLUSION

If the foregoing discussion of China's environmental policies and programs appears a little cynical in spots, it is only in contrast to the often exaggerated accomplishments claimed by Peking, especially when addressing an international audience. By placing under the umbrella of "environment" a great variety of programs pursued for the past 25 years, China can now sneer at the bumbling and stumbling efforts of industrialized nations to correct conditions stemming from decades of environmental neglect. But to question the image China is attempting to project is not to belittle her successes. She may not have found all the solutions to human and industrial pollution, but because of considerable wisdom and a measure of luck, she does have much of which to be proud.

China has been wise because very early on Mao Tse-tung recognized that the long-term success of economic development required that the people be protected from the hazards of environment and that the environment be protected from uncontrolled abuse. This resolve was based on very practical rather than strictly ecological considerations—long before the limited fad for ecology grew into a major international concern. Mao firmly believed that the basic physical needs of the population—good health, good water, adequate food—were prerequisites to any and all other national goals; hence, the early policies to improve sanitation and health and to make the land more productive. Only after these needs were largely achieved could environmental concern turn to some of the important but relatively less pressing problems stemming from industrial pollution.

Luck becomes a factor only at the implementation stage. China is "lucky" that more than four-fifths of her population is located in rural areas, where lower densities and essentially agricultural pursuits make environmental problems easier to manage. She is "lucky" that she does not have an economy of abundance which is so damaging to the environment, but rather an economy of frugality in which the "do not waste" ethic is relatively easy

To know the road ahead, ask those coming back.
CHINESE PROVERB

to enforce since it is a matter of necessity. China is "lucky" that her industrial pollution problems are less serious than those of highly industrialized, highly urbanized countries, and "lucky" that she can take advantage (even if surreptitiously) of the experience and technical and scientific know-how of those countries in selecting priorities and measures most suitable to China's needs and economic capabilities.

There is no such thing as a safe prediction for China, but it does seem that because of this combination of wisdom and "luck" China will not experience the type of environmental degradation that has occurred in most of the world's industrial nations. At present the problems are still numerous and nowhere near to being solved, but perhaps the peak has been reached and problems of environment and pollution will take a gradual downturn. The mass drive started in the early 1970's will probably have relatively limited immediate benefit, but it is an important educational effort, making both workers and cadres conscious of the economic and health consequences of uncontrolled pollution. It establishes a foundation for future policies which undoubtedly will place major emphasis on "prevention" to be implemented as the economy is developed, rather than on the more difficult "cure" for existing pollution problems.

Total mobilization is a vital prerequisite. In the United States it is only an abstract rallying call; in the People's Republic of China it is the key to virtually every major accomplishment.

NOTES

1 For a detailed discussion of developments in health and sanitation, see L. Orleans, *Health Policies and Services in China, 1974,* report to the Subcommittee on Health, U.S. Senate, Mar. 1974.

2 New China News Agency, 12 Feb. 1958, hereafter referred to as NCNA. Unless otherwise specified, all NCNA reports are from *Daily Report: China,* published by the Foreign Broadcast Information Service, hereafter referred to as FBIS.

3 NCNA, 1 June 1956; in *Current Background,* no. 405, 26 July 1956.

4 *Jenmin Jih-pao (People's Daily).* 5 Nov. 1965, hereafter referred to as *JMJP;* in *Joint Publications Research Service,* No. 5104, 27 July 1960, hereafter referred to as *JPRS.*

5 A number of articles on this subject were translated in *Union Research Service,* Hong Kong; see, e.g., "The Vigorous Movement of Accumulating Manure," 9 Mar. 1973 and "Accumulation of Manure in Rice-Producing Areas of China," 10 May 1974.

6 G. Winfield, *China: The Land and the People* (New York, William Sloane Associates, Inc., 1948).

7 *China Reconstructs,* Peking, June 1965, hereafter referred to as *CR.*

8 *Jen-min Pao-chien (People's Health),* no. 3, 1960, hereafter referred to as *JMPC; JPRS.* no. 2976, 15 July 1960.

9 NCNA, 15 Apr. 1966.

10 *CR,* Nov. 1971.

11 NCNA, 25 Feb. 1972.

12 NCNA, 3 Jan. 1972.

13 NCNA, 28 Oct. 1972. 1 mou = 1/15 hectare = 1/6 acre.

14 *Shui-li Yu Tien-li (Water Conservation and Electric Power),* no. 3, 5 Feb. 1966; in *Survey of China Mainland Magazines,* no. 566, 6 Mar. 1967.

15 See *CR,* Jan. 1973, at 13.

16 *CR,* Jan. 1973, at 19.

17 NCNA, 13 Mar. 1966.

18 *Peking Review,* no. 8, 22 Apr. 1958, at 15, hereafter referred to as *PR.*

19 NCNA, 24 Apr. 1966.

20 NCNA, 15 May 1966.

21 See O. Dawson, *Communist China's Agriculture* (New York: Praeger, 1970), at 109.

22 *Chih-wo-hsueh Tsa-chih (Botanical Journal)*, v. 1, no. 2, May 1974.

23 *Hua-hsueh T'ung-pao (Bulletin of Chemistry)* 23 Jan. 1974; in *JPRS*, no. 62,812, 26 Sept. 1974.

24 *PR*, no. 3, 21 Jan. 1972.

25 *JMPC*, v. 1, no. 10, Oct. 1959; in *JPRS*, no. 2745, 10 June 1960.

26 NCNA, 14 Sept. 1972.

27 *Wen-hui Pao (Wen-hui Daily)*, 22 Sept. 1972; in FBIS, 27 Sept. 1972.

28 *CR*, May 1971.

29 *CR*, June 1972.

30 As early as 1959 it was reported that most cities use the chlorine-ammonia sterilization method and that continuous testing and analysis is performed to assure the water is safe. (*JMPC*, v. 1, No. 10, Oct. 1959; in *JPRS*, no. 2745, 10 June 1960)

31 *K'o-hsueh Shih-yen (Scientific Experiment)*, no. 2, Feb. 1974; in *JPRS*, no. 62,812, 26 Sept. 1974.

32 *Ching-chi Tao-pao (Economic Bulletin)*, Hong Kong, 30 Aug. 1972; in *JPRS*, no. 57,290, 18 Oct. 1972.

33 *Ta-kung Pao (Impartial Daily)*, 11 May 1972; in *JPRS*, No. 56,741. 10 Aug. 1972.

34 *Hua-hsueh T'ung-pao (Chemical Bulletin)*, no. 1, 23 Jan. 1974; in *JPRS*, no. 62,812, 26 Sept. 1974.

35 Although in the past China completely ignored ocean pollution as an issue (as did most other nations), at the "Law of the Sea" conference in Caracas (Venezuela) in the summer of 1974 she strongly supported measures to protect the marine environment.

36 NCNA, 17 Nov. 1970.

37 NCNA, 17 Nov. 1970.

38 *JMJP*, 31 Jan. 1973.

39 *Chung-kuo Hsin-wen (China News)*, 9 Nov. 1970; in *JPRS*, no. 52,527, 3 Mar. 1971.

40 *PR*, no. 29, 20 July 1973.

41 *PR*, no. 24, 16 June 1972.

42 For a detailed discussion of China's current population policies, see L. Orleans, *China's Experience in Population Control: The Elusive Model*, report for the Comm. on Foreign Affairs, U.S. House of Representatives, Sept. 1974.

43 For a discussion of urban policies, problems and numbers, see L. Orleans, *Every Fifth Child: The Population of China* (Stanford Univ. Press, 1972), ch. V.

44 *Christian Science Monitor*, 22 Oct. 1974.

45 *Hung-ch'i (Red Flag)*, no. 12, 1 Dec. 1973, in *Survey of China Mainland Magazines*, 17–28 Dec. 1973.

24 They Shall Inherit the Earth

Malcolm Somerville

Malcolm Somerville, a professional seismologist, currently works with a consulting firm that specializes in structural engineering. Acknowledgments: Clyde Wahrhaftig, professor of geology at the University of California, Berkeley, contributed much of the requisite research for this article. Andre Lehré, a doctoral student in geology, assisted in the research.

The livelihood, both material and spiritual, of the Indochinese people has historically been their agriculture and their countryside. The destruction of much of the countryside, as well as devastating the morale of the people, has inevitably been accompanied by severe damage to the soils and fields which have sustained their societies through the centuries. The agricultural potential of Vietnam, as estimated for example by Moorman[1] in 1961, has probably been reduced drastically. The separation of the people from their countryside, caused by the war, may well continue long after the end of the struggle.

The extent of the damage, and the time required for recovery, where possible, cannot be accurately estimated in the absence of lengthy field studies. Some of the effects occur over a time span of years; for example, the permanent hardening of exposed soils to form the brick-like rock, laterite. The soils of the natural tropical forest are maintained by the forest itself, and are impoverished by the destruction of the forest. The time required for regeneration, if possible, of defoliated forests and mangroves, and the natural cycle of unwelcome bamboo and other pests, are measured in decades.

A consideration of the extent of the war's assault of the landscape and its danger to the soils suggests a drastic setback in the agricultural capability. Vagueness in the numerical and descriptive details of the soil damage indicates our general ignorance of the effects of the war technology.

Defoliating chemicals have been sprayed over at least 5 million acres, 12 percent of South Vietnam's area.[2] Orians and Pfeiffer[3] estimate that 20 to 25 percent of the forest lands have been treated with defoliants two or more times; an estimate by the Provisional Revolutionary Government is much higher, 44 percent.[4] Perhaps half of the total acreage under intensive cultivation has been destroyed. . . .

"According to Pentagon sources, aerial bombardment of Indochina from 1965 through 1969 reached 4½ million tons, nine times the tonnage in the entire Pacific theater in World War II. This is about half of the total ordnance expended."[5]

Further hundreds of thousands of acres in South Vietnam have been completely stripped of vegetation by the Rome Plow.[6]

These assaults result in soil impoverishment, erosion, or hardening to form laterite rock, and critical damage to cropland and estuary.

In some instances there is evidence that devastation of the soils is included in the military strategy. Referring to Operations Sherwood Forest and Pink Rose, Thomas Whiteside wrote:

> . . . the ultimate folly in our defoliation operations in Vietnam was possibly achieved during 1965 and 1966, when the military made large-scale efforts in two defoliated areas to create fire storms—that is, fires so huge that all the oxygen in those areas would be exhausted. The apparent intention was to render the soil barren. . . . Neither of the projects, in which tons of napalm were thrown down on top of the residue of tons of sprayed 2,4,5-T, succeeded in creating the desired effect.[7]

However, there is evidence that defoliation in itself is a disastrously effective weapon against the soils. The tropical forest is characterized by a complex interdependence of the soil, climate, plants and living creatures; the forest environment is unusually vulnerable to disturbances to any sector of the ecosystem. Most of the mineral nutrients of the humid tropical forest are carried in the standing crop or in the surface layers of the soil.[8] These nutrients include nitrates, phosphates, calcium, potassium, magnesium, sulphur, and other minor elements.[2] The heavy rainfall and high temperatures, among other minor factors, cause rapid washing away, or leaching, of soil nutrients. Humus, the layer of decomposing vegetation on the forest floor, is not accumulated, but is just maintained at a level sufficient for the survival of the forest. Fungi provide the role of trapping nutrients released from decaying plants and making them available to the roots of living vegetation.[2] The nutrients cannot be retained by the vegetation in this manner when defoliation accelerates leaching of the soil. Defoliation is responsible for harsher conditions of rainfall, temperature and wind on the forest floor, and thereby causes rapid decomposition of the humus. The nutrients are quickly washed from the soil and lost altogether from the forest.

Living creatures play an important role in sustaining the tropical forest. The forest is pollinated by birds, bats and insects, rather than by the wind, and its seeds are dispersed to new clearings by animals and birds.

> These complex plant-animal relations have reached their greatest intricacy in tropical forests because of the mild and predictable climate. Animals can be active the year around because many flowering and fruiting trees provide food continuously. Massive defoliation means an end to this reliable food supply and death for those animals that are most important to the survival of the forest plants.[2]
>
> . . . tropical forests hold the maximum number of individuals of most species that the resources will support. Reduction of forest habitats will decrease the populations of forest animals by an equivalent amount.[3]

Extinction of the fauna effectively prevents regrowth and spreading of the defoliated forest. Another factor inhibiting tropical forest regrowth, and hence retention of soil nutrients, is the tendency of pest species, such as grasses, shrubs and bamboo, to invade the forestland. This is a direct result of reduced soil fertility and the lack of seeds of the original forest plants. Giant bamboo is common in areas that have been cleared of trees for agricultural purposes, and other varieties are found as natural members of tropical forests in Vietnam.[9] The bamboo, growing in dense thickets, succeeds the natural tropical forest: consequently much of the fine hardwood forest now being destroyed by defoliation in Vietnam and Laos may be incapable of regeneration.

Orians and Pfeiffer, reporting on a visit to South Vietnam in March, 1969, wrote: "Two or three spray applications may kill approximately 50 percent of commercially valuable timber in such [upland] forests . . . possibly 20 to 25 percent of the forests of the country have been sprayed more than once."[3] As noted previously, this may be a conservative estimate of the extent of defoliation. The impoverishment of the tropical forest soil and the resulting succession by pest species make future agricultural use of the land very difficult. Defoliating, burning and cutting bamboo all fail to remove it once its hardy underground stems are established. The natural life cycle of bamboo is thought to be about thirty to fifty years.[9]

The directness of the plant-soil interdependence has been summed up by Leitenberg: "If the soil is changed the forest will change. If the forest is changed the soil will change."[10]

Angkor Wat, among other ancient shrines and monuments in Indochina, was constructed primarily of a durable, brick-like rock known as laterite. This rock actually forms from various soil types known as lateritic soils under certain tropical conditions. Lateritic soils develop in a strong rainy season alternating with a pronounced dry season of several months' duration.[11,12,13] These soils, characterized by a high iron oxide and aluminum content, occur over about half of Vietnam's terrain.[6,12] The hardening of lateritic soils to form laterite rock is thought to occur under the following conditions:[14] ample drainage of the soil, gentle topography, and thorough drying (facilitated by increased exposure to sun and wind), possible over a period of several years. The climatic conditions which develop the lateritic soil also play a part in its conversion to laterite; there is alternate wetting and drying, and the water table fluctuates accordingly.

Millions of acres of bulldozed, bomb-cratered and defoliated land have been exposed to elevated temperatures and wind speeds, following the destruction of vegetation. The Mekong Delta area is susceptible to laterization because of its gentle topography and alluvial soils, which may be well drained in some areas. The dry season is especially pronounced in the Annamite mountain chain to the north of Saigon, and in much of the plain of Cambodia. In these regions the dry season is often a complete drought lasting three months, sometimes occurring for a few years in succession.[15] The Bolovens Plateau in southern Laos would be particularly susceptible to laterization.

Although laterization may be prevented by the steepness of the terrain in some areas, the war weapons are subjecting much of the land to the same conditions as have been responsible for the widespread occurrence of laterite in the tropics, for example in Thailand. . . .[16,17]

The Midwest Research Institute in its final report to the Department of Defense outlined the dangers of laterization:

> This end result [laterization] is never reached under a cover of rain forest. Lateritic soils may develop under grassland, savannah, or forest. Under the forest, however, the concentration of iron and aluminum compounds occurs at some depth from the surface, while

> under grassland it occurs at the surface. Clear-cutting the forest and converting the area to grassland, though, may result in accelerating the hardening of lateritic soils into laterite rock, a process which greatly lowers the productive capacity of the site. . . .[6]

Laterization of the soil would most likely occur in bombed and bulldozed areas along roadways, damaged cropland, cleared areas around base camps and in the vast forest areas which have been completely stripped of vegetation by the Rome Plow.

> The Rome Plow has become one of the most surprisingly effective "weapons" of the Vietnamese war. This plow . . . is a sharpened 2,500-pound bulldozer blade long used commercially in ground-clearing operations. In Vietnam, Army engineers driving Caterpillar tractors with plows attached are cleaning hundreds of thousands of acres of jungle. . . . In the entire III Corps area about 102,000 acres were cleared between last July 1 (1967) and Dec. 3. Normally one plow, pushed by a Caterpillar tractor, can clear about 10,000 feet an hour of trees, up to 18 inches in diameter and of secondary growth. A heavy spike on one end of the blade, called a stinger, is used to split and weaken trees from 18 to 36 inches in diameter before they are cut off and pushed over . . . For rubber trees, which grow on plantations in orderly rows, two tractors dragging a chain between them can fell whole rows quickly.[6]

A similar area of land has been totally devastated by bombing. As mentioned previously, in 1967–1968 alone, 3½ million 500- to 750-pound bombs were dropped in Vietnam, leaving almost 100,000 acres occupied by bomb craters and excavating more than 2½ billion cubic yards of earth.[2] The bombing of Indochina has amounted to perhaps the most massive excavation project in mankind's history. It dwarfs the Suez Canal and Panama Canal projects, both involving the excavation of about a quarter of a billion cubic yards of earth. The total cratered area in Indochina exceeds the area of the State of Connecticut, 5,000 square miles. This estimate is based on Pentagon figures[5] for the aerial bombardment from 1965 through 1969, and on Orians and Pfeiffer's report that a 500- to 750-pound bomb creates a crater as large as 45 feet across and 30 feet deep.[3]

In some areas the landscape has been altered almost beyond recognition. An aerial view of the section of Vietnam near the Parrot's Beak was reported by T. D. Allman: ". . . the land is pitted by literally hundreds of thousands of bomb and shell craters. In some cases the years of day-and-night bombing have changed the contours of the land, and little streams form into lakes as they fill up mile after square mile of craters."[18]

Much of the newly exposed terrain is subject to increased soil erosion. Large areas of the Annamite Mountains, now (or recently) covered by dense forest, possess clayey red and yellow soils characterized by a deep weathering profile, and sandier brown soils in the drier parts.[19] These soils on steep slopes would be highly susceptible to gullying and other forms of accelerated erosion following the destruction of vegetation or extensive bomb-cratering.[20] Rapid gullying and landsliding would lead to floods of silt and sand in valleys and on downstream watercourses. Large-scale exposure of unweathered bedrock would cause rapid rainwater runoff, causing higher floods and more prolonged droughts than previously.

As evidenced by . . . photographs, bombing has been used to destroy rice paddies and their irrigation systems. This can cause extensive erosion of paddy fields cut into steep or sloping country. The restoration of irrigation schemes and filling of bomb craters in the paddies which can be salvaged will be a monumental task. In flatter areas, dried-out paddy fields may well become pavements of laterite.

A 1967 report of the Agronomy section of the Japan Science Council claimed that "anti-crop attacks have ruined 3,800,000 acres of arable land in South Vietnam."[21]

This would amount to about half of the total acreage under intensive cultivation.[2]

In western Ca Mau Peninsula there is a large area of peat underlain by an acid and possibly saline soil. The area is not yet developed for cultivation but has considerable agricultural potential for rice production with a moderate amount of fertilization.[1] This potential could be greatly reduced if deep bomb cratering destroys the protective peat cover and mixes the acid subsoil with the peaty surface soil.[20]

Destruction of mangrove forests bordering tidal estuaries permits bank erosion and salt-water intrusion of rice fields, with possible catastrophic reduction of rice yields. The Mekong Delta has suffered heavy bombing and defoliation. Tschirley, after visiting South Vietnam in March and April of 1968 to assess the ecological consequences of the defoliation at the request of the U.S. Department of State, reported:

> The mangrove species seem to be almost uniformly susceptible to Orange and White, the herbicides used for their control in Vietnam. . . . Strips of mangrove on both sides of the Ong Doc River, sprayed with Orange in 1962, were of particular interest. The treated strips were still plainly visible. Thus, one must assume that the trees were not simply defoliated, but were killed.
>
> . . . 20 years may be a reasonable estimate of the time needed for this forest to return to its original condition. . . .[9]

However, this time would be longer if there were removal of soil, which occurs more readily in a dead forest.

Ancestor worshipers, the Vietnamese saw themselves as more than separate egos, as a part of this continuum of life. As they took life from the earth and from the ancestors, so they would find immortality in their children, who in their turn would take their place on the earth. To leave the land and the family forever was therefore to lose their place in the universe and to suffer a permanent, collective death.
FRANCES FITZGERALD

According to the Stanford Biology Study Group:

> They [mangroves] also provide a special habitat for key stages in the life cycles of economically important fish and shell fish. . . . There will undoubtedly be a drastic and long-lasting effect upon river fishing and upon the natural processes of delta formation along Vietnamese rivers.[2]

There is only one report of the use of soil sterilants in South Vietnam. The report, in September 1967, referred to the Demilitarized Zone: "Chemical soil killers will be used in South Vietnam warfare for the first time under the Pentagon's plan to create a barrier against infiltrating North Vietnamese troops."[22]

As far as is known, 2,4,5-T, Picloram, and 2,4-D, the most commonly used defoliation and herbicide agents, do not kill soil microorganisms. However, the effect of the decomposition products on microorganisms is not known. Nor do we know how long dangerous teratogenic residues will persist and threaten the lives of unborn children.

After more than a decade of highly technological warfare in Indochina, we have only preliminary evidence concerning the extent of the soil damage. However, current indications are of a disastrous reduction in agricultural capability. Desolate, inhospitable countryside awaits millions of refugee peasants. How many people can the land now sustain, and how many will return to attempt to revive their wasted land?

NOTES

1 F. R. Moorman, "The Soils of the Republic of Vietnam," Republic of Vietnam, Ministry of Agriculture, Directorate of Studies and Research in Agronomy, etc., 1961, 66 pp., Map, scale 1:1,000,000.

2 Stanford Biology Study Group, "The Destruction of Indochina," *California Tomorrow*, San Francisco, 1970, 9 pp.

3 Gordon H. Orians and E. W. Pfeiffer, "Ecological Effects of the War in Vietnam," *Science*, 168, May 1, 1970, pp. 544–554.

4 Statement by Mme. Nguyen Thi Binh, Chief of the Delegation of the Republic of Vietnam at the 55th Plenary Session of the Paris Conference on Vietnam, Feb. 19, 1970.

5 Noam Chomsky, "Cambodia," *New York Review of Books*, June 4, 1970, footnote 53, p. 48.

6 Midwest Research Institute, "Assessment of Ecological Effects of Extensive or Repeated Use of Herbicides," Final Report, Aug. 15 to Dec. 1, 1967, AD 824,314, pp. 281–284, 292.

7 Thomas Whiteside, *Defoliation*, Ballantine Books, New York, 1970, 168 pp.

8 G. Aubert, "Influence des divers types de végétation sur les caractères et l'évolution des sols en régions équatoriales et subéquatoriales ainsi que leurs bordures tropicales semi-humides," in Symposium on Tropical Soils and Vegetation, Abidjan, UNESCO, 1959, pp. 41–58.

9 Fred H. Tschirley, "Defoliation in Vietnam," *Science*, 163, Feb. 21, 1969, pp. 779–786.

10 Milton Leitenberg, Draft of an unpublished mimeographed paper, Jan. 5, 1970, "The Long-Term Effects of the Use of Chemical Herbicides," Stockholm, Sweden.

11 Mary McNeil, "Lateritic Soils," *Scientific American*, 211, May 1964, pp. 96–102.

12 R. Maignien, "Review of Research on Laterites," UNESCO 1966, 148 pp.

13 Lyle T. Alexander and John G. Cady, "Genesis and Hardening of Laterite in Soils," U.S. Dept of Agriculture Soil Conservation Service Technical Bulletin 1282, 1962.

14 J. F. Taranik and E. J. Cording, "Laterite and Its Engineering Properties," Mimeo Rep. 59th Engin. Detachment (Terrain) (U.S. Army, A.P.O. 96491, 1967).

15 Y. Henry, "Terres rouges et terres noires basaltiques d'Indochine," Hanoi, IDEO, 1931.

16 Robert L. Pendleton, "Laterite and Its Structural Uses in Thailand and Cambodia," *The Geographical Review*, 31, Feb. 1941, pp. 177–202.

17 Robert L. Pendleton, "Laterite or Sila Laeng, a Peculiar Soil Formation," *Thai Science Bulletin*, 3, Feb.-Mar. 1941, pp. 61–77, pl. 1–24.

18 T. D. Allman, *Far Eastern Economic Review*, Feb. 26, 1970.

19 R. Dudal and F. R. Moorman, "Major Soils of Southeast Asia," *Journal of Tropical Geography*, 18, Aug. 1964, pp. 54–80.

20 Clyde Wahrhaftig, unpublished mimeographed paper, June 1970, "Some Possible Effects of Military Activity on Soils of Southeast Asia," Berkeley, Calif.

21 S. M. Hersh, *Chemical and Biological Warfare*, Bobbs-Merrill, New York, 1968, 354 pp.

22 B. Horton, "Infiltration Barrier Planned, Soil Killer to Bare Viet Strip," (AP) *Minneapolis Star*, Sept. 8, 1967.

25 India's Agricultural Balance Sheet

M. S. Swaminathan

M. S. Swaminathan is director-general of the Indian Council of Agricultural Research and president of the Indian Science Congress Association.

Reprinted and abridged from *The Ecologist*, 5 (October 1975), pp. 272–81, by permission of *The Ecologist*.

. . . In our earth, appropriately compared to a spaceship by Buckminster Fuller, the first-class compartment, which the rich nations as well as the rich of the poor nations occupy, is consuming a greater and greater proportion of the available resources. . . . Those living in the first-class compartment . . . have forgotten the meaning of the prayer "God, give us our daily bread," while for the vast majority of humanity squeezed into the small second-class compartment, "God is bread," to use the words of Mahatma Gandhi. . . .

The state of food and agriculture in the world shows two contrasting trends. In one kind of agriculture, larger and larger farms are being farmed by fewer and fewer cultivators. Such farms are highly automated and capital-intensive. . . . In one study conducted during the mid-fifties, it was found that while in many countries of Asia and Africa about 2.5 to 10 work days were needed to produce one quintal of grain, the extent of labour time needed to produce the same quantity of grain was about 3 hours in parts of France and only 6 to 12 minutes in parts of the United States. The gap in the relative productivity of farm labour was therefore about 1 to 800 even over 15 years ago and this gap, which also represents the relative earning and purchasing power of farm labour, has been growing ever since.

In many of the poor nations, including our own, smaller and smaller farms have to be cultivated by the same or even larger number of farmers. The percentage of work force employed in farming in India was 72.1 in 1971 and it is anticipated that this percentage will remain practically unaltered in 1981. The relative productivity of small farms, however, varies widely in the world. Mr. R. S. McNamara, President of the World Bank Group, pointed out recently that "if Japan could produce 6,720 kg. of grain per hectare on very small farms in 1970, then Africa with its 1,270 kg., Asia with 1,750 kg. and Latin America with 2,060 kg. per hectare have an enormous potential for expanding productivity." . . .

Besides farm size and land and labour productivity, the other two major differences in the agriculture of the rich nations as compared to the poor ones, are in the pattern of agricultural growth and the nature of the food chain. The mechanised and low-labour consuming agriculture has achieved increased productivity largely on the basis of a high consumption of energy derived from the non-renewable resources of the earth. Thus, while in 1964, about 286 kilo calories of energy were needed in countries like India and Indonesia to produce one kg. of rice protein, 2,860 kilo calories of energy were needed to produce one kg. of wheat protein and over 65,000 kilo calories to produce one kg. of beef protein in the United States. Such a situation has now resulted in a widespread awareness . . . that any finite resource, if exploited in an exponential manner, . . . will some day or the other get exhausted, thereby bringing the pattern of growth based on its consumption to ruin. It has also become clear that the tools of modernisation of agriculture, . . . if indiscriminately used and excessively based on non-renewable resources, will end in crises. . . . For example, 96 per cent of the energy input in the United States in 1970 came from oil, gas and coal, while in the same year non-commercial fuels like dung, firewood and wastes provided 52 per cent of our energy needs. . . . Our agricultural production process is still predominantly based on the use of renewable resources but our current productivity is very low. Hence, there is urgent need for the development of technologies where the productivity of land can be continuously increased with diminishing dependence on non-renewable components of energy, by deploying recycling processes more and more effectively. A consequence of the agriculture of the high-energy-consumption and low-labour input pattern is the diversion of labour from agriculture to more industrial pursuits and a close linkage between farm and factory. . . .

A biological consequence of affluence has been a rise in the consumption of animal products. The poor and rich nations are hence characterised today by the former depending largely on the plant-to-man food chain and the latter, on the plant-animal-man food chain. Consequently, the average per capita consumption of grain is about one tonne per year in the developed nations, out of which only about 70 kg. are consumed directly in the form of . . . products made from flour. The remaining 930 kg. are used to feed animals whose products like meat, milk and eggs are consumed by man. In contrast, the per capita consumption of grain per year in the developing countries is about 190 kg., most of which is directly consumed. . . . Our need is an agricultural system where the benefits of a large human and animal population, robust soils, abundant sunlight, rich ecological diversity, availability of large quantities of organic wastes and a fairly extensive irrigation network are optimised in a manner such that productivity is continuously increased without damage to the long term production potential of the soil, stability is imparted to the production as well as prices of foodgrains and labour and land use diversified, so as to increase real income and purchasing power. The mode of achieving these goals was summed up beautifully by our President Shri V. V. Giri when he called upon us to make "every acre a pasture and every home a factory."

About 10 years ago, some foreign experts believed that India could never become self-sufficient in its food requirements. They predicted the outbreak of widespread famine and hunger in India on the basis of theoretical calculations of the year when the food needs of the poor nations would exceed the capacity of the rich nations to meet them. Paul and William Paddock, for example, fixed this year as 1975. Some others felt that such a contingency would arise only in 1985. Our failure to achieve during the nineteen fifties and early sixties the anticipated results with the Community Development and the Intensive Agricultural District Programmes fanned the sentiments of the pessimists. It was widely believed that in our rural areas developmental action and achievement were being impeded by the limited sights of our farming community. Experts wondered why we should find it so difficult to become self-reliant in our food needs, when there appeared to exist several easy pathways of attaining this goal. An expert on rodent control would thus advocate the control of rats to wipe out the deficit in our food budget, while the agronomist would plead for using good seeds and some fertilizer for achieving the same end. The fact remained, however, that progress in improving production was slow and largely took place through an increase in the cropped and irrigated area, rather than through any appreciable increase in productivity. The economic plight of the small farmer remained unchanged since his real income would rise only if productivity was improved and marketing organised so that the small producer got a fair share of the price paid by the consumer for a commodity. While Government was anxious to push up agricultural production and Jawaharlal Nehru gave expression to this desire by his oft-quoted statement "Everything else can wait but not agriculture," there was despair during the drought years of the mid-sixties concerning our agricultural destiny. . . .

Historically, famines and scarcities have been known in our country from the earliest times . . . but it was only during the famine of 1868–69 that it was clearly stated that "the object of Government was to save every life." The birth of agricultural departments in our States and of Famine Codes took place as a result of the recommendations of the Famine Commission of 1878. The Scarcity Manuals currently used by several Governments are largely based on the Famine Codes of the last century. B. G. Verghese's penetrating articles entitled *Beyond the Famine* and *A Blessing Code Named Famine* on the droughts of Bihar and Maharashtra have highlighted the advantages conferred on the helpless rural population by famines, since it is only during such crises that at least some food and employment are assured to those living in absolute poverty.

Meanwhile a silent revolution has been taking place in the minds of our farming community as a result of the new production technologies introduced through the High-Yielding Varieties Programmes, Intensive Cotton District Programme including hybrid cotton which was introduced into commerical cultivation for the first time in the world and the Cattle Cross-breeding projects. Obsession with destructive criticism has blinded us to the meaning and implications of the increase in wheat production from a little over 12 million tonnes to over 27 million tonnes in just six crop seasons beginning from 1967–68. There has not been much interest in studying how a small Government programme in wheat was converted into a mass movement by our farmers. It is not only in the Punjab or Haryana that wheat production went up dramatically but also in non-traditional wheat areas like West Bengal. While in the past it was difficult to induce farmers to take to a rat control operation, farmers took pains, in the wheat belt, to see that rats did not migrate from sugarcane fields to wheat. While in the past everyone knew that farmers were sitting over a large underground water resource in the Indo-Gangetic plains, it was only the introduction of high-yielding varieties of wheat and rice that provided the necessary motivation for them to take to the construction of tube wells—whether made of metal or of bamboo—on a large scale. . . .

Farmers who learnt the economic value of good management in wheat also took to better practices in rice, potato and other crops, with the result that agriculture as a whole started moving forward in such areas. The sudden and steep spurt in fertilizer demand is an eloquent testimony to the credibility of the new varieties and techniques. The concept that agricultural advance in India would suffer due to the limited vision of the farming community was thus disproved. The social tensions between those who had access to the inputs needed for adopting the new technology and those who did not have similar access, only underlined the fact that those who have not derived economic benefit from such technology are equally anxious to take to the technology. The desire to change farming methods thus fanned both joy and sorrow in our countryside. This in turn generated considerable thinking and action on the part of Government, resulting in programmes for marginal and small farmers, expansion of credit facilities and more recently in integrated farmers' service societies on the lines recommended by the National Commission on Agriculture. . . .

Out of our total geographical area of 328.05 million hectares, the net area sown during 1969–70 was about 139 million hectares. Forests occupied about 65 million hectares and uncultivated land about 101 million hectares. . . . At 1960–61 prices, the contribution of agriculture, forestry and fishing to net national product in 1970–71 was 44.4%. The contribution of agricultural products to the total export earning in the same year was 37.1%. . . .

The physiography of our country shows great diversity. . . . We have major mountain ranges [including] the Himalayas, at once one of the youngest [and] the mightiest of the world's mountain sytems. . . . Plateaus, ranging in elevation from 300 to 900 metres, constitute a prominent feature of our topography. . . . A very large part of the country consists of extensive plains watered by great rivers where a considerable proportion of humanity live. . . . Variations of the order of 1 to 300 in the mean monthly flows of [our peninsular] rivers are common.

Our climate shows equally great diversity, ranging from continental to oceanic, from extremes of heat to extremes of cold, from extreme aridity and negligible rainfall to excessive humidity and torrential rainfall. . . .

On an average, we receive an annual rainfall of about 370 million hectare-metres. It has been estimated that about 80 million hectare-metres seep into the soil of which about 43 million . . . remain in the top layers and contribute to soil moisture which supports crop growth. The ground water recharge available for utilisation may be of the order of 26.75 million hectare-metres, while the current utilisation is about 10 million. . . . The area under irrigation during 1969–70 was 30.3 million hectares. The Second Irrigation Commission has calculated that the ultimate potential for irrigation from conventional sources is 81.7 million hectares. . . .

. . . In the care and maintenance of soil fertility and productivity, . . . our record is dismal. . . .

We have four major groups of soils—alluvial, black, red and laterite, several other types like forest, desert, alkaline, saline and acidic soils occurring in smaller areas. . . .

Consequent on our varied soil and climatic endowments, we have over 20,000 plant species, [far more] than those found in countries with a larger land mass. . . . We have over 4,000 woody species of plants and among them about 50 are of major utility. The potential productivity of our forests has been estimated to be of the order of 490 million cubic metres of wood. The current annual harvests comprising 12.9 million cubic metres of fuel wood and 8.9 million cubic metres of industrial wood are, however, extremely low. . . .

Our wild plant wealth, used in various forms by people in tribal areas, includes about 500 species. . . . If we total up plants under cultivation on some scale, their number comes to about 250, excluding the ornamental trees, shrubs and herbs. . . .

. . . It is only recently that serious attempts have been made to domesticate some of the medicinal plants which are otherwise directly collected from the wild states and used.

An example of such a recent domestication is the release for cultivation in the Bangalore area of strains of *Dioscorea floribunda,* a plant native to Central America and which contains Diosgenin, a steroid used in oral contraceptive pills. . . . Equally impressive is our animal wealth. According to the livestock census of 1966, our total livestock population was 344 million including 176.0 million cattle and 53 million buffaloes. Both in cows and buffaloes, we have an impressive array of hardy and productive breeds. Exotic breeds like Holstein, Brown Swiss and Jersey are being increasingly used in cattle breeding programmes. More than 60 per cent of the total milk production comes from about 23.4 million buffalo cows. The best dairy animals in our country have, however, an average production efficiency of only 25% of their counterparts in Europe and North America.

We have about 4.2 percent of the world sheep population, numbering according to the 1966 census about 42.0 million. In addition to wool, mutton and milk, sheep contribute at present 15.5 million pieces of skins and 2.1 million tonnes of manure. . . .

Animals also contribute over 28 million H.P. of energy per day for agricultural operations. . . . It has been calculated that to produce a yield of 2 tonnes per hectare, the average power requirement would be about 0.75 H.P. per day. The current power availability including that provided by man, animals, tractors and power tillers comes to only 0.30 H.P. per hectare. Thus, inadequacy of power is one of the basic causes for our inability to improve the efficiency of farming through timely agricultural operations. . . .

Fisheries, both marine and fresh water, constitute one of our great assets. Over 3 million persons live on marine fisheries. . . . Our marine fish catch has increased from about [400,000] tonnes in 1947 to about 1.2 million tonnes in 1971. We are now the second biggest shrimp producing country in the world. . . . [An] index of the untapped marine fish resources we have is provided by comparative figures on catch in different oceans. The yield per sq. km. is about 233 kg. in the Atlantic Ocean and 196 kg. in the Pacific Ocean, in contrast to 37 kg. in the Indian Ocean. Even out of this low yield, we catch very little. For example, the total tuna catch from the Indian Ocean is about 175,000 tonnes, out of which our share is about 5,000 tonnes. The rest is caught mainly by Japanese, Korean and Taiwanese vessels.

Though the current contribution of inland fisheries including both capture and culture fishes is low and amounts only to about 690,000 tonnes per year, there is vast scope for improvement through modern aquaculture techniques. Water pollution and water weeds could become some of the greatest threats to fish culture and hence deserve serious attention. Domestic and municipal wastes contribute more in many areas to pollution than industrial effluents.

Any account of our agricultural assets will be incomplete without a reference to the excellent network of agricultural universities, research institutes and demonstration-cum-training centres which we fortunately have. A recent analysis at the Yale University has shown that while investment in agricultural research has been very small in many parts of India in relation to other sectors of research, the payoff from investment on agricultural research in our country has been one of the greatest in the world. Among factors which have contributed to the effectiveness of our agricultural research system is the development by the Indian Council of Agricultural Research of a national grid of co-operative experiments conducted by scientists belonging to all the relevant disciplines and institutions. Such all-India coordinated research projects now number over 70 and cover all the major areas of crop, animal and fish improvement. The data collected in such projects are discussed at all-India workshops and decisions on recommendation to development agencies and farmers are made collectively by all the concerned scientists. Another important strength of our research system is the direct linkages which have been established between the research centres and the farmer through National and mini-kit demonstrations, Krishi Melas and travelling seminars. This feedback relationship assists in the fine tuning of the research apparatus to the needs of its clients. Though impressive in terms of contributions, our research efforts are still very small in relation to the magnitude and diversity of problems facing us. . . .

The low productivity of our agricultural systems is well-known and comparative statistics place our country in the bottom group with reference to the yield per hectare of many economic plants like rice, wheat, *Jowar,* maize, pulses, oilseeds and cotton. An important reason for our relatively poor agricultural productivity is the vast area under the important crops, which includes considerable marginal lands. Historically, cropping systems have evolved partially due to ecological and pest and disease compulsions but more importantly, on the basis of the home needs of the farmer and his family. We now know that many low-yield environments for rice or wheat may constitute high-yield environments for some pulse or oilseed crops. An effective food distribution machinery can pave the way for re-adjustments in crop planning based on considerations of ecology and economics. If this is not done, our agriculture will become increasingly inefficient and expensive and our cost of production will increase, since wages tend to rise with or without co-incident improvement in productivity. For example, our average rice yield now is about 1170 kg. per hectare and if the tentative Fifth Plan projection of productivity improvement is all that we can achieve, the average yield will be 1375 kg/ha by 1979. In contrast, the average yield of rice even in 1971 was over 5000 kg. per hectare in the Arab Republic of Egypt, Japan, Italy, the United States and several other countries. Thus, after the inadequacy of farm power, a major handicap is defective land and water use resulting in low yields. . . .

I mentioned earlier that one of our great natural assets is variability in weather and soil conditions which fostered the domestication of a wide range of economic plants and animals. Aberrations in weather, however, also constitute a major handicap. The rainfall distribution is skewed, about 80% occurring during the south-west monsoon season. About 30% of our geographical area receives annually less than 75 cm. rainfall and the occurrence of drought, floods, breaks in monsoon, cyclonic storms, thunderstorms and duststorms are common in one part of the country or the other. Although the periodicity of such weather aberrations cannot be determined with precision, it is possible to work out anticipatory measures and cropping systems on the basis of probability estimates. This is yet to be done systematically.

The next major handicap is the growing loss of soil and the damage that is being done to soil health and fertility. For reasons which are not clear to me, there has been a great neglect of the soil in our country in comparison with China or Japan. While it takes anywhere between 100 to 400 years for one centimetre of top soil to be formed in nature, all this soil can be lost in just one year due to erosion. It has been estimated that nearly 80 million hectares out of the 139 million hectares under cultivation require attention from the soil conservation point of view. A wide range of factors such as denudation of forests and vegetation cover, inappropriate tillage and cropping techniques and practices like shifting cultivation, are causing a considerable loss of valuable soil through water and wind erosion.

Shifting cultivation, known as *Jhuming* in the north-eastern Himalayan region, involves cutting down all vegetation from hill slope, use of fire to clear the debris, growing a crop like a hill paddy, millets, sweet potato or beans, abandoning the land after a few years and restarting the cycle at another place. According to an FAO estimate, this form of cultivation dates back to the Neolithic period. While this practice has gradually tended to disappear from States like Bihar, Orissa, and Madhya Pradesh, the area cleared annually for *Jhuming* may be about 100,000 hectares in Assam and Meghalaya. Observations in Assam hills indicate that at least 10 centimetres of soil may be washed away even from moderate slopes in each *Jhuming* cycle. One of the factors influencing such indifference to soil care is the fact that the land cleared for *Jhuming* is not owned by the cultivator, whose interest in the land is co-terminous with the cropping cycle. Shifting cultivation as well as growing one crop once in two to three years were two of the ancient methods of overcoming the implications of the law of the diminishing of the soil in relation to crop yield. In areas with settled cultivation, application of organic wastes and the cultivation of pulse crops and other legumes which can fix atmospheric nitrogen in the soil were the common methods of restoring soil fertility adopted in the past. . . .

Soil erosion leads to an enormous loss of nutrients. Some calculations show that the annual loss of soil due to erosion is about 6,000 million tonnes and that of nutrients is 2.5 million tonnes of nitrogen, 3.3 million tonnes of phosphorous and 2.6 million tonnes of potash. A portion of these nutrients may get deposited elsewhere. In fact, such loss of nutrients may be the most important cause of water pollution in our country. A major approach to soil conservation has been the construction of contour bunds for the purpose of decreasing erosion, conserving water above the bund and increasing infiltration. We have so far treated, during the various Plan periods, about 13.3 million hectares of agricultural land and 1.2 million hectares of non-agricultural land with various soil conservation measures. While these programmes have been very valuable both for diminishing erosion and providing employment, the agricultural benefits from such programmes have not been commensurate with the effort or expenditure involved. Often bunding has neither been followed up with other measures like providing a vegetation cover nor been carried out with an understanding of crop production technology. For example, a recent survey by the Madhya Pradesh Department of Agriculture on the effectiveness of contour bunds in increasing *rabi* wheat yields has shown that on the whole the effect of bunding was strongly negative. Similarly, a study on the effect of bunding on the yield of *kharif* crops in the Bellary area indicated that contour bunding decreased the yields of *jowar*, cotton and safflower . . . as a result of water stagnation and delay in cultural operations. . . .

One has only to see how land is being used for brick making and road making in our country to understand the extent of indifference to the care of the soil. While planned use of land for brick making could help in building up permanent assets like tanks, unplanned use results both in the loss of good agricultural land and in erosion. Similarly, there is always a danger that natural drains may be plugged if a total view on soil and water conservation is not taken while planning the construction of roads and railway lines. Closing natural drains leads to floods and thereby to considerable soil erosion. Thus, deforestation, shifting cultivation, overgrazing and improper cropping of undulating lands, bunding without vegetation cover, plugging of natural drains and other kinds of poor land management are causing increased runoff, reduced ground water recharge and severe erosion resulting in the deterioration of soil, loss of valuable nutrients, lower yields, flooding of lowland areas, sedimentation of small tanks and large reservoirs and the wastage of precious water to the ocean. It is not hence surprising that in an article entitled *The Eleventh Commandment,* Dr. W. C. Lowdermilk stated a few years ago that "the use of land is a down-to-earth index of civilization, for land has been the silent partner in the rise and fall of civilizations."

We can continue to neglect our soil only at the peril of our future. The authors of *Limits to Growth* have calculated that every child born today would need 0.08 hectare of land for purposes like housing, roads, waste disposal, power supply and other uses and 0.4 hectare of land for producing the food he or she needs. On this formula, we will need at least 5 million hectares of additional land every year to cater to the needs of those added to our population. In contrast even in 1969–70, the availability of agricultural land was only 0.34 hectare per person. . . .

Improper and inefficient water use, inadequate tapping of sunlight, poor utilisation of biological nitrogen fixation, wasteful disposal of wastes, lack of understanding of recycling processes and poor integration of crop and animal husbandry on the one hand, and terrestrial and aquatic production systems, on the other, are some of our other major liabilities.

The slow pace of progress in getting the best out of our water resources is evident both from the relative stagnation in *kharif* crop production and the low intensity of farming. During the *kharif* season, when much of the rainfall is received, the production of foodgrains was 65.6 and 62.0 million tonnes in 1964–65 and 1971–72, respectively. In contrast, the *rabi* production during these two years was respectively 23.7 and 42.7 million tonnes. Thus, our major gains in production have come from the non-rainy season. With the development of irrigation facilities it should have been possible to bring more area under double and multiple cropping. The intensity of cropping is, however, still low, and has risen from 114 per cent in 1965–66 to only 117 per cent in 1969–70.

A low intensity of cropping when water is available also implies a poor utilisation of sunlight, since in the tropics and sub-tropics green plants can be made to produce food, feed and fodder continuously by photosynthesis from water and atmospheric carbon dioxide. The cellulose produced by photosynthesis on the earth is not only the chief basis of all fossil fuels, but is also the most abundant renewable raw material currently available. . . .

In the area of nutrient supply to crops, organic manures like farmyard manure, compost, green manure, various oil-cakes and various waste products of animal origin like dried blood, bones, fish manure and urine have been used in the past. With the growth in population, farmyard manure and other organic wastes have increasingly been diverted to fuel purposes. Though some work has been done on the generation of gas from such wastes in order to obtain both fuel and manure from the same material, such techniques have not come into use on any significant scale. Much of the urban wastes, sewage water, cattle and human urine and human excreta are not recycled in a manner that will promote productivity.

A great marvel of nature is the way in which micro-organisms fix atmospheric nitrogen in the soil largely through leguminous plants. While synthetic nitrogen-fertilizer production requires very high temperature and pressure for combining nitrogen, hydrogen and oxygen, the nitrogen-fixing organisms like *Azotobacter, Rhizobium,* and blue-green algae are able to do this at ordinary soil temperature and pressure, with the help of the enzyme nitrogenase. Nitrogen is now being introduced into the earth in fixed form at the rate of about 92 million tonnes per year, whereas the total amount being denitrified and returned to the atmosphere is only about 83 million tonnes per year. The difference of 9 million tonnes per year may represent the rate at which fixed nitrogen is building up in the soil, ground water, rivers, lakes and oceans. Unfortunately, studies show that while Australia, the United States and Soviet Union are adding every year substantial nitrogen to their soils, we have a negative nitrogen balance. . . .

Before I end this discussion on our agricultural assets and liabilities, I must refer to the most important factor which will determine our agricultural future, namely, our human population. The principal characteristics of our population are the predominance of youth, poverty, under-nutrition and illiteracy. According to the projections of the Registrar General, we will have about 657 million people by 1981; over 40 per cent of them will be below the age of 18. Nearly 80 per cent of our population lived in villages during 1971. We are thus mainly a land of youth and of rural people. Impressive statistics on poverty are available and there is also a growing awareness of the implications of protein-calorie malnutrition on national development. Many authorities now compare nutrition education respectively to the hard and soft ware components of technology. Although in terms of percentage, the number of literates rose from 24.03 in 1961 to 29.34 in 1971, the number of illiterates increased in absolute number during this period. Most of our agriculture is managed by illiterate peasantry and Mahatma Gandhi warned us forty years ago that unless there was a marriage between intellect and labour in rural India, there would be neither agrarian advance nor rural prosperity. Our agricultural future will hence depend to a great extent on how successful we are in involving youth and the illiterate, both men and women, in rural transformation. The urgency of population stabilisation hardly needs any emphasis, since all our efforts will have no

Where there are millions and millions of units of idle labor, it is no use thinking of labor-saving devices. . . .

Not that there is not enough land. . . . It is absurd to say India is overpopulated and the surplus population must die. . . . Only we have got to be industrious and make two blades of grass grow where one grows today.

The remedy is to identify ourselves with the poor villager and to help him make the land yield its plenty, help him produce what we need and confine ourselves to use what he produces, live as he lives and persuade him to take to more rational ways of diet and living.

GANDHI

effect in improving the quality of life, if population growth cannot be arrested and stabilised.

The second-class compartment of Spaceship Earth is getting smaller and smaller, while the number of those who have to live there is getting larger and larger. Affluence is becoming a major claimant of world food and energy resources. Between 1967 and 1971, the developed market economies increased their agricultural exports on an average by 11 per cent per year, while the developing countries lost ground by 1 per cent. This is the net result of what is called the U.N. Development Decade. It is time the people living in the first-class compartment realised that when the compartment of the poor bursts due to excessive pressure, the whole spaceship will crash. But the poor have to begin to help themselves.

During the last few years, we have started making real progress in improving our agricultural capability. . . .

. . . Our liabilities, apart from those caused by population growth and aberrant weather, mostly arise from an improper use and management of [our] resources. Attempts to promote synergy, which is the only mechanism which can lead to rapid progress from small resources, have been few and halting. Our urgent needs hence are first, to develop and introduce in each ecological area an agricultural production technology which will lead to increased productivity based predominantly on the use of renewable resources and on the wise husbanding of non-renewable resources. We will have to learn to produce more and more food from less and less land. Secondly, we need to develop and introduce educational tools which will help to impart the latest technical skills to illiterate peasantry and which will enable educated youth to become catalysts of rural change. Finally, we need to develop and spread management and organisational techniques which will help those living in absolute poverty to overcome their handicaps and obtain their share of the fruits of agricultural advance.

26 The Wail of Kashmir

Lee Merriam Talbot

Lee Merriam Talbot is an official on the President's Council on Environmental Quality. An eminent ecologist, he has written numerous books, monographs, and articles on wildlife and conservation, as well as an award-winning TV documentary film, "Man, Beast and the Land" (1968).

A striking example of man's growing impact on the land is the Great Thar Desert in western India. At the time of Christ, Indian rhinoceros roamed in grass jungles in the middle of what is now desert. And for the past 80 years the desert has been advancing into the rest of India at the rate of one-half mile a year along its whole long perimeter. That means that in 80 years, an estimated 56,000 square miles, or an area equal to that of Wisconsin, has been turned into shifting sand.

The mechanics of this land degradation seem clear. The starting point is the mature forest with its wildlife, fertile soil, and abundant water. The lumber is cut, often clear-cut with young growth destroyed. The land is then cultivated for a time, then grazed and overgrazed. There can be no replacement of trees or grass, for everything green is eaten by ravenous livestock. When there is nothing left for cattle, goats take over, and when the goats have left, nothing remains but sand or blowing dust. This story holds true, with the same plot and characters, but with different stage scenery and costumes, throughout much of the world.

To illustrate the effect of this land-use pattern on wilderness, let us consider Kashmir, the ancient Moguls' "paradise on earth." This is a lovely mountain land in northernmost India, lying at about the same latitude as San Francisco, and bordered by Tibet, China, and Pakistan. The British with the local maharajah set aside magnificent wild areas here. But when independence came, here as in most former colonies, the tendency during the first burst of nationalism was to reject all that smacked of the previous "imperialism" or "colonialism." Parks and wilderness areas were thought of as something kept away from the people by the former rulers rather than as a resource maintained for them, so the first reaction was to destroy them, to take "what was rightfully ours." On top of this came political and military unrest with a side effect of a large population suddenly armed but with little discipline. Among the results have been large-scale poaching leading to the virtual extermination of the Kashmir stag, heavy forest cutting, and overgrazing.

Through much of Kashmir up to and above timberline one runs into herders and livestock. In less than ten years much of this land changed from dense conifer forest or park lands like the best of our Sierra or Rockies, to what are approaching high-altitude deserts, with the vegetation pulled apart, cut, overgrazed, and burned out—and the soil, too.

Economic need, destructive land use, and destructive nationalism form a constantly recurring pattern deadly to wilderness. Until all three of these factors are somewhat ameliorated it is hard to be optimistic about the future of wilderness lands throughout the world.

As my last example I would like to mention an area not usually thought of as living wilderness—the Middle East. Much of it is arid desert, but when Moses led the children of Israel through the Sinai wilderness, it was a live wilderness with wildlife and trees. Today one can go for days through that country and never see a living thing. The mountains above the Promised Land were cloaked with dense forest, with pine, oak, and cedar; and in the more open areas, Asiatic lions stalked abundant wildlife. Today these mountains are largely dead stone skeletons, and the last small remnant of the Asiatic lions is to be found 3,000 miles to the east. There are still two more

> **When man obliterates wilderness, he repudiates the evolutionary force that put him on this planet. In a deeply terrifying sense, man is on his own.**
>
> **J. A. RUSH**

or less living wilderness areas left—in northern Lebanon and western Syria. Until recently, protected by inaccessibility and unsettled conditions, the forest here remained intact; but within the last few years, lumbering and cultivation have begun to move into these last forests. When the land's fertility has been cropped out and the trees have been cut off, the crops will give way to grazing. Once overgrazing has gone far enough, the starving animals preventing grass, brush, and tree reproduction, the area will assume the desert aspect of most of the Middle East.

This remnant biblical wilderness illustrates one of the very real economic values of wilderness that, perhaps, is not often thought of in our country. It would be easy to say, looking at most of the desert Middle Eastern lands, that this area never did support much life, or that the old records of forests and crops are wrong, or that if there were trees here once there has since been a climactic change. But in these remaining wild forest areas we have the living proof that this was not the case. North Lebanon and western Syria provide a point of reference by which one may judge the condition of the land as it was, see what man has done to the rest of the land, and therefore see what can be done with what land is left.

27 Combating Environmental Degradation in Arid Zones

Eskandar Firouz

Eskandar Firouz is undersecretary of the Iranian Ministry of Agriculture and National Resources, and director of the Department of Environmental Conservation.

In relation to many environmental criteria, the increase in the populations of the nations of the Near East and of the rest of southwestern Asia is perhaps as critical as that of the more densely populated countries of the world. Undoubtedly, one of the most destructive results of the burgeoning populations is the degradation of the arid zones and of their component ecosystems. In the past the alteration of a specific ecosystem might take millennia; today in certain areas the needs of a growing society produce radical transformations of the natural environment in a mere decade or less.

Arid by nature, enormous areas of the Near and Middle East progressively assume a more and more xeric appearance with the retrogression of plant life as a result of various human practices. Forest is transformed to steppe, steppe to semi-desert, and semi-desert ultimately to sterile wasteland that is virtually devoid of all vegetation. This tendency is not confined to the arid regions, but it is there that the problem is now assuming disastrous proportions: the prospect in the offing is an almost total loss of vegetation over wide areas. The question is not whether such a state, if matters are allowed to proceed unchecked, will come about, but merely how soon!

The mechanism of range-land destruction is well documented. As the range deteriorates, the herdsmen turn to more adaptable species of livestock; sheep are substituted for cattle, and then goats replace sheep more and more. The problem is often compounded by the persistent tendency to convert easily accessible parts of such areas to dry farming, notwithstanding the fact that these lands are invariably ill-suited or marginal for such purposes. Thus still less forage is left for the additional livestock now required to support a larger populace.

If the phenomenon were only dependent on livestock grazing, perhaps the law of diminishing returns would set a limit to the extent of retrogression: that is to say, at some point prior to complete vegetation destruction, grazing would prove uneconomic and thus be terminated. Unfortunately, this is generally not the case, for the desert is also mercilessly exploited as a source of fuel. Trees are cut, then shrubs, and finally perennial and annual forms of vegetation are also doomed. The death-knell is sounded with wind and water erosion, soil deflation, sandstorms, sand-dune accumulation, loss of wildlife, and climatic changes. It is only when villages are buried beneath the sands of the desert, when there is misery and suffering for man, that the phenomenon is understood and bewailed.

Any environmental degradation is undesirable, but the immediate and pressing issue here is to determine the minimum levels to which plant retrogression could be allowed to proceed in arid zones, and to insure that such criteria of deterioration are not exceeded.

It is always on the move. Whenever man's grip on the soil slackens and his defenses break down, the desert moves in. . . . In the deserts of Persia and on the edges are oases and fertile stretches which are menaced by wind-drift of salt and sand by the erosion of wind and water. In the area of Khorasan there was, in ancient times, a great forest of which not a vestige is left today.

RITCHIE CALDER

It is evident that the insufficiency of resource managers precludes the implementation of sound range-management practices throughout the region. Eventually, training programmes may provide personnel in adequate numbers to monitor these problems and to propose as well as enforce guidelines. What is proposed in the meantime is that the respective governments undertake immediate and drastic action to halt, and hopefully reverse, these trends in the more endangered areas before the damage is irreversible.

Where forage production does not exceed certain levels, grazing or other exploitation of the vegetation is not economically productive and becomes, in fact, counter-productive. To disregard this fact will ultimately lead to national disaster. For once a given level of degradation is reached, attempts at restoration, if not wholly outside the realm of possibility, become prohibitively expensive.

Serious attention must evidently be given to demographic trends. The location of human settlements must be studied in depth and on a long-range basis with a view to limiting, if not relocating, villages and towns (particularly those with inadequate access to suitable fuel sources) in arid zones. Where plant production has fallen to unacceptable levels of denudation, or where climatic and edaphic factors support vegetation only on a marginal basis, governments should consider a total ban on vegetation exploitation. At the same time, the prospect of supplying desert residents with fuel sources, whether fossil fuels or hydroelectric power, should be seriously examined. Until alternative programmes are developed, various countries might consider the provision of energy sources at subsidized rates.

Such emergency measures should be enforced in order to buy time until comprehensive management plans can be implemented. If the action proposed appears drastic, failure to apply it might prove disastrous. Clearly there is no single solution to these grave and complex problems, and closing certain areas to population will increase the pressures elsewhere.

As an example of action, Iran has (by law) prohibited the cutting, etc., of all desert vegetation. However, enforcing this has, to date, been problematical in many areas. But in our "Protected Regions" not only is this aspect rigorously enforced by strict wardening, but over large areas (totalling something like four million hectares) of steppe and desert land in these regions grazing as well has been partially, and over some 70 per cent of the area totally, banned. As we have described and illustrated in *Biological Conservation* (Vol. 3, No. 1, pp. 37–45, 1970), the resulting regeneration in these regions has been very encouraging. But in the long term it is most essential to tackle the problem of human population, which in Iran is increasing at the quite devastating rate of over 3 per cent per annum. Consequently, the Government has instituted a family planning service, which it is ardently hoped will in time lead to the rate of increase being reduced to 1 per cent.

AFRICA

28 The Ecology of Development

D. F. Owen

D. F. Owen is professor of animal ecology at the University of Lund, Sweden. He is also the author of *Animal Ecology in Tropical Africa* and *Tropical Butterflies.*

Two-thirds of the present population of the world are living in poverty while the remaining third, in the developed nations, are living in comparative luxury. The average person in a developed country uses up non-renewable resources at a rate of between thirty and fifty times that of a person in an under-developed country. For some resources the average rate of utilization is hundreds of times greater than what occurs in an under-developed country: on average an Englishman uses as much steel in a month as a Gambian uses in his lifetime. Hence despite the high rates of increase of the population in under-developed countries the population in developed countries, with a much lower rate of increase, has a much greater impact on the world's resources: the ecological impact of a birth in a developed country is potentially fifty times greater than one in an under-developed country.

All the countries of tropical Africa are officially designated as under-developed and all are striving, in most cases unsuccessfully, to become developed. Put differently, this means that the aim of under-developed countries is to use resources at a rate that is equivalent to that in a developed country. All the countries of tropical Africa produce raw materials, renewable and non-renewable, that are sold to and consumed in developed countries. Just how long this state of affairs can be sustained is anyone's guess. How long, for instance, can Sierra Leone rely upon the export of uncut diamonds and Zambia upon the export of copper ore? Both of these exports depend upon demands for them by developed countries, both resources are non-renewable, and there will come a time when diamonds and copper can no longer be produced in sufficient quantity to interest the business community of the developed countries. When this occurs the present unstable economies of Sierra Leone and Zambia will become even more unstable.

Consider, as a more specific example, the fate of iron ore mined in Africa and exported raw to Britain. The iron is converted into steel and is then used for a variety of industrial purposes, among them the manufacture of motor cars, an industry which keeps thousands of people in Britain in employment and which contributes substantially to the country's economy. The motor car industry is geared to the concept of rapid production and rapid consumption: cars are not built to last and within a short time they are replaced by new models. Old cars "decompose" and little of the steel is recycled and used to make new cars. If it were re-cycled there would not be the same demand for iron ore and the economy of the producer countries would suffer. On the other hand there is clearly a limit to the available resources of new iron ore and the time will come when re-cycling is a necessity. It would also be technically possible to manufacture cars that last a long time. This would be less demanding on the dwindling raw materials (including many in addition to iron) but would presumably create unemployment for car workers in Britain and miners in Africa, and would reduce the profits made by business and by the government. It would also mean a change in the social attitudes of people in developed countries who would have to buy one car to last a lifetime rather than replace the car with a new model at the first opportunity. Steel is also used in construction work and thus becomes "locked up" for long periods of time, chiefly in the developed countries where most construction occurs. This pattern of production and consumption applies to all other non-renewable resources, many of them produced by under-developed countries, and all of them consumed mainly by developed countries.

The most significant aspect of the above considerations is that the present economic structure of the world would collapse if the developed countries were unable to use the resources of the under-developed countries. This is most strikingly apparent with oil which is produced mostly in under-developed countries (although not a great quantity is produced in tropical Africa) and consumed largely in developed countries. It is equally significant that there would be economic collapse if the under-developed countries consumed raw materials at the same rate as the developed countries.

The present concern of the industrial countries with the harmful effects of environmental pollution is regarded as something of a threat to development by the under-developed countries. Cleaner industry and more recycling of raw materials, both of which are now being advocated in industrial countries, could easily make industrialization a more expensive process, and this might inhibit rather than promote development in Africa. It has been suggested that if the industrial countries are worried about the effects of pollution there should be an effort to spread industry out on a global basis: many people in tropical Africa would welcome the sight of chimneys belching smoke as this is equated with development and jobs. There is much to be said in favour of this point of view but there seems little prospect of it being implemented.

The present ecological predicament in tropical Africa can be defined in relatively simple terms. Human numbers are expanding extremely rapidly as a result of the decreased death rate among children. The decrease has been brought about by the suppression (to a large extent) of the effects of many of the important killing diseases and by the spread of cultivation. Everywhere there are understandable demands for a higher standard of living which in effect means a higher rate of consumption of raw materials. All plans for economic development are in one way or another intended to facilitate a better standard of living which in Africa, unless there is a radical change in attitude, will result in even higher rates of population increase. The increase in human numbers in Africa has taken place chiefly in the last few hundred years and is continuing at an accelerating pace. The savanna and forest ecosystems which have taken thousands of years to evolve their biological complexity and stability have been largely destroyed, and almost everywhere there are signs of cultivation, erosion, and a general degradation of the environment. Raw materials have been exploited and exploitation is continuing, chiefly to the benefit of the developed countries. The population is largely dependent on food that it can produce for itself and the food that it can acquire from other countries in exchange for raw materials or in the form of aid. There is virtually no industry and prospects for industrial development seem poor. Indeed even if industrialization were possible it could proceed little further than producing goods for home consumption as the developed nations are unlikely to require quantities of manufactured goods from Africa. This predicament is not of course unique to Africa: it applies to varying extents to almost all tropical countries, and the paradox is that the continued existence of developed countries is dependent on this unsatisfactory state of affairs.

Table 1 shows the world population in 1960 and the projected population in 1980 and 2000. The projections are based on estimated current birth and death rates and are therefore optimistic as it is assumed that there will be no rise in the rate of population increase. Other estimates give higher and lower

TABLE 1 *Present and projected population of the world in millions. From Taylor (1969).*

	1960	*1980*	*2000*
World	2998	4519	7522
Developed	976	1242	1580
Under-developed	2022	3277	5942
Regional estimates			
Developed			
Europe	425	496	571
U.S.S.R.	214	295	402
North America	199	272	388
Japan	93	114	127
Temperate South America	33	47	68
Oceania	16	22	33
Under-developed			
Mainland East Asia	654	942	1509
Remainder of East Asia	47	87	175
South Asia	865	1446	2702
Africa	273	458	860
Latin America (part)	180	340	688

figures for 1980 and 2000, depending on what assumptions are made about fertility trends and death rates. The figures in [the table] are divided into regions and these in turn are divided into developed and under-developed areas of the world. Almost all the under-developed areas are in the tropics. . . . There is much disagreement among economists as to what is meant by development and under-development, indeed doubt has been expressed as to whether it is legitimate to consider development in economic terms. Economists consider that literacy, income per head (which in effect means the capacity to consume raw materials), child mortality rates, housing, and so on, can be used to judge the degree of economic development in a country, but all of these present numerous difficulties of interpretation. How can one say, for instance, that a country that grows most of its own food and yet has a low *per capita* income is less developed than one with a higher *per capita* income which has to import most of its food? In [the table] development is defined in demographic terms: an under-developed country has a rate of population increase that is known or believed to be in excess of 2 per cent per year, while a developed country has a population growth rate of less than this figure. By adopting this definition it is assumed that the rate of population increase is the chief obstacle to economic improvement in an under-developed country, but the definition evades the question of whether economic improvement is necessarily a good thing or for how long the present inequality among the countries of the world can be sustained.

The projected figures in [the table] show that although the present population of Africa (including that of the area outside the tropics) is lower than that of Europe, it should become equal to that of Europe some time in the 1980s, and by the year 2000 there may be nearly 300 million more people in Africa than in Europe. The ecological impact of this projected rise in human numbers will be spectacular. Moreover the estimate of 300 million more may be on the conservative side. Thus recent reports from Rhodesia indicate that the African population is increasing at a rate of 3.6 per cent per year which means that the population will double itself in about eighteen years, a phenomenon which will have political as well as economic implications.

Most African countries are forced by circumstances to encourage economic investment from the developed countries. Some people see this as a form of aid but in reality it amounts to little more than a financial inducement for Africa to export more of its natural resources and to accept in return manufactured products from the developed countries. The motives of the European, American, Russian, and Chinese governments in offering aid to the newly independent nations of Africa are not so much associated with making profits but with gaining influence, which, of course, in time, can lead to making profits. The former colonial powers probably feel somewhat guilty about what they did not do in the past or they are embarrassed because they failed to predict the unexpected turns that political events have taken. There is also the curious view that Africans need help and advice more than other people which in its extreme form can lead to disastrous situations where anything African must not be criticized. This does no good, and it is important to appreciate that Africans, like other people, can make mistakes and can become corrupted, a point which is expanded at considerable length by Andreski (1968). But perhaps the most important motive in offering aid is the possibility of political influence. Whether in fact there is any long-term influence is another matter, but the big powers evidently feel that there

> **The danger to the African is that . . . he loses his past without gaining the future. That, grasping the glittering bauble of materialism, he loses his greatest talent—that of living, of survival.**
> STUART CLOETE

is some advantage to be gained if they can claim support for their international policies from under-developed countries. The present declining influence of the United Nations in determining world events will perhaps result in a decrease in aid for entirely political purposes; indeed there may be an increase in bilateral aid more closely associated with marketing and economic development.

It is probably true to say that since the days of independence no area in the world has suffered at the hands of outside experts more than tropical Africa. There are experts in education from UNESCO, in agriculture from FAO, in health from WHO; there are those that give advice on economic development, how to build airports, roads, universities, and television centres, and those that want to try and preserve wildlife and areas of land which they claim are of scientific and aesthetic interest to the rest of the world. All are anxious to offer advice and to propose programmes which, they urge, are related to development and to the well-being of the people. Individual proposals may conflict and rivalries often develop between organizations trying to achieve the same ends. Africa is remarkably receptive to the ideas of others and missionaries have for years exploited this, but the missionaries now have to compete with the economists, educationists, and conservationists for the attention of the people. But the advent of the expert into African society does not appear to have achieved a great deal. There are probably a number of reasons for this, not the least of which is that many experts are badly informed, not only on the subject they are supposed to know something about, but also on the environment on which the organization they represent wishes to make an impact. This applies especially to the experts from the specialized agencies of the United Nations who are often appointed to their highly paid jobs for political reasons (each member state is entitled to contribute its quota of experts) rather than because of their abilities in a particular field. Then there is the problem of implementation. African politicians are often more than willing to receive offers of assistance for their country, especially if there are no strings attached, but if it becomes necessary that the recipient country should also contribute money and manpower to a scheme, difficulties arise, and in many cases little or nothing is achieved. But the chief difficulty with outside offers of assistance is that only rarely are the long-term effects of a proposal evaluated. Thus a short-term proposal to eradicate a disease or to improve crop production very often leads to little more than a local increase in population.

In the developed countries there are numerous charitable organizations whose aim is to relieve hunger, malnutrition, and the effects of disease, and to improve the agriculture. Their funds are dispersed in so many different projects that virtually all their efforts are short-term. OXFAM is one of the best known of the charitable organizations and (to take an example) in the year beginning in October 1965 it allocated about £385000 to about 170 projects in 22 tropical African countries. About a third of the money was allocated to projects aimed at preventing disease and another third to projects associated with improving food production. Just over 8 per cent was allocated to help victims of civil wars and to feed refugees, most of whom had fled from one independent country to another. About 22 per cent was allocated to training programmes, mostly associated with rural health and agriculture, but only 1.2 per cent went to family planning, the one project that might be expected to have more than short-term benefits. Indeed in the year in question there were only two allocations under this heading, one to Kenya and one to Sierra Leone, and the total amount granted was only £4743.

Most of the people of tropical Africa are illiterate and with expanding human populations there seems little possibility that literacy will become widespread in the immediate future. There are campaigns aimed at providing more schools and offering facilities for adults to become literate but most efforts are frustrated by the rise in the population. Many of the existing schools, especially those in rural areas, are staffed by incompetent teachers, many of them expatriates from Europe, America, and India. One difficulty is in persuading African graduates from the universities to take up teaching as a career. Most students receiving diplomas and degrees look towards the civil service (or its equivalent) or to business enterprises for jobs which not only offer a better salary and allowances but also carry more status. School teaching is often regarded as a last resort, to be entertained only as a temporary measure until something better comes along. There are, of course, some dedicated teachers, but many teach only because they have failed elsewhere, and it is likely that the continuing recruitment of teachers from abroad is preventing a development of local interest in teaching as a profession.

The universities of tropical Africa have developed much more rapidly than the schools. In most countries there are well-equipped universities that not only provide instruction to a high standard, but are also actively involved in research. Most of the scientific research is of an applied nature, but there is some pure research, which is sometimes criticized by politicians as not being relevant to national development. Several African academics have argued persuasively that there is a place for pure research as it trains people in the art of problem solving and exercises the imagination. It has been said that university facilities have developed too quickly, especially in relation to the situation in the schools, but on the other hand a big expansion of school education, especially at the elementary level, would lead to the production of large numbers of semi-literate people for whom there are no jobs. Illiteracy may be a big problem, but large numbers of literate people could create a bigger one in the present environment.

It has been suggested that improving the agriculture and eradicating disease will not in themselves result in a significant improvement in the already unsatisfactory state of affairs. The agricultural experts in particular have misled Africa into believing that the solution to poverty is to grow more food. More food for what? one might ask. It does not matter whether agriculture becomes highly mechanized or whether the approach urged by Dumont (1966), in which the silent majority (the peasant farmers) make an all-out effort to grow more food by using relatively simple methods, is adopted, the result will be the same: more people. The so-called "green revolution," invented by the western countries, but also practised by the eastern countries, is a myth, although without doubt big business will benefit from the increased sales of pesticides, fertilizers, and improved varieties of crops.

Prospects and possibilities for industrial development in Africa on the scale that has been possible in Europe, America, and Japan are remote: the industrial nations already have all the advantages in the highly competitive world of commerce and industry, and it is unrealistic to suppose that there will be a change in this position. It is to me inconceivable that a tropical African nation will be able to develop its industry sufficiently to produce and sell items like motor cars and jet aircraft to the industrial nations or even in competition with them. There is perhaps a little hope if the African nations can offer items or services for sale that are impossible to obtain elsewhere. It is not easy to imagine what many of these might be, but one obvious one is tourism. In East Africa in particular there are numerous tourist attractions, including game parks, which if developed further could produce a substantial revenue, and the same applies to the numerous attractive beaches which are much sought after by sun-loving Europeans with money to spend. The advantage of tourism is that it offers something that cannot be had elsewhere and, moreover, it does not result in a flow of raw materials out of the country. The packaging and marketing of crop products instead of exporting them as raw materials also comes to mind, but many of these crops are also grown elsewhere in the tropics, a possible exception being cacao, most of the world's supply coming from a few countries in West Africa. Would it be possible for the cacao-growing countries to stop exporting the raw product and to develop industries that result in marketing the finished products? But these and similar possibilities will do no more than merely scratch the surface of the problem. None is likely to produce the effect that everyone is hoping for: an expanding economy and a rapidly rising standard of living for the people.

. . . Economic development in Africa is impeded by the phenomenal growth of the human population and by the necessity, imposed by the industrial nations of the world, for Africa to function as a producer of raw materials for consumption elsewhere. If there is to be a solution to the present predicament there must be a massive effort to slow down the rate of population growth and a policy on the part of the industrial nations to de-develop sufficiently to allow under-developed countries to catch up. Neither of these stands much chance of being implemented in the immediate future. Sustained economic growth is the goal of all governments, both left and right wing, in both developed and under-developed countries, and although a diverse array of people have come to realize that in order to survive growth must be controlled, there is at the moment no political party in the world that would adopt this as its manifesto and expect to come to power by either democratic or other means.

NOTES

Andreski, S. (1968) *The African predicament: A study in the pathology of modernization.* Michael Joseph, London.

Dumont, R. (1966) *False start in Africa.* Deutsch, London.

Taylor, W. (1969) Population prospects for regions of the world. *J. biosoc. Sci. Suppl.*, I: 107–17.

29 The Integration of Environmental and Development Planning for Ecological Crisis Areas in Africa

Simeon H. Ominde

Simeon H. Ominde is head of the Department of Geography at the University of Nairobi, Kenya.

Reprinted and abridged from *International Social Science Journal,* 27 (3) (1975), pp. 499–516, by permission of Unesco.

INTRODUCTION

In recent years we have witnessed the emergence of a global debate which will increasingly condition the pattern of development in Africa at three very important levels of awareness. The first is the increasing realization that man's action on the planet within sovereign boundaries or globally could injure the very basis of our survival beyond repair. The second is the stunning realization that the concept of exponential growth in consumption, the belief that with the rising GNP the poor within nation states and among nations will catch up, are no longer valid. The third is a global awareness that the nation states through which a concerted action is possible have increasingly become the greatest obstacles to the objective of narrowing the gap between the poorer and richer parts of the world.

. . . Whilst the concept of eco-development is receiving increasing support in the more developed parts of the world, it has not attracted the same attention in the developing African region. In some quarters the concern with ecology of development is considered as a luxury preoccupation in a situation where the priority development problems are seen in terms of a sectoral war against widespread malnutrition, rampant disease, the high infant mortality, low life-expectancy, high illiteracy levels, deepening unemployment and increasing concentration of fruits of development in the hands of a minority élite.

CONCEPTUAL FRAMEWORK OF DEVELOPMENT AND ENVIRONMENT

The challenge of integrated development in ecological crisis areas becomes that of controlling environmental problems resulting from poverty while, at the same time, adopting measures to prevent economic growth and industrial development from having an unfavourable impact on the society and the environment and thus cancelling benefits. The strategy must therefore be to devise new development styles aimed at harmonizing economic and social growth objectives with those of the rational management of the environment.

The need for integrated development arises from the knowledge that some ecosystems are flexible and resilient and may be modified by the agency of man without serious damage to the environment. Others are more delicate. But all over the globe there are limits beyond which the unregulated activity of man can and will cause serious damage. Knowledge on the structure and functioning of these ecosystems is needed. In these respects the scarcity of trained manpower and the limitation of resources in Africa constitute serious constraints on the new development approach, yet we may summarize the emergence of African crisis areas under two broad headings.

In the first place, Africans have used the rural environment in a variety of ways. These range from low-intensity hunting, fishing, and gathering, pastoralism and subsistence agriculture to intensive commercialized farming using imported technologies. In the transformation of rural systems crises have developed where diversity has been replaced by monocultural agricultural practices. New problems have arisen through efforts to raise agricultural productivity through fire, fertilizers and the introduction of new cultigens. At a very different level Africa has seen the introduction of large river-basin control schemes designed either to raise water for irrigation or for new forms of power.

Another set of environmental problems has emerged as a result of the widening impact of the urban-industrial systems. The emergence of the modern urban systems and associated industrial technologies combined with the pattern of population growth and change present African development planners with some of the most acute environmental problems. Within the urban areas immense problems of human existence are emerging as rural migrants are forced to transform their way of life in new and changing urban settings.

DEMOGRAPHIC COMPONENT OF ENVIRONMENTAL MANAGEMENT PROBLEMS

Between 1950 and 1970, the estimated increase in the population of Africa was from 217 million to 344 million. Forward projections for the continent indicate that the total population may rise to 800 million by the end of the century. Over the same twenty years the world population is estimated to have increased by 46 per cent as against 56 per cent for the less-developed regions. The population gain for Africa over the same period was 59 per cent. Africa's share accounted for 11.1 per cent of the world's increase.[1]

Apart from the impact of increasing numbers and consequent pressures on resources, Africa will maintain the youthful character of its population. Persons aged 0–14 will constitute approximately 40 per cent or more of the total. The dependency ratio is expected to remain high well into the twenty-first century.

Among the most important aspects of the demographic situation in the region is the spatial shift of population frequently referred to as the rural–urban population drift. Spatial redistribution of population and in particular the accelerating rate of urbanization are responsible for some of the most serious aspects of environmental disruption in Africa.

At the global level there has been concern with the environmental impact of rapid urbanization. During 1970 this was responsible for a demand for urban space for some 490 million people.[2] At the regional level, development planners have to devise integrated plans to accommodate an urban population which was expected to increase the share of African cities from just over 20 per cent in the late 1970s to about 40 per cent towards the end of the century.

However, the pattern of urbanization is known to differ significantly from one part of the region to another and among the different countries of Africa. In Southern Africa, the most urbanized of the subregions, it is expected that the proportion of urban population will rise from 37.61 per cent to 57.93 per cent in the year 2000, while in Northern Africa, the next most urbanized subregion, it is to rise from 26.64 per cent in 1950 to 52.92 per cent. The proportions for Western Africa are 11.42 per cent and 39.85 per cent respectively, for Middle Africa 6.89 per cent and 40.66 per cent, while Eastern Africa, the least-urbanized region, is expected to experience a fourfold increase in urban population from 5.39 per cent to 21.23 per cent between 1950 and the year 2000.

At the national level there are indications of accelerating urbanization and the associated growth of the primate cities. In Western Africa, the urbanized population of the Ivory Coast is expected to increase tenfold between 1950 and 1985. That of Nigeria is expected to increase fivefold while in Mali and Senegal the increases are expected to be fourfold. Available evidence indicates that Tanzania may expect a sevenfold increase of urban population whilst Kenya and Uganda each can expect a ninefold increase. It is only in the already more urbanized Northern and Southern subregions that national increases are expected to be more moderate.

These national rates of increase are the outcome of the increasing flight of rural populations to the few urban centres in response to accelerating rural population growth rates and consequent environmental crises. As living conditions deteriorate a growing volume of migrants will be forced off the land.

There has been a tendency for the accelerating drift of the rural population to result in the dominance of one or two large cities.[3] In the 1960s, countries such as Algeria, Egypt, Libya and Morocco each had approximately 70 per cent of their total populations in cities of 100,000. The 1969 census revealed that the total population of Nairobi and Mombasa in Kenya accounted for well over 70 per cent of the total urban population.

In 1950 there were only two cities in Northern Africa over the 1 million threshold, but by 1970 the number of such cities on the continent had increased to eight. In the meantime their total populations had risen from 3.4 million in 1950 to 15.6 million in 1970. It is forecast that in 1980 there will be fifteen cities of over 1 million inhabitants in Africa: two in Western Africa, two in Eastern Africa, two in Middle Africa, five in Northern Africa and four in Southern Africa. By 1985 the number is expected to rise to nineteen with a total population of approximately 47 million.[4]

In the absence of comprehensive development plans it is expected that the concentration of population in the few cities of 1 million could be even more spectacular in a continent where, in most cases, the city of 1 million is an ecologically novel phenomenon.

ENVIRONMENTAL MANAGEMENT PROBLEMS OF THE RURAL DOMAINS

One of the most important revolutions which have taken place since independence in Africa is the emphasis on rural development. The inherited colonial system and the urban bias of earlier development strategies are now being subjected to criticism and are in fact being replaced by a more holistic approach. This is backed by a realization that, despite the expected accelerated growth rate of urbanization, the vast majority of the African people will still be rural by the end of the century. In 1950 well over 80 per cent of the total population was rural, and with the expected increase in the rate of rural–urban migration some 60 per cent of the total population will still be living in the rural areas by the year 2000.

The general nature of the rural domain is one of great ecological complexity to which must be added the complexity brought about by the action of man within the different cultural settings. The earlier, urban-biased development led to neglect and, in some cases, to serious environmental deterioration, itself in part responsible for migration from the land. Environmental planning must take into account the role of the rural domain in the production of the basic necessities of the rural population and the resources on which the developing economy of Africa must be based.

Traditional agriculture in Africa is characterized by high labour inputs, self-reliance and independence from external factors other than natural.[5] In different regions of the continent it is hampered by an increasing population and inadequate technology. It is the growing pressure of population on the traditional sector that leads to environmental damage and hence to flight from the rural areas.

A large part of Africa suffers from variable rainfall and the unpredictability of water supply. Extensive areas are unsuitable for dry land agriculture and other forms of land usage requiring artificial augmentation of the water supply. Overstocking, flash flood hazards and famines are major problems which must be overcome by modernization of traditional agriculture in these ecological crisis areas if the basic requirements of the growing populations are to be met.

One of the most important ecological effects of agriculture is to be seen in the application of modern technology which may ignore the characteristics of ecosystems and lead to serious environmental damage, through the application of fertilizers and chemical weed control, mechanization, irrigation and new, high-yielding plant varieties.

In some parts of Africa, the use of insecticides and especially of chlorinated hydrocarbons have had serious effects on the fauna of the rivers and lakes. It has been noted that even the use of fertilizers may pose an ecological threat. The use of nitrate fertilizers may in the end lead to the poisoning of water. The runoff of fertilizers into water bodies and their subsequent eutrophication are among the potential dangers of modern agriculture which must be monitored and checked in different ecosystems.

Other forms of modern agriculture such as mechanization also have adverse side effects, for instance on the structure of the soil, while irrigation has been known to bring about problems of salinization and waterlogging. Attempts to use new seed varieties have introduced new diseases.

Some of these problems require improvement in management techniques. The use of chemical control methods may be supplemented by biological methods of control. There is need for a search for other methods including crop diversification and improved land-tenure practices.

Among the most important ecological modifications of the environment is river-basin development. Africa has some of the world's largest rivers, which have increasingly been affected by the creation of dam sites. Many of these technological developments have been carried out without a prior study of the ecological side effects. The Kamburu Dam ecological study is currently being conducted when the dam has already filled up.

The range of environmental problems include the proliferation of waterborne diseases, sedimentation of the reservoirs, the spread of aquatic weeds, diminishing fish resources, inundation of extensive agricultural land and displacement of human populations and wildlife. In a number of cases such as Kariba, the Aswan Dam and the Volta Dam there are indications of micro-climatic changes.

The tropical grasslands and savannah domains of Africa constitute some of the most extensive ecosystems supporting man on the globe. They are used not only for agriculture but for grazing or as sources of fuel, recreation, tourism and support wildlife. Their development poses a number of ecological problems, some similar to those in agriculture.

Most of these areas suffer from periodic water shortages. The recent sub-Saharan drought is a catastrophic reminder of the plight of millions of people and livestock occupying these areas which are threatened by overstocking and, in some cases, rural over-population. Among the major constraints to a fuller use of these ecosystems are the inadequacy of knowledge and general neglect. The importance of research as a basis for integrated development of these and other areas has been fully recognized in the Unesco programme on Man and the Biosphere.

Closely related to the ecological problems of these regions are those of the desert and semi-desert biomes. African deserts and semi-deserts belong to the hot climatic regions where moisture deficit and in particular excessive loss and lack of moisture are major constraints. The ecosystems include man-made deserts resulting from excessive grazing, soil erosion and fires.

From the development point of view, these regions are also characterized by general neglect, lack of development infrastructure such as roads and communication and sparsity of population. The raising of the low-carrying capacity requires controlled grazing, effective water-management techniques and improvement of the plant cover.

In addition to the tropical deserts and semi-deserts, the tropical forest and Mediterranean woodland biomes of Africa constitute some of the most important resource areas. These include forests which lack a pronounced dry season and those that have developed under conditions of seasonally dry climates. The forests include a wide range of ecosystems which in most cases are little known.

Traditionally the greater part of the tropical rain-forests of Africa have had low population densities. However, with increasing population there are pressures for a more intensive use of the forests. In marginal areas where these forests carry heavy population densities their conservation now constitutes an important policy objective in the face of demand for agricultural land. In general these forests constitute an important source of the world timber supply. There is need for new management techniques to ensure that the more valuable resources are not replaced by economically unproductive ones.

The Mediterranean woodland biomes of the southern tip of Africa and the Atlas lands of North Africa have been well described in the botanical literature. They are mostly hilly areas and mountains devoted to certain forms of horticulture and grazing, or unutilized because of the nature of their topography. Those that have been used for grazing have often suffered from serious environmental degradation, including erosion and land slippage. The long period of settlement and impacts of different civilizations have left these biomes severely degraded. The development challenge in these regions is one of reversing the degradation through the transformation of vegetation, the control of the effects of fire and greater emphasis on conservation measures.

The aquatic ecosystems of the continent include the highly fertile marshes and swamps and the inland lakes as well as the estuarine and nearshore areas. These aquatic ecosystems derive their importance from their use for water supply, irrigation, waste disposal, production of fish and wildlife and

In much of Africa, nature has defied man's attempts to settle and civilize the land. There are many mountains that disappear upward into drenching, uninhabitable cloud forest, and some . . . that soar above 16,500 feet, perpetually capped by snow and ice. The dark expanse of the equatorial forest . . . covers nearly a million square miles; its rain-leached soil offers a poor living to man. In the Sahara's broad ocean of trackless, burning dunes . . . nothing lives at all.
BASIL DAVIDSON

recreational facilities. In some parts, especially in East Africa, the aquatic ecosystems shared by a number of nations are threatened by industrial pollution and eutrophication. There is need for information on these ecosystems and integrated planning for their proper management, including international co-operation. Intoxication and other forms of contamination are environmental costs which must be prevented for the limited water resources of the rural areas.

Increasing tempo of development in the rural areas is invariably associated with increasing density of the rural transport infrastructure. Rural road construction has a disturbing effect on the natural ecosystem. Care is needed in the development of a transport infrastructure integrated with the forms of land usage, including potential recreational areas.

A number of activities damaging the environment are the result of increasing rural industrialization. The list of rural industries is short but with increasing emphasis on regional development it is clear that an important share of industrial establishments will be located in what are largely rural environments. In addition there are the raw-materials-based or agro-industries which of necessity must be situated in the farming belts. As Africa passes through the various stages of industrialization it will be necessary to pay increasing attention to the problems of waste generation.

ENVIRONMENTAL MANAGEMENT PROBLEMS OF THE URBAN AREAS

From the ecological point of view an urban region is a dynamic and open-ended system, which must be regarded as a component of the wider rural ecosystem. It transcends the boundaries of the natural ecosystems linking them together in a complex network of inter-relations and interaction. But a city as a dynamic entity has a character of its own.

The development of a city involves the interaction of both population size and its economic viability. The development objective of a city is that it must continue to provide for employment and increasing income, health, housing and educational needs. The preservation of environmental quality therefore becomes synonymous with social welfare. Thus, as a country urbanizes, both the economic and social indicators of development rise.[6]

However, the dynamic growth of a given city must at some point reach a threshold where hidden costs begin to emerge. It is at this point that environmental deterioration sets in. The nature of environmental costs differs at various stages and depends on the urban share of development resources. But a critical point always exists. In analysing the environmental problems of urban systems it is necessary to distinguish the effects of the rate at which urban regions are growing and the problems of growth within urban regions. The rate of growth of the regions is primarily determined by the impact of external forces including the national and international economic systems. We have already referred to the effect of rural-urban influx which results from the spatial imbalance of economic development and the rate of population growth. However, another category of problems of environmental degradation is largely the result of the internal forces of competing demands of urban transportation, housing, industrial production and recreation.

At the national level, the corridors of transportation linking the various urban centres are already experiencing problems of congestion. This problem is more evident in the case of motor transportation. Although a rising proportion of resources are being invested in permanent motorways, the congestion of national and the few international highways will become more acute as inter-African trade grows.

Many countries, hampered by limited resources and accelerating population growth, cannot find the necessary resources for expanding the national network of transportation. A number have begun to examine alternative forms of transportation. However, transportation as a problem of the urban centre is more important within the region. Many of the urban centres of Africa have inherited internal transportation networks that are seriously at variance with the flow pattern of population. In some towns the historic segregation of residential areas means that the majority of workers living in low-cost housing areas are located along extended lines of communication linking housing estates and places of work.

On the other hand the extended nature of the modern city in Africa leads to an increasing reliance on motor transportation, whether public or private. Increasing congestion on the highways and the rising rate of accidents are some of the hidden costs of urbanization. Motor transport is thus an aspect of environmental crisis the solution of which can only be found in integrated planning.

An important impact of urbanization in Africa arises from the demand for housing facilities. Frequently the location of work places takes no account of the pattern of residence. Inadequate housing in African cities is one of the fundamental causes of social disruption. In 1969 approximately 10 per cent of the total population of Dakar were living in slums and uncontrolled settlements. In 1967, Dar es Salaam had an estimated 34 per cent of its total population in slums and uncontrolled settlements. Lusaka recorded some 27 per cent of the population living in similar parts of the city.[7] In 1971, it was reported that approximately one-third of Nairobi's population lived in unauthorized housing and that a much larger proportion lived in public housing which was critically overcrowded and inadequately serviced.[8] Uganda has similarly reported the problems of poor housing in Kampala.[9] In 1971 at least 17,000 people were living in the "cardboard villages" or shanty towns of Khartoum North. One of the most seriously affected cities was Port Sudan where it was reported that almost half the population was living in squatter settlements.[10]

Allied to the problem of housing is that of waste disposal. Whilst a large proportion of the total urban population is outside the water-borne sewage and regular refuse-removal services, the sewage-disposal systems of African cities are grossly overloaded by demands of a rapidly rising urban population. Many of the local authorities are unable to raise the necessary financial and other resources for investment in these essential services.

Moreover, the negative aspects of accelerated urbanization and the concentration of the process in a few unplanned cities have social and psychological dimensions. Accelerated urbanization is perpetuating the ethnic heterogeneity of the African city. In many parts the rural–urban drift is transferring persons from familiar social environments to the impersonal environments of the large city. The problems of human adjustment to new environments and to the unfamiliar and less-integrated cultural settings are issues that will face planners for decades to come.

INTEGRATION OF ENVIRONMENTAL AND DEVELOPMENT PLANNING

Integrated Environmental and Development Strategies in the Rural Areas

Population growth is not the main cause of the development problems. It is in this context that African nations have underlined the view that concern with the problems of population growth should not lead to a neglect of the critical issue of development. It is realized that population policies alone cannot solve problems of development. However, as part of an integrated development policy they have an important role to play.[11] The need for such a policy is best summarized as follows:

> The growth of population is not always an obstacle to development and lower population growth does not automatically cause faster rate of development. Nevertheless, there is a consensus that very rapid population growth is usually an obstacle to development. Recent economic and social trends in the developing countries suggest that rapid population growth has not prevented socio-economic progress in most cases. On the other hand, in a number of countries population growth has outspaced economic growth and in certain individual sectors progress has been very slow in comparison with population, a case in point being agricultural and food production.[12]

The majority of African countries experience a very high rate of rural population growth. The rates may differ as between different regions of the national domain. In a number of cases socio-economic development programmes are already hampered by the fact that higher commitments to these areas face serious financial constraints because of the needs. Within individual countries regional disparities in population concentration and population growth constitute major problems. In Kenya the rural parts of

Central, Nyanza and the Western provinces show regional pressures on resources that can only be expected to increase at the expense of the quality of the environment.

However, while development planning, which includes the moderation of the rate of population growth, is an important alternative in a situation where population growth coupled with inadequate institutional and social structures threaten to lead to serious crises such as food and water shortages, hunger and increased rates of morbidity and mortality, rural-population policy must be interpreted more broadly.

The basic aim of a rural development policy is the elimination of poverty, which poses the need for an integrated development policy which takes into account the rural–urban continuum. Priority must be given to a regional development concept which is resource-oriented and has a demographic component. The currently increasing emphasis on rural development by a number of African countries to promote growth poles is a vital aspect of the drive to reduce the rural poverty. This strategy may in some cases include basic reforms involving redistribution of land to the landless, to lessen rural out-migration. However, it is not the reversal of rural–urban migration itself that is the basic objective, but a better balance in settlement patterns and a more rational use of resources including the adoption of new technologies.

Within this larger development framework, the urban centres and other points of economic growth cease to be parasitic on rural resources. On the other hand, the framework provides an opportunity for resource use within the terms of eco-development to take place while maintaining the quality of the environment.

Many rural development plans have foundered because local participation has been ignored. A policy of environmental protection which is not based on participative development has a weak foundation.

Integrated Urban Development Strategy

In considering urban growth strategies, it is necessary to underline the novelty of the urban eco-system over most of the continent. Through its historical origins and internal design, the urban system presents some of the most acute environmental problems of development.

An intense debate has centred round optimal city size and the desirable national population distribution, which has remained inconclusive. Even the relationship between city size and density and environmental costs such as air pollution or traffic congestions is far from clear. However, this is not to deny that some cities of the region are growing too fast. It is also possible to argue that while the large cities enjoy economies of scale, the fruits of investment may not benefit those for whom equalization policies are intended. The gains in most cases go to landowners. The conclusion reached is that it is not possible to alter city size. However, there is the need to intervene on the problem of the rural–urban influx and to give priority to policies for redistribution within cities.[13] Lloyd Rodwin (1973) sees matters as follows:

> For the next generation (15–20) years it is probably not feasible (nor desirable) in most developing countries to stop the spread or the growth of giant cities (with populations of 500,000 or more) but it may well be feasible to lower their rate of growth in relation to the growth rates of other urban centres.[14]

Similarly, the search for an optimal national pattern of settlement with reference to environmental damage has been inconclusive. Available information does not indicate that any particular settlement pattern would be less damaging to the environment. There is need for caution and flexibility regarding the problem of primacy and optimal national settlement policy.

But this should not be interpreted to imply that aggregate size needs no attention. There is room for co-ordination of policies at the sectoral, spatial, national and regional levels with a view to influencing urbanization trends, population change and distribution.

Many of the current problems of African cities arise not from the population size but from the internal organization and structure. They relate to internal strategies for controlling land use, the relative priorities of housing as against job opportunities and financial policies designed to benefit urban dwellers as a whole. The first area involves the manner in which urban land is allocated and used. There is need for adequate control either through ownership, or regulation and zoning to meet the needs of the majority, and not of land speculators.

There is also a divergence of views as to whether priority should go to housing or to job-creation. It has been stressed that the need of most urban dwellers is for work opportunities and essential services such as transport, water supply and waste disposal. Housing tends to be awarded low priority, and it is argued that most of the low-cost housing intended for the poor have benefited the wrong socioeconomic categories. In most African urban centres, one basic constraint is the unsatisfactory financial relationship between the local authorities and the central government. In some cases the national financial policy with respect to rates (local taxes) has operated against the healthy development of urban centres.

NOTES

1 United Nations, *Demographic Trends in the World and Its Major Regions, 1950–1970.* (E/CONF. 60/BP/1, 3 May 1973.)

2 S. H. Ominde, *Urban Growth in Africa.* Commonwealth Association of Planners, Africa Regional Conference, Nairobi, 10–16 February 1974.

3 United Nations Economic Commission for Africa, *Africa Social Situation,* p. 20. (E/CN.14/POP./38, October 1971.)

4 Ominde, *Urban Growth in Africa,* p. 12.

5 United Nations, *Environmental Costs and Priorities,* p. 7. (Working Paper No. 1. 4/Rev.1, 1 July 1971.)

6 Ibid., p. 21.

7 Ibid., table IV, p. 32.

8 Kenya, *National Report on the Human Environment in Kenya,* p. 16, Nairobi, Ministry of Natural Resources, June 1971.

9 Uganda, *National Report on the Human Environment,* p. 15.

10 Sudan, *National Report of the Democratic Republic of Sudan to the UN Conference on the Human Environment,* pp. 3–4. National Council for Research, March 1971.

11 United Nations, *Population Change and Social and Economic Development,* p. 60, World Population Conference. (E/CONF.60/4, Bucharest 1974.)

12 Ibid., para. 125, p. 62.

13 United Nations, *Report of the Symposium on Population, Resources and Environment,* pp. 25–29. (E/CONF.60/CBP/3.)

14 L. Rodwin, *Population and Settlement: Location Policies Relevant to Urban Development.* (E/CONF. 60/SYM.III/10, Stockholm 1973.)

30

Tourism and Wildlife Conservation in East Africa

Anthony Netboy

Anthony Netboy, a free-lance writer and editor, has written two major works on the salmon: *The Atlantic Salmon: A Vanishing Species?* (1968) and *The Salmon: Their Fight for Survival* (1974). He is also the co-author, with Bernard Frank, of *Water, Land, and People* (1950). In the past ten years he has published more than a hundred magazine articles on fish, wildlife, parks, forests, and other natural resources.

Reprinted from *American Forests*, August 1975, pp. 24–27, by permission of the author and the American Forestry Association.

You have to get up early to see the game in the savannahs of East Africa. The animals begin to stir as the sun rises. The lions are finishing off their nightly kills while jackals and vultures wait patiently to devour the remains left behind. The noble cheetah stretches his long, lithe body and looks around for a chance to hunt. "The antelope herds," as Norman Myers notes in his fascinating book *The Long African Day*, "out on the plains where they have spent the night away from brushwood and ambush, are beginning to make their way to the rivers and water holes, [while] the bat-eared foxes sit outside their holes sunning themselves. The light sparkles on Mt. Kilimanjaro's eastern glacier. Lion cubs sport in the soft warmth. The elephants as usual are eating." The day begins.

This spectacle occurs daily in the national parks and game reserves of Kenya, Tanzania, Rhodesia, Uganda, Zaire, and other African countries. Some of these areas are very large. The Serengeti National Park in Tanzania is a fifth larger than the state of Connecticut, and the Ngorongoro Crater Conservation Area, also in Tanzania, is twice as large as Rhode Island.

The most famous wildlife areas in Kenya are Tsavo National Park, covering some 8,000 square miles, noted for its elephants, rhinos and hippos; Amboseli Game Reserve, 1260 square miles, dominated by Mt. Kilimanjaro, Africa's highest mountain which rises 19,300 feet above sea level; Masai Mara Game Reserve, 700 square miles, of which 500 square miles are set aside for the Masai nomads, a tall, handsome, cocoa-colored people who pasture cattle, sheep and goats, and shun civilization; Nairobi National Park, only 44 square miles, situated four miles outside the city, swarming with game; and Nakuru National Park, harboring about a million flamingos as well as 370 other bird species.

In addition to the Serengeti and Ngorongoro, Tanzania is noted for Lake Manyara National Park lying at the foot of the escarpment in the Great Rift Valley. Here one can see not only wildebeest, waterbuck, giraffes and zebras, but also leopards and lions (often in trees), and flocks of flamingos, white and pink-backed pelicans, saddle-billed storks and goliath herons; 5,000 square-mile Ruaha National Park on the Great Ruaha River; the Tarangire Game Reserve of 525 square miles; and a few smaller reserves.

The parks and game reserves of East and South Africa harbor the earth's last great wildlife treasures, but they contain only a small fraction of the animal populations of a century and a half ago, before the white man colonized most of Africa, slaughtering the game as he advanced, in order to clear the land for farming or cattle and sheep raising, just as he did in North America. The Africans, of course, slaughtered the wildlife too, but not on such a scale as the white colonists.

There is a difference between national parks and game reserves. Both of them, incidentally, owe their origin to the European nations who controlled nearly all of Africa until very recent times, and they took their cue from the United States which established the world's first national park, Yellowstone, in 1872, and wildlife refuges later. The first African national park was created in the Belgian Congo in the late 19th century.

In the African national parks there are no human habitations except those of park personnel and their facilities and tourist lodges; there is no farming or grazing of livestock. The land belongs to the wildlife and as Perez Olindo, director of Kenya's national parks, says, "Visitors are admitted as a privilege." In the game reserves there are core areas for wildlife and around them areas for human habitation where farming or cattle grazing is permitted but only by the indigenous people, such as the Masai, who were there before they were designated as reserves.

Hunting is permitted in special hunting blocks. While a good deal of big game shooting by wealthy sportsmen still occurs (it costs $700 to obtain a license to kill an elephant in Kenya), adverse public opinion is potent discouragement. As a matter of fact, one gets the impression that the day of the swashbuckling big-game hunters like Theodore Roosevelt and Ernest

Hemingway is about over. Many professional hunters and sportsmen have laid aside their rifles and taken to the camera.

The variety and number of animals one sees on a visit to East Africa is staggering. The spectacle that unfolds on the Serengeti plain, or on the floor of Ngorongoro Crater, is what much of the planet looked like a million years ago when human beings were scarce and the earth belonged to the quadrupeds and associated bird life. The fauna are interwoven in biotic communities where each species has its place, from crocodiles basking along the shores of a river to the great flocks of flamingos feeding in the alkaline waters of shallow lakes. Mostly big game inhabit the savannahs or open plains, tall and short grass areas interspersed with deciduous trees like acacias, euphorbia and other species and dissected by riverine forests and thickets along the watercourses. Outcroppings of rocks, called "kopjes," and termite hills up to fifteen feet high, fill out the landscape.

The plains support grasses of extraordinary vigor and reproductive capacity. Some animals, like the zebra, eat the coarse top of the grasses; others, like the shaggy wildebeest (also called gnu) and blue-flanked topi, eat the leafy center, while gazelles graze on high-protein seeds and young shoots at ground level. The 18-foot giraffes, tallest animals on earth, eat the leaves of the umbrella-like thorn trees (a species of acacia) which nature has thoughtfully provided for them. Elephants browse as well as graze, and are specially fond of twigs which they digest easily. They uproot and pull over trees, sometimes destroying forest glades, and by digging for water with their tusks, often through a sandy river bed, enable other animals like the rhinos to survive.

The most numerous species on the savannahs are the wildebeest, the dreamy-looking red hartebeest, zebras, and fleet-footed gazelles. Less numerous are buffalo, impala, topi, warthog, eland, hippos, and rhinos. Numbers fluctuate with climatic conditions. When grass is abundant populations increase, while in dry periods they decline. Many animals then die of starvation, as we saw on our visit in March, 1974. All the major predators—lion, cheetah, leopard, wild dog—inhabit the plains along with the scavengers: hyenas, jackals, serval cats, vultures, and others.

The Serengeti alone harbors some 350,000 wildebeest, 180,000 zebra, and over a half million gazelle, as well as smaller numbers of ostrich, eland, topi, hartebeest, and the predators who follow these herds. The wildebeest move en masse in the dry season to a distant region of greener pastures, a sight one can never forget: the landscape is filled with these animals as far as the eye can see, from horizon to horizon. They move slowly, but when they come to a river they may jump into it in panic, and many are drowned. The migration of the animals is nature's way of permitting the vegetation to recover, with the help of naturally-occurring fire. In contrast, domesticated animals often overgraze the land—large areas of East and South Africa are badly overgrazed. The carnivores, who prey mostly on the young and sick animals, help to keep populations in check.

At the entrance of every national park in Kenya there is a plaque with the statement of President Jomo Kenyatta:

> The natural resources of this country—its wildlife which offers such an attraction to visitors from all over the world, the beautiful places in which these animals live, the mighty forests which guard the water catchment areas so vital to the survival of man and beast—are a priceless heritage of the future. The government of Kenya, fully realizing the value of its natural resources, pledges itself to conserve them for posterity with all the means at its disposal.

In accordance with this pledge, management policies are designed to preserve the wilderness and protect the flora and fauna. The main intention is to allow visitors to enjoy the animals, the spectacular scenery, the wide open spaces and the diversity of cultures without any distractions. Kenya's

> **Perhaps some of the tremendous renewal of energy one experiences in East Africa comes from being put back in one's place in the universe, as an animal alongside other animals—one of the many miracles of life on earth, not the only miracle.**
>
> ANNE MORROW LINDBERGH

national park system and reserves cover most of the country's representative geographical zones. There are moorland and mountain forest parks, like Masai Mara; freshwater and alkaline water parks like Nakuru; bird sanctuaries, tropical forest reserves, and marine parks on the coast where during the migration season one can watch the colorful shore birds from the mudflats and mangrove swamps, see the living coral from glass-bottomed boats, or swim in the warm (but slightly polluted) waters of the Indian Ocean.

"The tourist [coming] to Africa," says Park Director Perez Olindo, "needs to be educated to appreciate and respect African values in the entire spectrum of activities. In this regard we want to make it abundantly clear that we have no intention of importing systems alien to the African environment and any attempts in that direction we feel should be vigorously discouraged."

Within their own limited means, and with considerable help from other countries and from private organizations like the World Wildlife Fund, the Ford Foundation, the East African Wildlife Society, and the African Wildlife Leadership Foundation (an American organization), some of the countries have embarked on programs of needed research, park expansion and construction of tourist facilities.

Tourism is a specialized industry. Travellers go to far-off places to see strange and unusual sights. African countries have a built-in advantage over other nations because they are the last great homes of wild animals in the world. The Pleistocene (Ice Age) did not cause the extermination of a great majority of mammal species in Africa as it did in Europe and North America. More and more people come to see these wonderful animals. In Kenya, tourism is now earning more foreign exchange than coffee, the major export. In 1974 they had over 400,000 visitors.

Governments are stimulating tourism by building roads and working in partnership with private investors to construct hotels and lodges to meet growing needs. Nairobi, the jumping off place for East African safaris, is a clean, modern city situated at 5400 feet above sea level, and while close to the equator, has one of the most equable and pleasant climates of any city in the world. Safaris in Kenya and Tanzania are as pleasant as western people could desire; modern lodges and game-viewing facilities are increasing rapidly, and service is generally first-rate.

Tourism, however, while stimulating the economy, has certain booby traps. An excess of tourists can destroy the very attractions they come to visit—witness some of our own national parks. In Africa one can already see certain undesirable impacts on the delicate habitats and ecosystems, as in Amboseli, Nairobi National Park, and even in the vast area of the Serengeti. Visitors are accustomed to driving all over the terrain in minibuses and private vehicles in search of the less abundant species which they are most desirous of seeing and photographing—lions, cheetah, leopard, elephants, rhinos, and hippos.

The tendency to congregate around some captivating animal has the effect of temporarily urbanizing the wilderness, as when a dozen minibuses surround a nervous cheetah or a lion sitting in a tree. Recently steps have been taken in some parks in Kenya to forbid vehicles from leaving the roads and tracks and roaming all over the landscape. And it is even contemplated to limit the number of cars and buses that will be allowed to enter a park at one time. Roads, incidentally, are mostly unpaved and are laid out so as not to disrupt habitats and ecosystems.

The maintenance of wildlife on such a scale as in the dedicated areas of Africa merely for the pleasure of sightseers is always subject to controversy. There is constant pressure to reduce the size of the parks in order to give the natives more land for farming and grazing their livestock, and from time to time governments succumb to these demands. The people must be shown, as the politicians are aware, that protection of wild animals and their habitats, including the cropping of surplus numbers, are to the advantage of the nation as a whole. They bring foreign tourists and foreign exchange, and also provide thousands of jobs in the parks themselves and in hotels, lodges, and other service industries.

One of the biggest threats to the parks is widespread poaching, an endemic practice hard to root out. The African Wildlife Foundation reported in 1973 that elephant poaching had reached such alarming proportions that it was feared the species might become extinct in the near future. At least a thousand elephants, and possibly more, were being slaughtered illegally every month, as the trade in ivory had increased rapidly. The tusks of one small animal could bring as much as $2,000. On September 1, 1973, Kenya banned all elephant hunting and dealing in ivory; Zaire (formerly the Congo) and Tanzania had banned it earlier.

Unlicensed hunters kill hundreds of thousands of animals every year for their meat or skins, especially leopard, zebra (a ring that was exporting 25,000 zebra skins annually through Kenya was recently broken up), baboons, and other species. In many cases, as cattle herds expand, they usurp wildlife ranges and big game is squeezed into national park enclaves. When they wander outside park boundaries they may be shot by licensed hunters or killed by poachers.

In a recent television interview with Bill Moyers, Maurice Strong, director of the United Nations Environment Program, headquartered in Nairobi, asserted that "The rest of the world should help support the wildlife of East Africa." One can readily agree with him, for they belong to mankind in general rather than to any nation or group of nations. Already greatly reduced, they must be protected from further depredations. They give joy to all those fortunate enough to behold them in their natural setting, and are among the world's most lasting and priceless heritages.

OCEANIA

31

Strategy for Ecodevelopment of an Island Community: A Case Study of Nissan Island, Bougainville, New Guinea

George Chan and Balwant Singh Saini

George Chan, a professor of engineering in New Guinea, and Balwant Singh Saini, a professor of architecture in Australia, were asked by the government of Papua New Guinea to prepare a plan for the future of Nissan Island. This article is drawn from their report.

Reprinted from *Ekistics,* 40 (239) (October 1975), pp. 232-40, by permission of the authors and the publisher. Published by the Athens Center of Ekistics, Athens, Greece.

Following the trends in most parts of the developing world, there are two major factors which are known to have made the maximum impact on the physical environment of the islands in Southeast Asia and the Pacific. They are the population expansion and the crisis caused by urbanization. No doubt, there are other factors which are equally important, but these two seem to be the most significant.

They have followed in the wake of changes which have taken place in the old colonial picture. The emergence of nationalism has created considerable impatience in many quarters to make these islands self-reliant by rapidly building their economies. An obvious and important result of this trend is evident in the development of mineral resources and (to an increasing degree) tourism, both of which have focused our attention on the need for conservation of the environment with all its ecological complications.

Population figures in the islands may look impressive against those provided by larger countries in Asia and America, but when their size and geographic limits are considered, it is not difficult to see how fast they are approaching a crisis. Most exhibit a high rate of growth with annual average expansion of over 2.6 percent, a rate 0.6 percent higher than the world average.

The rising populations are beginning to strain the islands' existing resources resulting in a continual drift away from the stagnant rural economies to a few urban centers. One answer to this problem may lie in not only strengthening and improving the economies in the existing rural settlements, but also deliberately creating or enlarging a selected group of rural service centers somewhat on the lines of those established in remote areas of Israel. Essentially, rural areas of the islands require an eco-development program which integrates plantation and agricultural production, local institutions, housing, and other sectors into all-round development projects. Such projects need economic and physical planning, as well as the conservation of natural resources.

PAPUA NEW GUINEA'S PROGRAM

Papua New Guinea is an emerging country with most of its population in remote rural and island territories (fig. 1). [It achieved] independence in 1975, [and] the basic objectives of the Papua New Guinea government closely follow those outlined by UNEP/UNCTAD in October last year in Mexico and set out in a document known as the Cocoyoc Declaration on the subject of self-reliance.

It states:

> It [self-reliance] implies mutual benefits from trade and cooperation and a fairer redistribution of resources satisfying the basic needs. It does mean self-confidence, reliance primarily on one's own resources, human and natural, and the capacity for autonomous goal setting and decision making. It excludes dependence on outside influences and power that can be converted into political pressure. It excludes exploitative trade patterns depriving countries of their natural resources for their own development. There is obviously a scope for transfer of technology, but the thrust should be on adaptation and the generation of local technology. It implies decentralization of the world economy, and sometimes also of the national economy to enhance the sense of personal participation. But is also implies increased

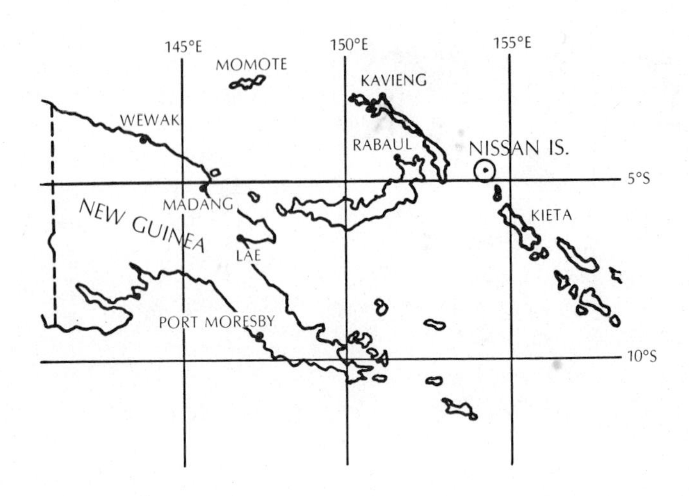

FIGURE 1 *Papua New Guinea*

international cooperation for collective self-reliance. Above all, it means trust in people and nations, reliance on the capacity of people themselves to invent and generate new resources and techniques, to increase their capacity to absorb them, to put them to socially beneficial use, to take a measure of command over the economy, and to generate their own way of life.

Emphasizing the need for a new approach to development styles, the Declaration calls for:

> Imaginative research into alternative consumption patterns, technological styles, land use strategies as well as the institutional framework and the educational requirements to sustain them. Resource-absorbing and waste-creating overconsumption should be restrained while production of essentials for the poorest sections of the population is stepped up. Low waste and clean technologies should replace the environmentally disruptive ones. More harmonious networks of human settlements could be evolved to avoid further congestion of metropolitan areas and marginalization of the countryside. In many developing countries the new development styles could imply a much more rational use of the available labor-force to implement programs aimed at the conservation of natural resources, enhancement of

environment, creation of the necessary infrastructure and services to grow more food as well as the strengthening of domestic industrial capacity to turn out commodities satisfying basic needs.

The government of Papua New Guinea drew its own set of principles in an Eight-Point Improvement Plan which is based on equality, self-reliance and rural development. Essentially the plan aims at increasing the control of the economy by Papuans and New Guineans, and achieving more equality in the distribution of economic benefits and services throughout the nation. It means revitalization and improvement of the rural areas where 90 percent of the people live and work. The main emphasis is to develop a self-reliant economy, less dependent for its needs on imported goods and services and better able to meet the needs of its people through local production.

Ten thousand islands lie scattered over the face of Oceania, ranging from tiny atoll islets barely visible above pounding surf to continental Australia, three million square miles large. Every conceivable kind of physical setting is to be found. Almost within sight of the snow fields which cap New Guinea's central mountains are sweltering equatorial swamps. And the traveler need not voyage from Australia's desert to rain-soaked Hawaii to compare climatic extremes: he can find nearly as great contrast on single islands.

DOUGLAS L. OLIVER

ECO-PLAN FOR NISSAN ISLAND

We applied these principles in drawing up a plan for Nissan Island.

Nissan, which was completely overrun first by the Japanese and then by the US Defense Forces during the Second World War, is still fairly isolated from the rest of the world. It is approximately 30 square miles in area. Most of the four thousand people in its dozen odd villages are primarily dependent upon coconuts, taro, sweet potatoes, and bananas as well as fish, oysters, prawns, and crayfish.

Nissan not only epitomizes the problems of dangerous decades ahead for people of the Pacific region but it has indeed relevant implications for the whole world. In a miniature and simple form, this island highlights a global problem faced by all of mankind with populations increasing at an enormous speed and with limited resources likely to dictate the extent to which we can grow.

Our plan is based on the concept of the Integrated System of Farming (IFS) developed by George Chan during the last few years. It involves utilization of sun, sea, sand, sewerage, sludge, and sanitation for water supply, waste disposal, and water pollution control. There is also a proposal to produce gas for cooking, growth of algae for animal feed, fish culture, vegetable gardens, and the use of sun to produce low-cost energy for lighting, cooking, hot water, and even refrigeration (fig. 2).

The essential approach is to recycle everything from waste to garbage using the sun as the only external source of energy to produce animal feed, fertilizer and food at very little cost, with the least amount of effort, and without polluting the area, soil, or sea.

Water Supply

Fresh water is limited to rain water and this can only be collected during the rainy season. Since both roof catchment and storage tanks are usually inadequate, contain impurities such as rat droppings, dead insects, leaves, and dust, and can be a breeding place for mosquitos, it is important to consider alternative sources.

Groundwater above mean sea level is practically fresh, although the salinity increases with depth. Shallow wells with waterproof bottoms and infiltration galleries are practical. Water can be distilled from the sea by means of a solar still consisting of little more than a plastic-lined shallow basin with glass covers, using the sun as the source of energy.

Water Disposal

There is no such thing as waste, only people who do not know how to make use of human and animal excreta, food remains, and garbage in a scientific and sanitary manner. When these are left to rot they are not only "wastes" but also nuisances and health hazards.

An integrated system isolates human and animal excreta for reuse. Latrines and animal pens are flushed into a digester. The digester allows retention of the wastes for twenty-four hours under quiescent, anaerobic conditions. About 69 percent of the suspended solids will settle, and decomposition of the organic matter by bacteria will reduce the volume and change its characteristics considerably.

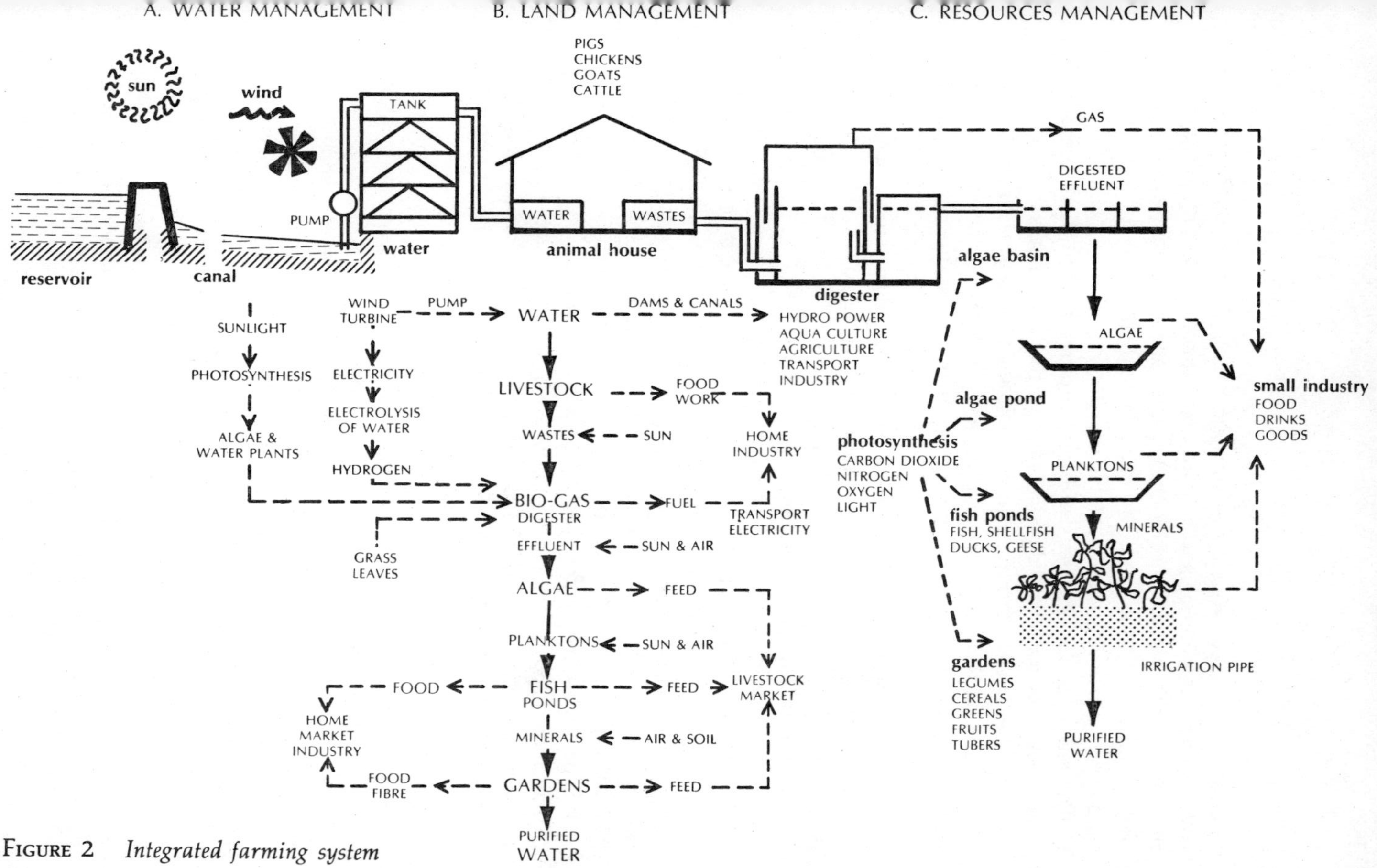

A. WATER MANAGEMENT
B. LAND MANAGEMENT
C. RESOURCES MANAGEMENT

sun
wind
reservoir
canal
PUMP
TANK
water
PIGS
CHICKENS
GOATS
CATTLE
WATER
WASTES
animal house
digester
GAS
DIGESTED EFFLUENT
algae basin
ALGAE
algae pond
PLANKTONS
fish ponds
FISH, SHELLFISH
DUCKS, GEESE
MINERALS
IRRIGATION PIPE
gardens
LEGUMES
CEREALS
GREENS
FRUITS
TUBERS
PURIFIED WATER
small industry
FOOD
DRINKS
GOODS
photosynthesis
CARBON DIOXIDE
NITROGEN
OXYGEN
LIGHT

SUNLIGHT
PHOTOSYNTHESIS
ALGAE & WATER PLANTS
WIND TURBINE
PUMP
ELECTRICITY
ELECTROLYSIS OF WATER
HYDROGEN
GRASS LEAVES
WATER
DAMS & CANALS
HYDRO POWER
AQUA CULTURE
AGRICULTURE
TRANSPORT
INDUSTRY
LIVESTOCK
FOOD WORK
HOME INDUSTRY
WASTES
SUN
BIO-GAS DIGESTER
FUEL
TRANSPORT ELECTRICITY
EFFLUENT
SUN & AIR
ALGAE
FEED
PLANKTONS
SUN & AIR
FISH PONDS
FEED
LIVESTOCK MARKET
FOOD
HOME MARKET INDUSTRY
MINERALS
AIR & SOIL
GARDENS
FEED
FOOD FIBRE
PURIFIED WATER

FIGURE 2 *Integrated farming system*

Gases are formed (60 to 70 percent methane), and can be stored if a gas cover with a water and oil seal is fitted to the digester. Painting the gas cover black allows the maximum solar energy to be absorbed and heats the anaerobic liquid inside to activate the digestion. At the same time, pathogenic organisms are for the most part destroyed.

The only way the digester can function properly is to discharge a fresh load of waste into it every day. This means that the scheme can only work if pig and/or chicken farming is done to provide the wastes which, together with natural bacteria, natural plant life, and the elements in the air (oxygen, nitrogen, and carbon dioxide) are the raw materials for the system. The pig pen and chicken house should have concrete floors and plastic drains respectively, so that the wastes can be easily washed into the digester. The objection that can be raised by the traditional farmer to fencing the animals is that he will have to find or buy feed for them. This feed is produced at little or no cost in the same scheme.

As with any other domestic gas, methane should be treated with due respect. When it burns, it combines with oxygen to form carbon dioxide which should be allowed to escape through open windows. For intermittent use of the gas, as in cooking and boiling water, it can be used directly. But for continuous use, as in refrigeration or even lighting, it is better to use the gas to run an electric generator outside the building.

It is perhaps worth noting that the digester gas, with as high as 30 percent carbon dioxide, is safer than ordinary town gas which contains 9 percent carbon monoxide. But we should always remember that air containing between 5 and 15 percent methane is explosive; so there should be no naked gas light when any repair is being made to the digester.

A digester of 300 gallon capacity (costing about US $300) is adequate for twenty-five to thirty pigs and the same number of chickens, and willl produce enough fuel to meet the needs of an average family. One of 50 gallon capacity, costing about $60 to $70, can cater for five to ten pigs and produce gas for cooking purposes only. The advantage of this is that the year-round production of gas will end the need to collect hard-to-find firewood or to import expensive kerosene.

Since we intend to use the end products of our treatment system for other useful purposes, it is worth having an extension to the digester to retain the effluent for another twenty-four hours in a tank or pipe. More settling will take place, with further digestion by anaerobic bacteria and destruction of pathogens (bacteria, helminths, viruses). A better effluent is also obtained for subsequent purification by oxygen. This is important, because the lower number of germs and the longer retention period increase the efficiency of germ removal.

Animal Feed

The main objection from the islanders to fencing their animals is that they have to find or buy the feed, so they prefer to let their pigs and chickens roam and forage freely. As a result the animals are underfed and produce less meat and fewer eggs.

The effluent from the digester is used to grow animal feed. Algal growth is encouraged by discharging the effluent from the settling tank into long, shallow channels, each formed in a V-shape with a depth of not more than 3 feet to allow sunlight to penetrate down to the bottom, and lined with puddle clay, or in rectangular shape about 12 inches deep built in concrete.

The algae, particularly the blue-green type, is a good source of protein and vitamins that can be used to enrich animal feed. As a result, the animals will grow and breed faster, thus increasing the meat supply.

It is worth pointing out that the algae is *not* part of the original waste, but is a simple natural plant than can change carbon dioxide from the air or water into food in the presence of sunlight by a biochemical process known as photosynthesis. So this type of feeding is totally different from that practiced in some places where pigs feed on raw human excreta or even their own; or from what is done on a large scale in highly developed countries where animal excrement and chemicals such as urea are mixed with animal feed to supplement the protein content.

The oxygen produced during the fixing of carbon dioxide by the algae not only purifies the effluent by converting the organic matter into minerals, but it also kills the pathogens that have come out of the digester and settling tank unharmed.

Fish Culture

The effluent from the algae ponds contains nutrients that can be used in fish culture, particularly for tilapia and carp that grow and reproduce well. This fish is another source of minerals and another source of protein for animal feed. Provided that the pigs are not fed with fish two weeks before they are slaughtered, the pork will not have a fishy taste that may be objectionable to consumers.

Such fish would be perfectly safe for human consumption as they feed on algae and protozoa, not excreta. In any case, any enteric microorganisms will long since have been killed by the digestion and oxidation processes. Fish from the effluent ponds of this nature would be safer than several species available in the markets, which feed on raw sewage from sea outfalls in many parts of the world, or which are toxic.

It has been found that ducks kept enclosed within the fish pond will not compete for food with the fish, but will eat what the fish do not want, thus helping to keep the pond clean. Between them they also keep mosquito larvae and weeds under control.

Shellfish can be collected from the shore when they are small and placed in plastic tubes of 4 to 5 inches in diameter, and the purified effluent from the algae or fish pond circulated through them by means of a small pump, thus supplying algae and oxygen to the shellfish. If a short length of tube is replaced by some transparent material, it will be possible to watch the shellfish grow.

If the same effluent is circulated in tanks where seed oysters are placed, the oysters will grow fat and succulent on the algae. Perhaps it should be pointed out that there is far less fear of contamination in this sytem than exists in the present growing of oysters and other shellfish in open sea beds, with the increasing use of the sea as a sewer for domestic and industrial wastes.

Vegetable Garden

The overflow from the fish ponds contains the end products of the decomposition of the organic matter in the sewage, and these are minerals that are suitable for use as fertilizer. The stabilized sludge at the bottom of the digester is pumped out periodically and dried in a shallow bed in the ground to kill any pathogen present, before it is used as humus to improve the physical fertility of the soil. Without the use of chemicals, except for some trace elements for certain cultures, the purified effluent of the fish ponds can make vegetable gardening a very profitable proposition.

As with the fish and duck combination, it has been found that geese enclosed in a taro patch will eat the unwelcome weeds but leave the taro alone. In this case, the geese still have to be fed other food, but they do the back-breaking work of weeding and leave the farmers free for other more pleasant or profitable occupations.

The vegetable will provide the bulk of the animal feed required in this project, but where it can be sold at a high price for human consumption, it is preferable to buy the cheaper animal feed, thus providing additional income for the rural family.

Application of such a program requires the active collaboration of architects and planners, public health specialists, engineers, agriculturalists, and others; but once the program is set up it can easily be maintained.

SETTLEMENTS

Apart from proposals for improvement of land, air, and sea transport facilities and upgrading of utility services generally, the planning of Nissan essentially concentrates on an integrated program of rural development based on the integrated farming system and the establishment of a town center as a focus of urban activities for the island as a whole.

Nissan has considerable existing resources. Its 30 square miles of land have a number of well-developed plantations and are studded with coconut palms. A variety of agricultural crops is already grown and its miles of coastline make sea products accessible.

There is a good road system threading through the island and there is considerable possibility for improving the existing jetties and establishing and developing improved harbor facilities. The existing airstrip could be improved and better terminal facilities could be provided. A number of missions have already established educational institutions and a vocational school could be set up for those who wish to develop trade and other skills (fig. 3).

Agriculture forms perhaps the most important basis for eco-development of the people of Nissan. The most economic farm unit recommended is a cluster of eight families (or multiples thereof) each of which could be serviced with farm land of approximately 2 to 2½ acres. The villages are generally located either close to the beach or in areas where there is room for expansion in all directions.

In cases where existing villages have more than eight families, it is possible to locate farms in an extended form with those at long distances made accessible by paths acting as rights of way.

Each farm unit has been suggested to service a family of six (two parents and four children). It is to include a 112 square foot piggery for ten sows, a 150 square foot digester, an algae basin and pond of 1,200 square feet, as well as land for a fish pond and vegetable garden. Pigs, chickens, and ducks

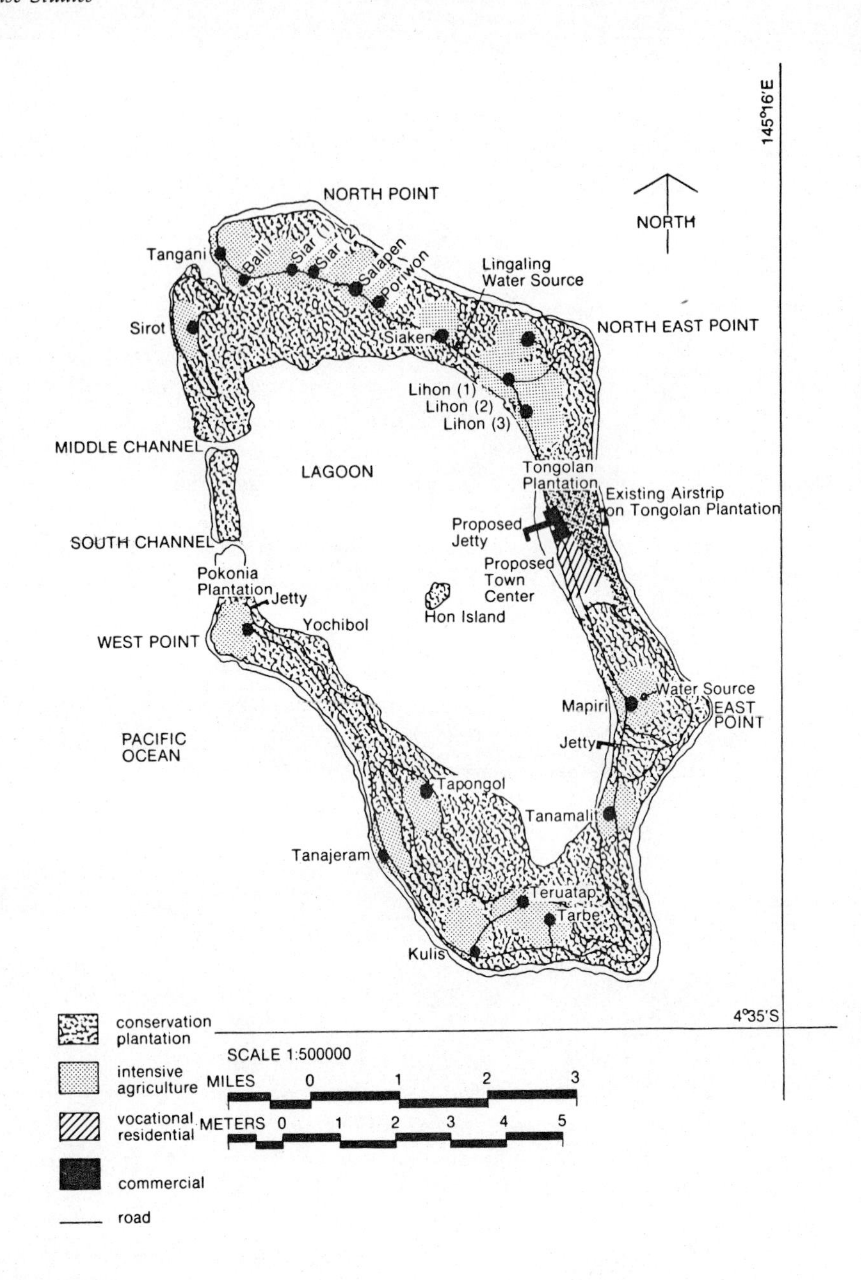

FIGURE 3 *Master plan for Nissan*

should not be allowed to roam freely but should be fenced in and given proper nutrition and husbandry. Chickens for both meat and eggs should have proper cages; and ducks should have suitable ponds.

The extensive coastline facing the lagoon and the Pacific Ocean would be a reserve belonging to the whole community. Land exceeding 25 percent (that is, dependent upon heavy vegetation to control erosion), foreshore, and mangrove forests, and land unsuitable for development such as swamps would not be used. In addition, there are areas of special recreational, traditional, and scenic value, all of which must be carefully preserved for future generations.

Housing

The people of Nissan largely live in villages located alongside the existing road network that circles the entire island. They have cleared ground amidst coconut groves and mixed growth of forest and coconut palms. There is abundant land for village residential development within the island's 30 square mile area.

Our plan recommends no change in the traditional sanctioned pattern of village residential development. Inspection of existing village houses indicated to us the remarkable ability of the people of Nissan to use their initiative for exploiting local materials and manpower, all of which has been achieved by self-help and cooperative methods. In our view, it is important that these qualities are retained and, in fact, active steps should be taken to encourage them.

In many cases, people have used scrap metal and other materials left by the Japanese and US forces during the Second World War as lining for the walls and roofs of their houses. A number of local Catholic missions have experimented with the use of stabilized coronous blocks made from crushing local coral and cement. Although the layout and designs of these buildings leave much to be desired, their durability and comparative maintenance-free qualities suggest their increased use in local construction.

Town Center

An island such as Nissan with a population of some four thousand, which is likely to increase in the foreseeable future, deserves a carefully devised program for a town center. Our plan recommends allocation of land for this purpose close to the airport and the harbor.

Apart from offering the island administrative, institutional, and commercial services, this center could also act as an exit point for island products to other parts of New Guinea. A regular farmer's market, cooperative store, and a variety of small shops could be built there. The whole development will not only meet the needs of the residents but also those who decide to live near the town center. It will complement the commercial activity of small trade stores close to the villages in other parts of the island. The services of a post office and a bank could also be made available.

As a long range project, a complex of village Government Council offices and a community center in the form of community hall and associated facilities have been recommended for building on a prominent site in the town center.

CONCLUSION

The eco-development plan endeavors to give direction to the growth of Nissan. With this growth it will be necessary to encourage individual enthusiasm to improve and maintain the general appearance of the island by designing all buildings in keeping with its rural character. We have strongly recommended that no buildings on the island should be constructed higher than the nearest palm tree. We must preserve Nissan's attractive beaches from encroaching development and prevent them from becoming littered.

Most of all, a small island such as Nissan must at all cost try and keep itself free from the onslaught of tourism and preserve itself mainly for the local people.

The Nissan Plan is only one of many attempts to help make people of Papua New Guinea self-reliant. Its success will lie in the way it is implemented. Knowing the qualities of present leadership whose own roots are firmly planted in the villages of their country, we have every confidence in the success of these proposals.

ANTARCTICA

32 Conservation in the Antarctic

M. W. Holdgate

M. W. Holdgate is director-general of research in the Department of the Environment and of Transport, London. He has been director of the Central Unit of Environmental Pollution in Britain, director of the Institute of Terrestrial Ecology, and deputy director for research for the Nature Conservancy, and during 1963-66 was senior biologist with the British Antarctic Survey.

Abridged from *Antarctic Ecology,* Volume 2, edited by M. W. Holdgate (published for the Scientific Committee on Antarctic Research by Academic Press, London, New York, San Francisco). Reprinted by permission of the author and the publisher.

THE OBJECTIVES OF CONSERVATION IN THE ANTARCTIC

The objectives of conservation in the Antarctic regions have never been set out clearly, but it is evident from the records of the Antarctic Treaty Consultative Meetings (Anderson, 1968; Carrick, 1964), the SCAR Working Group on Biology, the International Whaling Commission and other bodies that the following have been generally recognized:

1. The general protection of the scenic beauty of the Antarctic regions south of 60°S latitude, and of their birds, mammals, and terrestrial and freshwater life.

2. The protection of the remaining undisturbed ecosystems of the Subantarctic and temperate oceanic islands north of 60°S latitude, and as far as possible the restoration or stabilization of those island ecosystems that have been disrupted by man and by the alien species he has imported (Holdgate, 1968).

3. The wise management of the biological resources of the southern ocean, so that a protein crop can be taken without irreversible damage to the ecosystem or the undue depletion of populations of the cropped species (Carrick, 1964; Murphy, 1964; Gulland, 1970; this symposium*).

No policy has been spelled out for the use of the resources of the Antarctic land, largely because although major mineral deposits occur there, including the world's largest coalfield (Adie, 1970; this symposium), their exploitation must involve great political and economic difficulties. An explicit, positive policy for the planned development of the recreational use of Antarctica is also lacking, although tourist visits to the region are increasing.

THE PRESENT STATE OF ANTARCTIC ECOSYSTEMS

Land and Fresh Water South of 60°S

Except in areas formerly exploited by sealers and in the immediate vicinity of present stations and sites of intensive research, the scenic and biotic resources of Antarctica remain in a substantially natural condition. The continent is the only great tract of land on earth of which this can be said. None the less, man has had, and is still having some effect on the ecosystem.

In the past, sealers overcropped the breeding stocks of fur seals (*Arctocephalus tropicalis gazella*) in the South Shetland, South Orkney and South Sandwich Islands so drastically that an initial population of about a million animals was virtually eliminated by 1830. Elephant seal (*Mirounga leonina*) were also taken, and penguin populations probably suffered through the killing of birds for food if not for oil. Roberts (1958) lists the known Antarctic sealing expeditions and makes the magnitude of this impact evident. Although these depredations ceased long ago fur seals are only now beginning to recolonise their former range (Aguayo, 1970: this symposium; Bonner, 1968). More recently, cropping of seals for dog food may have led to local reductions in Weddell Seal (*Leptonychotes weddelli*) populations, even

---

**Editor's note* "This symposium" refers to both volumes of *Antarctic Ecology* (see permissions note above).

though the totals taken are trivial as a proportion of the Antarctic stock as a whole.

Today, the most obvious and publicised threats to the Antarctic land ecosystem arise from the construction of stations, with their associated fuel storage, power generating, aircraft landing and ship docking facilities. These must involve local disruption of the ecosystem, and because many stations are placed on rocky sites in coastal areas where the biota is richest, may have a relatively more serious impact than might be expected from their small area. Such stations also contaminate the ecosystem locally with waste, disposal of which is difficult in a region where burial is generally impossible, and decomposition slow: at most stations refuse and sewage are dumped into the sea with noticeable small-scale effects.

These local influences are, of course, negligible when considered in the total context of Antarctica. A more significant threat is posed by the movement and activities of men (especially scientists) in the field. Vehicular, and even pedestrian travel can break up vegetation and desert pavements and

On nineteen days out of twenty [Antarctica] is a terrain as frozen and desolate as the land God gave to Cain. Yet it can be transformed within minutes to an Eden of ethereal beauty. Quite suddenly the wind will drop, the sky will clear, the light will strengthen until mountains 300 miles away can be seen by the naked eye, and the ice will glow with colours so brilliant and be encompassed by a stillness so absolute that they have to be experienced to be believed.
IAN CAMERON

cause erosion. Sampling of soil and moss carpets for botanical and zoological studies leaves gaps in the ground cover which can also be widened by erosion. Two factors combine to aggravate this danger. First, Antarctic vegetation even in the Maritime zone grows slowly and scars take years to heal. Secondly, although the Antarctic regions are of vast extent, only a tiny proportion of the land surface is snow free and supports vegetation or soil and this comes under a disproportionate pressure both from scientists and travellers. In the neighbourhood of stations there is a real danger that trampling and sampling will ultimately come to hamper future research.

Disturbance of bird and seal colonies is another localized threat. Helicopters are known to have caused diminished breeding success among penguins in the "show" colony at Cape Royds (Stonehouse, 1970: this symposium) and the relative disturbance due to flights at different altitudes is described by Sladen (1970: this symposium). In areas where helicopters and aircraft are used intensively, there is a distinct risk that the validity of data on bird breeding success will be reduced.

Land Areas North of 60°S

The impact of men on the oceanic islands of the southern circumpolar belt has been reviewed by Holdgate and Wace (1961), Dorst and Milon (1964) and Holdgate (1968) and is summarized in Table 1 (based on the latter paper). These islands are particularly vulnerable to human impact because these ecosystems have been developed within the isolation of an ocean barrier, and are species-poor, lacking many groups and species of plant and animal that are dominant in continental situations. Mammalian predators and herbivores are generally absent: when they are imported, the island vegetation is often drastically changed by grazing and their vast seabird breeding colonies cannot withstand the impact of such animals as rats or cats. Disturbance of the vegetation by burning, grazing or construction aids the spread of alien plants (especially ruderal weeds) which are less able to invade closed communities (Wace, 1968). Imported invertebrates often spread readily into niches that have no native occupant, especially in areas where the native vegetation is also disturbed and alien plants are spreading.

Like the Antarctic islands, those of the southern temperate zone were the scene of intensive sealing from about 1780 to 1830 and again between 1860 and 1880. At South Georgia about 1,200,000 fur seals were killed in the former period. Archipel de Kerguelen, Macquarie Island, Iles Crozet, Prince Edward Islands, Gough Island, Tristan da Cunha, Beauchêne Island, the Cape Horn archipelago and the New Zealand shelf islands all supported considerable stocks and were visited many times. On many of them elephant seal were also killed for blubber, as were King penguins at Macquarie Island. Male elephant seal continued to be cropped on a sustained yield basis at South Georgia until 1964/65 and an industry continues at Archipel de Kerguelen. . . .

MARINE ECOSYSTEMS

The Antarctic oceanic ecosystem has been disturbed by the removal from it of a high proportion of its original baleen whale population (Zenkovich,

TABLE 1 *Human settlement, feral mammals and degree of damage on the Islands of the Circum-Antarctic Seas*

Islands	*Zone*	*Human population*	*Feral vertebrates*	*Condition*
South Sandwich Islands (eleven islands)	Maritime Antarctic	Formerly sealers, for short periods. Small scientific station for a few summers on one island. A few scientific parties for short periods	None	Undamaged
Bouvetoya	Maritime Antarctic	None (weather station planned)	None	Undamaged
South Georgia (with outliers)	Subantarctic	Sealers from 1796 and many whalers from about 1910–65. Now only small scientific station	Reindeer (two herds) Rats	Local damage around stations and where reindeer grazing is heavy. Outlying islands undamaged
Marion Island	Subantarctic	Sealers formerly active. Now only weather station	Cats Formerly a few sheep	Vegetation undamaged, but cats damaging avifauna
Prince Edward Island	Subantarctic	None	None	Undamaged
Iles Crozet	Subantarctic	Sealers formerly active. Weather station since 1963	Pigs on Ile aux Cochons, 1820–60, now extinct. Rabbits, goats said to have been present but now also extinct. Rats (?)	Vegetation substantially undamaged, especially probably on Ile de l'Est. Surveys required
Archipel de Kerguelen	Subantarctic	Sealers formerly active (1800–30). Weather station since 1949, with own farm	Rabbits (abundant), sheep, reindeer, pigs (local), cats, rats	Vegetation of main island damaged by rabbits. Local damage by sheep. Avifauna on main island damaged by cats and rats. Outlying small islands undamaged

Table 1 continued

Islands	*Zone*	*Human population*	*Feral vertebrates*	*Condition*
Heard Island	Subantarctic	Sealers formerly. Scientific station 1947–54	None	Undamaged
Macquarie Island	Subantarctic	Sealers active 1820–80. Scientific station since 1945	Rabbits, imported in 1880, now abundant. A few sheep. Formerly a few goats and horses now extinct. Cats, rats, and predatory ground-living bird *Gallirallus* numerous	Widespread severe damage to vegetation by rabbits, and to smaller birds by predators
Beauchêne Island	Temperate	Sealers formerly.	None	Undamaged
Falkland Islands	Temperate	Settled since eighteenth century. Now used for sheep ranching	Sheep farmed over much of land. Cattle locally feral in interior. Guanaco, feral on one island. Cats, rats, mice widespread	Vegetation substantially altered by grazing except on smaller off-lying islands. Bird fauna probably substantially affected but remains rich
Gough Island, Inaccessible Island, Nightingale Island	Temperate	Sealers formerly. Weather station on Gough Island since 1955. Visits by Tristan Islanders to other two islands since 1810	Mice (on Gough Island)	Substantially undamaged
Tristan da Cunha	Temperate	Small settlement since 1810	Goats formerly (now extinct). Domestic stock maintained on lower grazings: cattle and sheep semi-wild in some areas. Cats, rats, mice widespread	Vegetation altered in heavily grazed areas, and bird fauna reduced by predators and man

Ile St Paul	Temperate	Fishing base, 1843–1914, and intermittently since. No present inhabitants	Rabbits abundant until 1957: later rare or ? absent. Cats, rats, mice	Rabbit grazing caused former disturbance of vegetation: condition now uncertain. Bird fauna probably reduced by predation
Ile Amsterdam	Temperate	Fishery base, 1843–53: farmed, 1971. Weather station since 1950	Cattle (numerous), formerly goats, sheep (now extinct). Cats, rats, mice	Native vegetation greatly altered by grazing and avifauna severely reduced by predation
Campbell Island	Temperate	Sheep farming, 1890–1927. Whaling station, 1908–14. Weather station since 1941	Sheep (now greatly reduced). Some cattle, formerly goats, pigs, guinea fowl, game birds. Cats, rats present	Grazing formerly altered vegetation. Management now aiming at removing sheep and aiding recovery. Populations of smaller sea birds reportedly reduced by predators
Auckland Islands	Temperate	Settled by Maoris, 1841. Whaling station 1849–52. Evacuated 1856: subsequent spasmodic farming. Weather station 1941–45	Pigs, goats and cats on main island. Rabbits, cattle on Rose and Enderby Islands. None on Adams Island	Adams Island undamaged. Vegetation and fauna elsewhere variously affected
Snares Islands, Bounty Islands, Antipodes Islands	Temperate	Small summer field station on Snares Islands. None elsewhere		Substantially undamaged

1970: this symposium; Mackintosh, 1970: this symposium). It is not clear how much this has been followed by increases in the populations of penguins, seals or fishes, all of which feed on the krill once taken by the whales. Sladen (1964) has suggested an increase in the chinstrap penguin, *Pygoscelis antarctica,* correlated with the reduction in whale numbers, but the data are insufficient to establish the position for bird populations as a whole (Holdgate, 1967). *Euphausia superba* has a central place in so many food chains that it seems improbable that a "surplus" of it, created by the elimination of one of its many consumers, will remain uncropped for long, and if this is so, the whaling industry must have set in motion a complex readjustment of the Antarctic marine ecosystem as a whole. The killing of fur seal, elephant seal and penguins on some of the southern islands must have caused similar local changes in balance. There is nothing unusual in this, for all the oceans of the world have been affected by exploitation and their present "equilibrium" (if there is one) is different from that preceding the development of fisheries, and must be adjusting continually to the increasing skill of man as a predator.

Pollution is the other major disturbance caused by man to oceanic ecosystems. In Antarctic waters it is unquestionably trivial compared with northern seas. Yet organochlorine pesticide residues have now been found in penguins at McMurdo and in euphausiids, penguins and seals at Signy Island. . . .

CONSERVATION MEASURES IN ANTARCTIC AND SUBANTARCTIC LAND AREAS

The Antarctic Treaty (H.M.S.O., 1965), which came into force on 23 June 1961, is in conservationist's terms a Management Plan. It lays down certain rules which govern the management of the resources of the Antarctic south of 60°S latitude, for example prohibiting warlike activities and contamination with radioactive waste and opening the whole area for purposes of peaceful scientific exploration. In this, it assumes that the most useful "crops" available from the continent today are scientific knowledge and international harmony, and establishes a framework for their attainment. Moreover, like other enlightened conservation organizations, it provides for the periodic review of the management plan at regular consultative meetings of all the governments concerned.

Wildlife Conservation on the Antarctic Land

Wildlife conservation in the Antarctic has been provided for under the Antarctic Treaty by detailed conservation measures, termed the Agreed Measures for the Conservation of Antarctic Fauna and Flora. These are based on scientific advice provided by SCAR through its Working Group on Biology, so that SCAR, in this respect, acts as scientific adviser to the Governments signatory to the Treaty and participating in the Consultative Meetings. The Agreed Measures apply in the same area as the Antarctic Treaty itself, namely the entire area south of 60°S latitude, but because states rights on the high seas are reserved, fishing is not affected and it is likely that nations also retain the right to take whales, seals and birds at sea. Consequently the Agreed Measure has full force only on land and on shelf ice, which is explicitly included with the land in the text. The legislative principles involved in the Agreed Measures and the ways they are being implemented by different countries have been reviewed by Anderson (1968) and will not be considered further here. . . .

The Agreed Measures provide for:

(*a*) overall protection from killing, wounding, capture or molestation for all native mammals (except whales), and native birds in the Antarctic, at all stages in the life cycle (including eggs);

(*b*) lifting of this protection by Governments, who may issue permits to allow selected individuals to kill or capture birds and mammals:

   (i) to provide indispensable food for men and dogs, in the Treaty Area, in limited quantities;

   (ii) to provide specimens for study, for Zoological gardens, for museums and for similar scientific, educational and cultural purposes.

Information on the numbers of each species taken under permit is required and these data are to be exchanged between the participating Governments. It is laid down that the numbers of a species taken under permit should not ordinarily exceed that which the population is capable of making good during the next breeding season, and that activities likely to disturb the balance of natural ecological systems should not be permitted. . . .

Besides these general measures, there are provisions for:

(i) The special protection of species which can be shown to require this because of their rarity, vulnerability or some other good reason. At present this special protection applies to the Ross seal, *Ommatophoca rossi,* and all fur seals (genus *Arctocephalus*). Specially protected species may be killed only for a compelling scientific purpose or in an emergency.

(ii) The special protection of areas. This provision is intended to safeguard outstandingly interesting or unique samples of vegetation, soil or habitat. Within these areas, plant collection and the collection of birds and mammals is permitted only for compelling scientific purposes that cannot be served elsewhere, or in an emergency. There is also a prohibition on driving vehicles across the areas. . . . .

(iii) Reduction to a minimum of harmful interference with living conditions of the Antarctic fauna, for example by preventing dogs running free, or low flying helicopters over penguin colonies. The same Article of the Agreed Measures requires reasonable steps to be taken to minimize pollution of inshore waters, e.g. by the pumping of ships bilges.

(iv) Prohibition of the import into the Antarctic of non-indigenous species of animals and plants, except under permit. Only laboratory animals and plants, domestic livestock (excluding poultry) and sledge dogs may be imported. Poultry are prohibited because of the danger of introducing bird diseases that might spread to the native avifauna. Dogs must be inoculated against specified diseases two months before import. All introduced animals and plants have to be kept under controlled conditions and eventually removed or destroyed.

Future Needs for Conservation on the Antarctic Land

The Agreed Measures provide so comprehensive a frame-work for Antarctic wildlife conservation that they are most unlikely to require significant amendment. They are indeed more complete than any other measures applying to a large region of the world. Future action is required largely in three fields: perfection of their application, especially to specially protected species and areas, development of a positive scheme for management, especially of specially protected areas, and adoption of "educational" means to make all personnel visiting the Antarctic, whether scientists or tourists, aware of the need for conservation.

Specially Protected Species Ross seals and fur seals have been given special protection under the Agreed Measures. All present evidence points to the Ross Seal being more numerous than previously supposed (Ray, 1970; this symposium). Its habits, so far as these are known, rarely if ever bring it within the area of application of the Agreed Measures, since it lives in the pack ice and rarely hauls out on Antarctic land or shelf ice. For this latter reason, its inclusion as a Specially Protected Species under the Agreed Measures is not altogether easy to justify, especially now that there is substantial agreement on the regulation of pelagic sealing in the Antarctic, under which the species receives protection in its main habitat.

Conversely, fur seals breed on land and are vulnerable when doing so. They yield a valuable product and are therefore attractive to sealers. They are rare in the Antarctic zone where they are establishing small nuclei of repopulation (Aguayo, 1970; this symposium). Given special protection now, they may be expected to re-establish themselves securely in their former range.

Three criteria have generally been used by the Working Group on Biology of SCAR in bringing forward cases for special protection. These are: (*a*) rarity, either in the world or in the Antarctic zone, (*b*) vulnerability, (*c*) capacity to benefit from special protection. . . .

What is apparent . . . is that before a species is proposed for special protection, a careful assessment of its numbers, biology, and vulnerability is required, and that the schedule will need periodic revision as new data become available. Such ecological research as a basis for conservation was urged by Carrick (1964) and remains highly relevant.

Specially Protected Areas The Article of the Agreed Measures governing the special protection of areas establishes two criteria, outstanding scientific interest and uniqueness (rarity) of the ecological systems, represented there. These criteria can be further refined:

(*a*) the series of specially protected areas should include representative samples of the major Antarctic land and freshwater ecological systems, and of the variations they display in relation to edaphic, climatic, or geographic variables;

(*b*) areas with unique complexes of species should receive special consideration, as should any areas which are the type, or only known, habitats for plant or invertebrate species, or contain outstandingly interesting breeding colonies of birds and mammals;

(*c*) areas which have been the scene of intensive scientific study and thus provide a baseline for long-term investigations should be eligible for special protection.

Criteria (*b*) and (*c*) are clearly compatible with those laid down in the text of the Agreed Measures. For criterion (*a*) to be acceptable, the sites proposed must have a certain "quality": they must be outstandingly good examples of the ecosystems they represent. The criteria are interrelated, for areas that satisfy (*a*) and (*b*) are likely also to be good research sites.

If this argument is valid, there are two logical conclusions. First, the series of specially protected areas envisaged under criterion (*a*) can only be chosen after careful field surveys. Secondly, a classification of the land and freshwater ecosystems of the Antarctic is required, against which surveys can be judged. At present, such a classification is available for vegetation in the Maritime Antarctic (Gimingham and Smith, 1970: this symposium), and to some extent for soils (Ugolini, 1970: this symposium; Allen and Heal, 1970: this symposium). The breeding ranges of bird and mammal species are also well enough known for the significance of particular colonies to be judged. But there is no adequate classification of invertebrate habitats or of lakes.

It is not surprising that the series of fifteen Specially Protected Areas so far established is not comprehensive. It contains four areas selected primarily for botanical purposes, seven areas selected primarily to protect breeding colonies of birds and mammals, and four areas selected as samples of ecological systems as a whole [see Table 2].

However, the size of most of [the breeding] colonies is quite unknown and we have no baseline data such as are desirable if these areas are to serve in any way as a long-term control.

Representative samples of most maritime Antarctic vegetation types are probably included. . . . No proper botanical or invertebrate surveys have been done, nor has any attempt been made to protect a series of representative freshwater bodies. No area of "desert pavement" soil, with its associated sparse biota, as in the Victoria Land dry valleys, features in the list. None of the intensively studied lakes on Signy Island or in Victoria Land receives special protection.

This problem is not confined to Antarctica. Throughout the world, nature reserves (which is what the specially protected areas really are) have been selected subjectively, because of known interest, and a more objective approach seeking to protect representative series of sites has followed. During the International Biological Programme such an objective survey is being mounted by the CT section on a world wide basis (Nicholson, 1968; Peterken, 1968) and it would obviously be appropriate to use a modification of the IBP Check Sheet and classification in a careful survey of possible Antarctic specially protected areas.

Management Management is an essential component of conservation, although its intensity varies. It does not always imply interference with the ecosystem, but does invariably mean that a positive plan is drawn up for conservation in a region or country as a whole, and for each reserve or special area established there.

In the Antarctic, the need for continued scientific consideration of conservation policies has been recognized by the Governments who have welcomed SCAR's interest and encouraged its continuation. It remains for SCAR to consider its scientific objectives and to establish a framework for their attainment. . . .

Inevitably, these tasks will require international planning but must be implemented at national level. They cannot be undertaken at once over the whole Antarctic. All can, however, be started in selected areas, and . . . monitoring . . . , involving repeated censuses or surveys of key breeding colonies or sites, will have scientific interest in its own right as well as value in conservation. Returns of species killed or taken under permit should be analysed in relation to these long-term studies of species abundance and distribution.

For each specially protected area, it is desirable that a management plan is drawn up, preferably under the supervision of a member of the SCAR Working Group on Biology. . . . .

Such plans should be kept up to date and copies should be exchanged between all SCAR nations. When activities take place in the area under permit a record of them should be kept as an annex to the plan. This record should include the location of collections, the numbers of each species taken, and any structural alterations caused to the habitat.

These proposals may seem needlessly formal, complex and restrictive under present conditions in Antarctica. None the less, experience elsewhere

TABLE 2 *Specially protected areas and the reasons for their designation.*

Area	*Latitude*	*Longitude*	*Reason for establishment*
Taylor Rookery, Mac. Robertson Land	67°26′S	60°50′E	Protection of largest known colony of Emperor Penguin (*Aptenodytes forsters*) breeding on land
Rookery Islands, Holme Bay	67°37′S	62°33′E	Protection of breeding colonies and habitat of all six bird species resident in Mawson area, of which Giant Petrel (*Macronectes giganteus*) and Cape pigeon (*Daption capensis*) breed nowhere else in region
Ardeny Island and Odbert Island, Budd Coast	66°22′S	110°25′E	Protection of breeding colonies and habitat of Antarctic petrel (*Thalassoica antarctica*) and Antarctic fulmar (*Fulmarus glacialoides*), together with other bird species
Sabrina Island, Balleny Islands	66°54′S	163°20′E	Protection of biologically richest sample of Balleny Islands group, including only known breeding site of chinstrap penguin (*Pygoscelsis antarctica,*) in Ross Sea sector
Beaufort Island, Ross Sea	76°58′S	167°03′E	Substantial and varied avifauna, representative of coastal Ross Sea area
Cape Crozier, Ross Island	77°32′S	169°19′E	Rich bird and mammal fauna, microfauna and microflora, with habitat of considerable interest. Includes emperor penguin and Adélie penguin (*Pygoscelsis adeliae*) colonies, skuas (*Catharacta skua maccormicki*), and Weddell seal (*Leptonychotes weddelli*)
Cape Hallett, Victoria Land	72°18′S	170°19′E	Ecosystem of outstanding interest, including large Adélie penguin colony, skuas, and small area of unusually rich and diverse vegetation dominated by bryophytes, and supporting a variety of terrestrial invertebrates
Dion Islands, Marguerite Bay	67°52′S	68°43′W	Protection of the only known breeding colony of emperor penguins on the west side of the Antarctic peninsula

Area	*Latitude*	*Longitude*	*Reason for establishment*
Green Island, Berthelot Islands	65°19′S	64°10′W	Protection of an exceptionally luxuriant bryophyte vegetation locally overlying up to 2 m of peat, with its associated terrestrial fauna
Byers Peninsula, Livingston Island	62°38′S	61°05′W	Diverse plant and animal life, with vegetation representative of western South Shetland Islands. Substantial population of elephant seal (*Mirounga leonina*) and small breeding populations of fur seal (*Arctocephalus tropicalis gazella*)
Cape Shirreff, Livingston Island	62°28′S	60°45′W	Diverse plant and animal life, substantial numbers of elephant seals and small colonies of fur seals
Fildes Peninsula, King George Island	62°12′S	58°58′W	Protection of a sample of a large, biologically diverse area with lakes that are ice free in summer. As amended in 1968, this Specially Protected Area includes only one large lake with the surrounding land
Moe Island, South Orkney Islands	60°45′S	54°41′W	Diverse vegetation, with samples of most of the main communities of the Maritime Antarctic, also convenient as a control in case intensive research alters the ecosystems on the adjacent Signy Island. Substantial bird population, including Adélie and chinstrap penguins
Lynch Island, South Orkney Islands	60°40′S	45°38′W	Protection of one of the most extensive areas of grass (*Deschampsia antarctica*) known in the Treaty Area, with associated soil and invertebrate fauna
Southern Powell Island and adjacent islands, South Orkney Islands	60°45′S	45°02′W	Protection of a range of Maritime Antarctic vegetation types, a considerable bird and mammal fauna including breeding Weddell seal, large numbers of elephant seal in summer, and a small colony of fur seal. A representative sample of the South Orkney Island ecosystem

confirms the need to compile an accurate dossier of information about strictly reserved areas and of the changes which take place there. In the long term, when human impact on the Antarctic land will undoubtedly increase and when the cumulative disturbance of decades of scientific activity must become a local hindrance to research, these procedures will not appear so irrelevant.

Information From the earliest stages, those concerned with the drafting of Antarctic conservation measures recognized the need to secure the willing support of all personnel (especially non-scientists) working in the region. Now that Antarctic tourism is justifiably expanding, the need for information about Antarctic wildlife, its interest, its vulnerability, and the conservation measures drafted for its protection is even more acute. . . .

Conservation on the Southern Islands

The land areas north of 60°S and within the area of interest of SCAR are not subject to the same universal conservation measures, and some indeed have no conservation legislation at all, despite their high scientific interest. Those listed as "substantially undamaged" in Table 1, extending from the young Antarctic volcanoes of the South Sandwich arc to the temperate islands of the Tristan da Cunha group and the New Zealand shelf form a series of high scientific interest, and the three smaller members of the Tristan group are probably the least modified group of temperate oceanic islands in the world. Little or no economic crop is likely from these islands, apart from fishery in some inshore waters and the harvesting of the surplus production of male fur seals once stocks have risen to an appropriate level. Conversely, the potential return to scientific knowledge is very great. It may be assumed that for these land areas, as for the land south of 60°S, conservation should be directed largely to the protection of this scientific interest.

The conservation measures required for such islands are not unduly complex. They include:

(*a*) prohibition of the killing, capturing or molestation of native birds and mammals except under permit for scientific and analogous purposes, at a level which will not cause serious disturbance of the populations, and regulation by permit of the collecting of other animals and plants;

(*b*) prohibition of the importation of all alien mammals, because of the vulnerability of island vegetation to grazing and bird populations to predation by mammals, both of which are absent from the natural situation;

(*c*) prohibition of the importation of alien plants and invertebrates, as far as this is possible;

(*d*) prohibition of any avoidable disturbance of the island habitat, because alien plants, with their associated invertebrates, have been shown to spread chiefly in places where the native vegetation has been disrupted by grazing, burning or trampling. Such alien species invade undisturbed vegetation much less readily.

Of these measures, (*b*) is the most important. Native bird and mammal populations can withstand a substantial level of culling by man, but are rapidly and irreversibly reduced by predation by rats, cats, or dogs. Island vegetation, likewise, is well able to withstand a normal level of scientific collection, but liable to irreversible disruption, even with consequent erosion of soil, following grazing by rabbits, sheep or cattle.

It is evident that the general form of the Antarctic Agreed Measures could, with slight modification, likewise be applied to the uninhabited southern temperate islands, and South Africa has done this for Prince Edward and Marion Islands. It is desirable that nations exercising sovereignty over the other islands consider a similar step. The problem is, of course, a global one, and in this context it is significant that the CT section of the IBP is sponsoring a world-wide survey of oceanic islands of scientific interest, in which SCAR is being asked to participate. . . .

Management plans should be drawn up for these islands, as for Antarctic Specially Protected Areas. Such plans will be more complex than for the latter. On many oceanic islands there are weather stations that must be resupplied and whose members need access to the island as a whole and facilities for scientific work. On some islands also, feral mammals have already become established. While these populations have real scientific interest, it is not easy to justify the disruption of a unique oceanic island ecosystem in order to study the behaviour of feral domestic stock, and where possible consideration must be given to their removal. New Zealand biologists are already eliminating sheep from Campbell Island: the wild cattle of Ile Amsterdam and the reindeer of South Georgia are likewise targets for control, if not elimination. All such proposals should, of course, follow a biological survey and assessment of the trends the island ecosystem is following, and the desirability and possibility of modifying them.

THE CONSERVATION OF THE SOUTHERN SEAS

Unlike the islands and mainland of Antarctica, the southern oceans contain valuable natural resources which have been exploited by man for a long period. The whaling industry, after several decades of overcropping and consequent reduction of the resource, has now fixed harvesting levels below the sustained yield figure, and a slow recovery of the stocks is predicted (Mackintosh, 1970: this symposium; Gulland, 1970: this symposium). Pelagic sealing has only recently started in Antarctic waters and guide lines for its regulation have been drafted in advance. Fish and krill have not yet been exploited commercially, but Moiseev (1970, this symposium) computed that the latter may have an available yield of about 45×10^6 metric tons/yr which would double world fishery landings, and many of the technical problems for such a fishery are being overcome.

Guide Lines for the Voluntary Regulation of Antarctic Pelagic Sealing

SCAR . . . has so far been active only in one field of marine conservation, the preparation of guide lines for the conservation of seal stocks and the regulation of Antarctic pelagic sealing. . . .

In advising on conservation measures, the SCAR Working Group on Biology was hampered by a lack of reliable data on the numbers and distribution of these species, and of the age structure of their populations. The interim guide lines proposed in 1966 and extended in 1968 emphasize this and provide for adjustment of sealing levels as knowledge improves. The main principles of the guide lines are (Polar Record, 1967):

1. The Antarctic seal stocks are a resource of potential value, which should be used in a rational way.
2. Harvesting of seal stocks should be held at or below the maximum sustainable yield.
3. Sealing should be regulated at a level at which the natural ecological ecosystems are not seriously disturbed.
4. To this end, the following precise conservation measures should be adopted:

   (*a*) figures should be set, in an Annex to the guide lines, which represent the best estimates of maximum sustainable yield for the time being;

(*b*) the Antarctic should be divided into a series of zones (the same as those used by the whaling industry) and these should be closed to sealing in rotation;

(*c*) certain areas where seal populations are the subject of scientific study should be designated as seal reserves, and commercial sealing excluded;

(*d*) a sealing season and a "close season" should be established, and for the Weddell seal the latter should protect the species while it is breeding in inshore waters, where local populations could readily be killed out by sealers;

(*e*) certain species (Ross and fur seals) should be protected at all times;

(*f*) animals should not be killed by sealers when in the water (because of the high rate of loss);

(*g*) governments should keep the exchange records of the numbers of adult males, pregnant females and pups of each species killed, and should also encourage biological research on seals, so as to improve the data on which quotas and sealing zones are based;

(*h*) if at any time the total harvest of seals in the area south of 60°S approaches the maximum sustained yield, or is disturbing the ecological system in any area, the Consultative Parties should consult together to plan a meeting to discuss steps to remedy the situation.

Like the Agreed Measures, these guide lines thus provide for the special protection of some species and areas, the proper record of numbers killed, and the regulation of an activity on the basis of scientific knowledge. There is a similar problem, because the guide lines had to be prepared in advance of really adequate information, and a similar need for new research.

The Need for Further Measures

The Guide Lines for the Voluntary Regulation of Pelagic Sealing in the Antarctic go as far as is reasonable at present, when knowledge is inadequate and an industry has yet to begin. Ultimately, they may require extension in two ways: in area of application so as to cover pack ice north of 60°S and in form of expression so as to become internationally binding rather than voluntary. More fundamental, however, is the need to apply scientific knowledge in a similar manner to the wise management of the other resources of the Antarctic ocean—krill and fish—and indeed to the management of oceanic resources generally (Gulland, 1970: this symposium). Krill lies at the centre of Antarctic food chains, and there is scarcely a species of bird or seal in the region whose population would not be liable to adjustment were krill greatly reduced in abundance by cropping. Perhaps even more seriously, excessive harvesting of krill could disturb the general balance of the Antarctic marine ecosystem. A critical scientific study of this situation, and the drafting of guide lines for the management of an industry based on krill, may be the most pressing task in Antarctic conservation today.

CONCLUSION

One theme recurs through this whole field, as it does everywhere in conservation. This is the need for legislation to be a tool of management and for the management to be based on sound scientific knowledge and directed toward the attainment of set objectives. Generally speaking, conservation in the Antarctic is well advanced, and if it is to be consolidated it needs to be supported by more comprehensive research, and implemented by more precise planning. The landscape and wildlife of the Antarctic land are a resource of great aesthetic and scientific value, and the life of the ocean has great economic as well as scientific importance. The international agreements now secured demand the support of scientists, through SCAR, to ensure that policies continue to be guided by the best knowledge available.

REFERENCES

Adie, R. J. (1970). Past environments and climates of Antarctica. This symposium, 7–14.

Aguayo, L. A. (1970). Census of Pinnipedia in the South Shetland Islands. This symposium, 395–97.

Allen, S. E. and Heal, O. W. (1970). Soils in the Maritime Antarctic Region. This symposium, 693–96.

Anderson, D. (1968). The conservation of wild life under the Antarctic Treaty. *Polar Rec.* 14, No. 88, pp. 25–32.

Bonner, W. N. (1968). The fur seal of South Georgia. *Br. Antarct. Surv. Sci. Rep.*, No. 56.

Carrick, R. (1964). Problems of conservation in and around the Southern ocean. *In* "Biologie Antarctique: Antarctic Biology" (Carrick, R., Holdgate, M. W. and Prévost, J. eds). Hermann, Paris.

Dorst, J. and Milon, P. (1964). Acclimatation et conservation de la nature dans les îles subantarctiques françaises. *In* "Biologie Antarctique: Antarctic Biology" (Carrick, R., Holdgate, M. W. and Prévost, J., eds). Hermann, Paris.

Gimingham, C. H. and Smith, R. I. L. (1970). Bryophyte and lichen communities in the Maritime Antarctic. This symposium, 752–85.

Gulland, J. A. (1970). The development of the resources of the Antarctic seas. This symposium, 217–23.

H.M.S.O. (1965). The Antarctic Treaty and the Recommendations of subsequent Consultative Meetings held at Canberra, 1961, Buenos Aires, 1962 and Brussels, 1964. Miscellaneous No. 23 (1965), Cmnd 2822.

Holdgate, M. W. (1967). The Antarctic Ecosystem. *Phil. Trans. R. Soc.* B 252, 777, pp. 363–83.

Holdgate, M. W. (1968). The influence of introduced species on the ecosystems of temperate oceanic islands. *Proceedings, I.U.C.N. 10th Technical Meeting, I.U.C.N. Publications,* New Series, No. 9, pp. 151–76.

Holdgate, M. W. and Wace, N. M. (1961). The influence of man on the floras and faunas of southern islands. *Polar Rec.* 10, No. 68, pp. 475–93.

Mackintosh, N. A. (1970). Whales and Krill in the 20th century. This symposium, 195–212.

Moiseev, P. A. (1970). Some aspects of the commercial use of the Krill resources of the Antarctic seas. This symposium, 213–16.

Murphy, R. C. (1964). Conservation of the Antarctic fauna. *In* "Biologie Antarctique: Antarctic Biology" (Carrick, R., Holdgate, M. W. and Prévost, J., eds). Hermann, Paris.

Nicholson, E. M. (1968). "Handbook to the Conservation Section of the International Biological Programme." I.B.P. Handbook No. 5. Blackwell, Oxford.

Peterken, G. F. (1968). "Guide to the Check Sheet for IBP Areas." I.B.P. Handbook No. 4. Blackwell, Oxford.

Polar Record (1967). Report of fourth Antarctic Treaty Consultative Meeting, Santiago, Chile, 1966. *Polar Rec.* 13, No. 86, pp. 629–49.

Ray, C. (1970). Population ecology of Antarctic seals. This symposium, 398–414.

Roberts, B. B. (1958). A chronological list of Antarctic expeditions. *Polar Rec.* 9, No. 59, pp. 97–134 and No. 60, pp. 191–239.

Sladen, W. J. L. (1964). The distribution of the Adélie and chinstrap penguins. *In* "Biologie Antarctique: Antarctic Biology" (Carrick, R., Holdgate, M. W. and Prévost, J., eds). Hermann, Paris.

Sladen, W. J. L. and Le Resche, R. E. (1970). New and developing techniques in Antarctic ornithology. This symposium, 585–96.

Stonehouse, B. (1970). Adaptation in polar and subpolar penguins (Spheniscidae). This symposium, 527–41.

Ugolini, F. (1970). Antarctic soils and their ecology. This symposium, 673–92.

Wace, N. M. (1968). Alien plants in the Tristan da Cunha Islands. *Proceedings, I.U.C.N. 10th Technical Meeting, I.U.C.N. Publications,* New Series, No. 9.

Zenkovich, B. (1970). Whales and plankton in Antarctic waters. This symposium, 183–85.

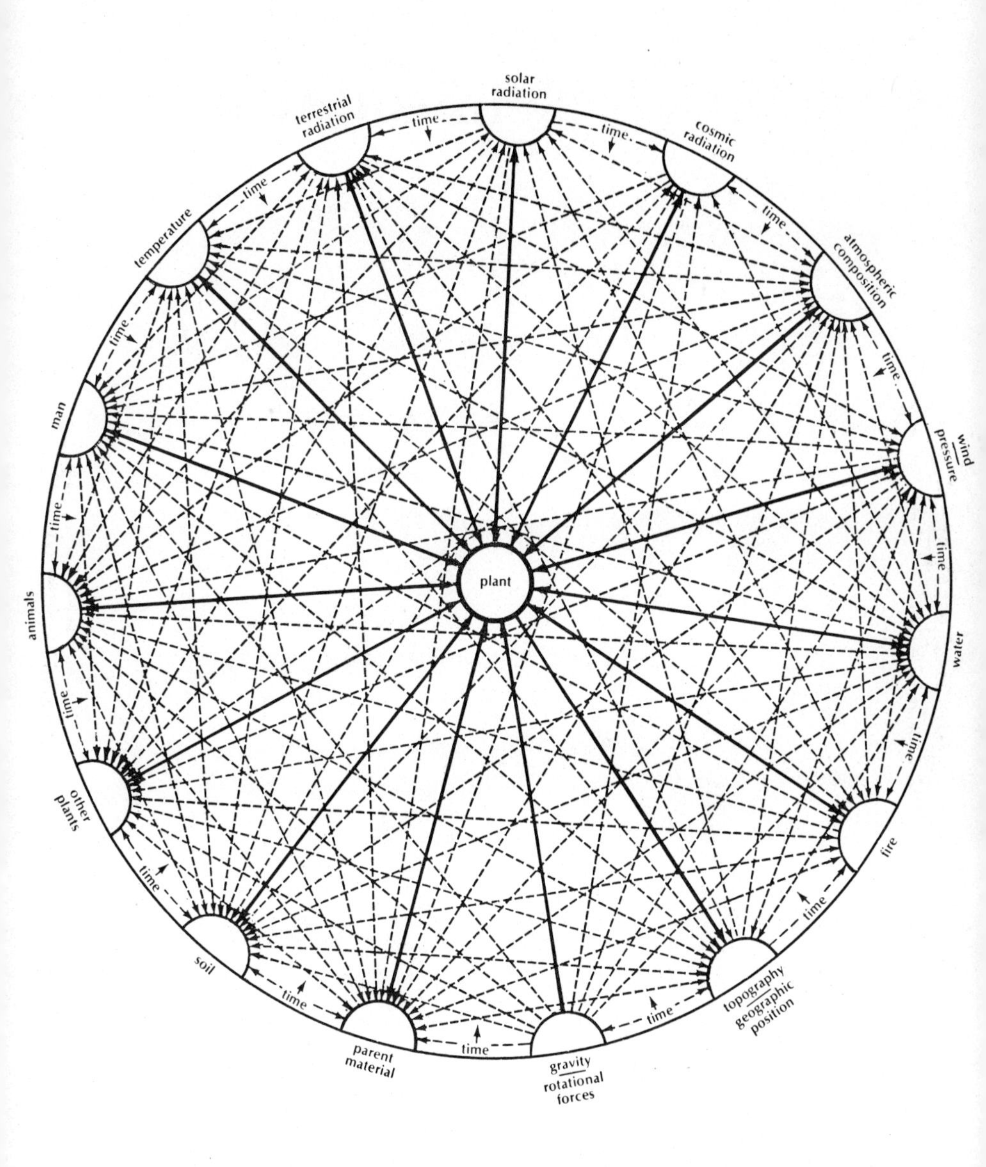

Part III The Future: Divergent Views

Introduction to

Part III

An ex-coroner once wrote an ode to his mother that included this line: "If perchance the inevitable should come. . . ." On reading the tangled predictions concerning man's survival, or even the earth's survival, it is possible to fall into the same beautifully absurd error. Is the demise of man as a species imminent? Has impatient man bedeviled a patient planet beyond endurance? Technological optimists foresee an ever greater proliferation of gadgetry, a rise in life expectancies, improved wheat strains, and the acceleration of human knowledge; Malthusians see a future that ranges from the merely appalling to the unthinkable. In the last selections in this book there is considerable gloom and a smattering of euphoria—and a confoundment of conflicting, provocative viewpoints. But they all demonstrate that neither overconfidence nor hopelessness is appropriate in pondering the future.

To Maurice Strong, the key question is not whether the planet will survive but whether it will remain fit for humans to live on. As he points out in Selection 33, our technological civilization has led to a web of interdependencies from which no nation can extricate itself and from which, ultimately, even such issues as national security, war, and peace cannot be separated. He explains why only a cooperative global approach can deal with this reality. Calling for forceful, rapid, even radical action to avert disaster, he outlines a six-point planetary policy for environmental security.[1]

In Selection 34, John Maddox resolutely takes the antitechnologists and prophets of doom to task, and dissects the U.S. environmental movement along the way. His argument is that, both for advanced and developing communities, science and technology are indispensable aids to the improvement of economic resources and to the solution of serious social problems.

Until a little more than a century ago it took millions of years for the human population to reach the 1 billion mark; today it takes only 15; and by the year 2000 it could take only nine. The simple message: people should multiply in their head before they decide to multiply in bed. Malthus may not have been wrong after all. As Garrett Hardin scoffed in 1972:

> Malthus has been buried again. (This is the 174th year in which that redoubtable economist has been interred. We may take it as certain that anyone who has to be buried 174 times cannot be wholly dead.)[2]

But Malthus was in error on one pivotal point: populations do not inevitably increase to absorb available resources. That is, at a certain level of wealth, health, literacy, and modification of religious beliefs, populations begin to stabilize at "zero growth." The catch is that under present political, economic, and social conditions, this "certain level" is unattainable for a large majority of the world's people. In Selection 35, Jonathan Spivak reports on the ominous future that is forecast for a world whose population has outstripped expectations—expectations that were alarming to begin with. The one hopeful note is that China, the most populous nation on earth, has a population-control program that appears to be working.

Are thousands upon thousands of people doomed to face endless hunger or starvation? Is food scarcity inevitable? Absolutely not, say Frances Moore Lappé and Joseph Collins in Selection 36. There would be enough food to go around if the agricultural resources of both the developing and the developed nations were used rationally and the gross inequities in world commercial and economic arrangements were acknowledged and rectified. Cogently arguing that people can feed themselves from their own resources in any country on earth, the authors first dismantle six myths about the world's food supply. They then offer six principles as the basis for an international campaign to remove the barriers that thwart people from becoming self-reliant in food production.

E. F. Schumacher, the author of Selection 37, is an unorthodox Western economist who believes that the economic system should serve man, and not the other way around; unlike most modern economists, he does not regard material growth as a primary measure of social progress. The economic system he advocates is derived largely from Buddhist economics, the keynote of which is simplicity and nonviolence and whose aim is to obtain the maximum of well-being with a minimum of consumption. In Schumacher's view, man can best meet his needs by scaling down his wants: small is thus not only beautiful but usually cheaper and more richly satisfying as well. As Schumacher indicates in Selection 37, he is not opposed to machines; he pleads merely for a redirected technology, one that is scaled to the human needs of a given community and enhances the beauty and integrity of the local environment. This "technology with a human face, . . . instead of making human hands and brains redundant," can lead to unparalleled productivity. Schumacher is not just spouting theories. The Intermediate Technology Development Group that he mentions toward the end of the selection is, in fact, an organization that he himself established. It does research, puts out a journal, designs products, and has, over the past decade, helped many developing nations devise technologies compatible with their cash-poor, labor-intensive circumstances.

Murray Bookchin is another proponent of an eco-technology that will help society live in harmony with the natural world. In Selection 38 he derides pollution-control programs that treat the symptoms rather than eliminate the causes of environmental decay. The roots of the ecological crisis, in his view, do not lie in technology and/or population growth but in a modern society that sets man against nature and man against man, a society in which man, like every aspect of nature, is turned into a commodity, and the work of eons of organic evolution is undone in the wild race to produce and consume. To prevent the ultimate ecological backlash—the early annihilation of man as a viable animal species—Bookchin calls for nothing less than revolutionary change and a utopian reconstruction of society.

Robert Heilbroner, the author of the last selection in this book—"What Has Posterity Ever Done for Me?"—may well have been inspired in the choice of his title by the statement of the chairman of the board of a major American popular magazine, who not long ago publicly asked that egocentric question. Heilbroner acknowledges that reason cannot offer a compelling argument to care desperately for posterity, but he believes that even if commitment to life's continuance entails sacrifices, they are preferable to the terrible anguish we must feel on imagining ourselves as "the executioners of mankind." It is his hope that a survivalist ethic, based on personal responsibility, will gain prominence as coming generations experience the egregious consequences of our ecological sins.

"When one reaches the edge of an abyss," writes Laurence Peter, "the only truly progressive move is first to step backward—and then change course."[3] And to change course means to change our attitudes. Granted, the struggle for power and wealth will go relentlessly on, but we cannot wait for the millennium to look at things differently and to take action. We can honor the advice of Aldo Leopold:

> Examine each question in terms of what is ethically and esthetically right, as well as what is economically expedient. A thing is right when it tends

> to preserve the integrity, stability, and beauty of the biotic community. It is wrong when it tends otherwise.[4]

As individuals we can adopt a new set of values to replace institutionalized witlessness, although this will call for degrees of sacrifice, sensitivity, and generosity rarely encountered in man's relation with his environment. Such a value system would recognize man's place in the eternal flow of things. It would have at its core the ancient, almost forgotten knowledge that all life is the manifestation of a single mystery.

The Chinese ideogram for "crisis" has a double meaning: danger and opportunity. The ecological crisis is dangerous in the extreme, but it also offers boundless opportunity for growth in different and better directions and for unprecedented domestic and international cooperation.

To a large extent, the future is in our hands. The very least each of us can do is to try to shape it in ecologically creative ways, whether for ourselves, for posterity, or even more altruistically for the beautiful planet we inhabit. In the words of novelist Antoine de Saint-Exupéry: "As for the future, your task is not to foresee, but to enable it."

NOTES

1 Thousands of organizations throughout the world are working to combat environmental degradation. They are listed in *World Directory of Environmental Organizations,* issued periodically by the Sierra Club, San Francisco, since 1973. *Conservation Directory,* a list of nationally active groups in North America, is compiled periodically by the National Wildlife Federation, Washington, D.C.; and the Onyx Group, Inc., has compiled *Environment U.S.A.: A Guide to Agencies, People, and Resources* (New York: R. R. Bowker Co., 1974). Scores of books are available as guides for individual and group ecological living. Among those with excellent down-to-earth instructions are: Robert S. de Ropp, *Eco-Tech: The Whole-earther's Guide to the Alternate Society* (New York: Delacorte Press/Seymour Lawrence, 1975); Philip Nobile and John Deedy, eds., *The Complete Ecology Fact Book* (Garden City, N.Y.: Doubleday & Co., 1972); and Garrett de Bel, ed., *The Environmental Handbook* (New York: Ballantine Books, 1970). On a more metaphysical plane is Tom Bender's *Environmental Design Primer* (New York: Schocken Books, 1973).

2 Garrett Hardin, in Garrett Hardin and R. Stephen Berry, "Limits to Growth–Two Views," *Bulletin of the Atomic Scientists,* November 1972, p. 23.

3 Laurence J. Peter, *The Peter Plan: A Proposal for Survival* (New York: William Morrow & Co., 1976), p. 167.

4 Aldo Leopold, *A Sand County Almanac,* enl. ed. (New York: Oxford University Press, 1966), p. 240.

33 A Global Imperative for the Environment

Maurice F. Strong

Maurice F. Strong is director of the United Nations Environment Program (UNEP), headquartered in Nairobi. He was secretary-general of the United Nations Conference on the Human Environment held in Stockholm in 1972.

A former United States secretary of state was widely criticized for his geopolitical "brinkmanship." Some of the current debate on the likelihood of environmental disaster suggests that the whole world is playing brinkmanship with the systems of nature in the belief that somehow, at the last moment, we will stop at the edge of the precipice. Even if this course were a good idea—which it obviously isn't—we would not be able to stop in time. There are too many facets in the interaction between man and nature, between our technological society and the biosphere. It is not, however, the single climactic global disaster in which everything, everywhere, simultaneously falls apart that we have to fear most. What we face in dealing with the environment is a series of problems that will become regional crises at different times and places if we do not act quickly enough, and that could escalate into disasters of sufficient magnitude to have political and moral implications for the whole world community if we do not act at all. The key question is thus, not whether planet Earth will survive, but whether it will remain fit for human habitation.

The human impact on the physical elements of the biosphere stems entirely from man's economic and social behavior. Our increasing awareness of physical constraints is therefore reflected first in accelerating pressures on our political and social systems, on the values and behavior patterns those systems reflect, and on the institutions through which we seek to manage our affairs.

Our institutional structures at all levels of government are already showing signs of severe strain. National governments seem to be increasingly overwhelmed by the sheer magnitude and complexity of unprecedented problems in decision making brought on by the cumulative and interacting effects of population growth, industrialization, urbanization, resource depletion, environmental degradation, and unrestrained consumerism. Qualitatively new problems of choice seem to lead to a hardening of the arteries in traditional decision-making procedures that threatens to result in institutional paralysis. There may be a blind insistence on continuing to deal with issues rendered obsolete by the course of events or, alternatively, panic may set in and lead to overreaction and the initiation of crash programs all too likely to be technically wrong or politically repugnant or both.

The United States is an example of the lack of long-range planning followed by what some might regard as near panic. In a forty-year period the country has moved from being the world's leading exporter of petroleum to its leading importer, and is beginning to reshape its attitudes and policies only in the face of the current energy crisis. The same response is apparent in the country's rapid transition from a philosophy based on an abundance

The reason why the world lacks unity and lies broken in heaps is because man is disunited with himself through failing to look at the world with new eyes.
THOREAU

of food grains, with its accompanying emphasis on exports, to a philosophy of scarcity and export restrictions. Both the energy crisis and the food crisis were eminently foreseeable—although they were clearly not foreseen by decision-makers at the highest political levels.

At the level of the world community, the international structure is no better equipped to deal with today's overriding issues. These are no longer national matters of security, ideology, the deployment of arms, or the competitive pursuit of illusory advantages for the prestige of individual states. The self-interests of all nations today are inevitably interwoven in a web of interdependencies characteristic of our present technological civilization. This reality requires a cooperative approach to the management of the interacting relationship between the development and distribution of resources, the minimal needs for sustaining decent standards of human life, and the protection of the environment on which that life depends. These interlocking concerns must be moved to the top of the world's agenda for thought and action. They are ultimately inseparable from the issues of war, peace, and national security that have preoccupied us up to now. The potential for conflict in such issues as competition for resources, lopsided deployment of technological power, pursuit by nations of activities environmentally harmful to others, and unequal pressures created by population growth is as great as the incentives they provide for cooperative measures to deal with them.

It is still possible to avoid the worst implications of the limits of our resources and the strain on our environment, but we need time to gain some measure of control over those human activities that are at odds with a viable natural system. We need time to adapt scientific and technological priorities to the imperatives of contemporary life; time to reform institutional structures and lifestyles. Unless there is strong, rapid, even radical action now, we may be headed for disasters almost too dreadful to contemplate. For if we fail to act decisively, we shall surely slip into a pattern of decisions and nondecisions, actions and nonactions, whose outcome will determine which individuals and which nations will survive and which will not. If we respond only to immediate pressures, fail to plan ahead, and compete instead of cooperate, we will break the bonds of interdependence and revert blindly to nationalism, protectionism, and parochialism. The fabric of world community, so painfully constructed, would begin to disintegrate into pockets of affluence and privilege struggling to resist the encroachment of a mounting tide of poverty and despair. Some collapse of the social order would probably take place well before the world reached the theoretical physical outer limits of its finite natural system.

In the face of potentially serious breakdowns in specific geographic areas, with tragic consequences for large numbers of people, the ad hoc responses that have characterized our reactions to these environmental emergencies in the past can no longer be tolerated. In our own self-interest we should consider the creation of disaster-prevention programs on a global scale. Just as schools have fire drills and ships have lifeboat drills, spaceship Earth needs emergency survival systems.

We must evolve a strategy for global environmental security—a planetary policy to avoid disaster and provide a greater sense of direction in human affairs. Such a policy could be built around six basic elements.

First, population must be stabilized. This can probably be achieved most rapidly in the industrialized countries, since they are already moving in that direction. The goal will be more difficult to achieve in the developing countries, where population growth is greatest and the new demands it creates on resources are most difficult to satisfy. Some of these countries are already facing the prospect of population limitation through the procrustean means of famine and disease; such forbidding prospects will undoubtedly accelerate if efforts to limit population in more humane ways prove too little or too late. Population control can only be brought about by making improved techniques of family planning universally available and by providing incentives for the reduction of family size largely through a rise in the standard of living.

In an era when national borders are closed to large-scale immigration, the population growth of each country is first and foremost a matter of its own concern. Thus, the first task of the international community should be to help all nations evaluate the trade-offs that only they can make between population growth and distribution, available resources, and desired standards of living. The population levels of individual nations become a world concern only when they begin to impinge on the resources and rights of other states. This is already happening in some instances and is clearly going to take place in others. To say that population is primarily a national responsibility does not relieve us of the need to see it also as a global concern.

Second, there must be a worldwide program to conserve scarce resources. The ethic of limitless abundance must give way to the ethics of scarcity and conservation. A rise in the cost of natural resources will provide incentives for the development of more efficient methods of using and reusing materials. Greater emphasis must be placed on the development of technologies and patterns of consumption that are less energy-intensive than those presently in use, on closed-system production methods, and on techniques for recycling. Higher resource costs may also result in differential price structures with lower prices for raw materials used to sustain life and higher prices on those employed for less essential purposes. Reserve stocks of vital resources should be maintained in quantities sufficient to meet foreseeable emergencies. This should be done under international control—as the United Nations Food and Agriculture Organization is currently proposing in respect to stocks of food.

Everybody likes to hear about a man laying down his life for his country—but nobody wants to hear about a country giving her shirt for her planet.
E. B. WHITE

Third, new models for economic and social progress should be elaborated and adopted. In the industrial world, this would involve life-styles oriented to nonmaterial satisfactions and changes in patterns of consumption from quantity to quality. It would require new criteria for decision making that would assign appropriate values to social factors, to amenities, and to the preservation of such public resources as water and air. In the developing world new directions for growth should be designed to avoid the errors and excesses made by already industrialized countries. Progress should focus on labor-intensive activities and the sound ecological development of indigenous resources, with particular emphasis given to rural areas. The need to avoid overburdening the environment in any particular region would motivate a wider distribution of industrial capacity, both within nations and internationally, than has been past practice. Environmental considerations also provide new reasons to seek less wasteful uses for natural products and a more equitable, international division of labor. These precepts, together with the growing demand for natural resources, could produce a significant shift of comparative economic advantage toward the developing countries. That should be encouraged despite the transitional problems such a readjustment would create in more industrialized parts of the world.

Fourth, there must be a much larger flow of resources between rich and poor countries, with heavy emphasis on providing basic social services to the poorest sectors. This policy should be seen, not as charity, but as a prerequisite for the basic economic and social security that is essential to the health and stability of functioning world community. Programs of direct foreign aid between individual nations must give way to more objective and automatic means of transferring resources. For example, a disproportionate share of the Special Drawing Rights created by the International Monetary Fund might be made available to developing countries. They might also be assisted by a toll on the use of the oceans and the atmosphere for transport and waste disposal. In effect, we must extend into international life the same principles of distributive justice and minimum opportunities for all that are accepted as the basis for relations between rich and poor in our national societies.

Fifth, science and technology must be mobilized on a worldwide scale to reduce our ignorance about environment, resources, and population, and to help devise new ways of improving the human condition. This effort will involve a large-scale redirection of research and development expenditures and the reorientation of the priorities of science and technology. It will involve reshaping the policies of industrial corporations, the leading practitioners of the world's technology, to assure that they are compatible with the social goals of society.

A more effective relationship must also be developed between scientists and technologists on one hand and political decision-makers on the other. Specifically, the scientific community must provide the politicians with environmental advice on risks and opportunities before they make decisions that will shape our future. The Earth Watch monitoring system now being set up as part of the United Nations Environment Program is designed to help meet this need. When fully functioning, it will provide an objective assessment of the condition of the environment and give early warning of impending risks.

Sixth, the resources of the oceans beyond national jurisdictions must be put under international control to assure that the principal beneficiaries of their exploitation are those living in the developing world. This great new source of wealth must not become the object of fierce competition among the wealthy countries having the technology to exploit it, or the basis of a new imperialism in which the wealthy gain a disproportionate share of the benefits at the expense of the poor. Even before exploitation begins, all nations should agree that the a priori consideration must be the preservation of the marine environment and its vital, life-sustaining processes. Many of these issues will be discussed at the forthcoming United Nations Conference on the Law of the Sea.

Seventy percent of the earth's surface is ocean beyond national jurisdictions. Bringing this vast expanse under the rule of law is one of the greatest challenges facing this generation. The way we meet that challenge will largely determine the future of rich–poor relations on this planet.

It is evident that the measures suggested here require new dimensions of international cooperation and an expanding role for the organizations through which this cooperation must be carried out. Much of the program could, of course, be accomplished through bilateral and regional arrangements but even these must, to an increasing extent, take place within a larger global framework. Such a "new internationalism" is not a utopian dream but an objective necessity—and one that is well within our reach. But to achieve it, the community of nations, particularly the countries of the industrialized world, will have to exercise a much greater degree of enlightened political wisdom than they have yet evidenced.

In the final analysis, nations will have to recognize that there can be no basic or enduring conflict between their national interests and the interests of the whole human community. The same compelling pressures of broader self-interest that induced man to form ever larger social and political units—first the family, then the tribe, the village, the town, the city, the city-state, and finally the nation-state—must inevitably impel mankind toward a planetary society. Loyalty to this planetary society will modify but need not negate an individual's loyalty to his nation, any more than loyalty to the nation negates loyalty to the family, the tribe, or the city.

The environment issue has helped make us aware that we do indeed have only "One Earth" and that all humans share both its dangers and its possibilities. Perhaps this awareness will help give birth to a new sense of loyalty to the planetary society. If so, environmental awareness may be the key to a new internationalism that will enable the whole human community to realize the promise offered by technological society.

34 What Can Be Done?

John Maddox

John Maddox, one of England's leading scientists, is editor of *Nature* magazine (London).

In the doomsday literature, it is common to deride the contributions which science and technology can make to the abatement of environmental problems. Dr. René Dubos, in *Reason Awake,* complains at the foolishness of relying on what he calls "technological fixes." In reality, however, to the extent that the environmental problems of which he and others write are realities, there is powerful evidence that research and development have an important contribution to make to the creation of a society which is at once humane and able to take charge of its own affairs. Nobody pretends that technology by itself could be a sufficient cure for present ills, for legal and administrative procedures, national and international, are indispensable.

The truth is, however, that in the modern world, science and technology have become essential components of the resourcefulness and ingenuity which the human race must exercise if it is to continue to survive. If the discovery of metal tools made possible the preservation of neolithic society, the development of the instruments of the new technology will help to ensure the survival of contemporary society. Until recently at least, so much has been taken for granted. One of the abiding flaws of the doomsday literature is that it ignores the contributions which science and technology are at present making to the solution of contemporary problems.

Medical research, which has admittedly made possible the abrupt reduction of death rates in developing countries since the Second World War, may yet contribute in important ways to the reduction of fertility. In the long run, reducing infant mortality in developing countries may be as important as the provision of still more convenient techniques of contraception. Is that to be scorned? And can it sensibly be held that the impending success in the treatment of presently intractable forms of cancer, and the further decrease of mortality among the middle aged in advanced societies, will be unwelcome, even if one side effect may be an increase in the size of some populations? The children who are by these means helped not to be orphans might be asked for their opinions.

To the extent that many of the social problems of developing countries are consequences of poor communications and the incarceration of too many people in agricultural communities, there are even important benefits to be won in the immediate future from the development of telephones, television and other devices in telecommunications. In advanced societies, at least, the experience of the past few years has shown quite clearly that these have important contributions to make to education and to social mobility, which in turn are among the more important prerequisites of stability, demographic and otherwise. Is it not mistaken to suppose that these developments, valued as they are in advanced societies, have no value elsewhere in the world? Is that not an assumption calculated still further to increase the sense of deprivation with which the less fortunate nations are saddled?

In the years ahead, as in the present and in antiquity, science and technology are indispensable instruments in the improvement of resources. The agricultural revolution in Europe in the eighteenth century, like the green revolution now under way in Asia, was made possible by a better understanding of the ways in which plants can be made to grow and yield good crops. In agriculturally advanced communities, in North America and Western Europe for example, these developments have brought unexpected side effects—it is necessary, for example, for the plant breeders to keep at least one step ahead of the plant diseases which are unavoidable in intensive agriculture. But to describe such a balance between practice and discovery as a technological fix is no more meaningful than to complain that the animal husbandry of the early neolithic societies was mistaken because it created a need for veterinary science. And at present the prospect is that scientific techniques of plant breeding and agriculture will make possible not merely still further increases of the yield of crops in Asia but the development of crops which provide a better diet and will prepare the foundations for productive fish farming in fresh-water ponds as well. In short, scientific research and the promise which it holds for the years ahead has made the threat of famine go away.

In exactly the same way, the contributions which science and technology have made in recent years to the improvement of natural resources have meant that nations no longer need to fear that their survival will be threatened by a lack of essential raw materials. . . . Research and development have also helped not merely to make it possible to exploit metal ores which

Pessimism in our time is infinitely more respectable than optimism: the man who foresees peace, prosperity, and a decline in juvenile delinquency is a negligent and vacuous fellow. The man who foresees catastrophe has a gift of insight which insures that he will become a radio commentator, an editor of *Time* or go to Congress.

JOHN KENNETH GALBRAITH

would previously have been unworkable but have made possible the substitution of comparatively plentiful materials for those previously in short supply. But this is a continuing and accelerating process. Even if supplies of some materials, petroleum for example, are dramatically much smaller a century from now than they are at present, in gloomy forebodings about impending scarcity it is surely appropriate to give at least some weight to the way in which scientific research and development have provided the human race with more freedom and flexibility than in the past.

For advanced and developing communities alike, technology as it is at present and the innovations now in prospect also promise a continuing improvement in economic resources. Although Dr. Commoner complains that technology means industry and industry means pollution, the creation of egalitarian and socially progressive societies in the developing nations will be possible only if there are ways in which whole populations can be provided with food, housing, health care and education. The necessary pooling of common tasks is possible only if science and technology are used as ways of increasing the productive capacity of human hands and muscles. The machine, in other words, however it may be hated, is no less important now and in the decades immediately ahead than has been the wheel. And who will call that a technological fix?

But science and technology have done and will do more than lighten labor and avoid drudgery. The temper of modern society owes much to the temper of modern science and in particular to the precept that understanding is the most secure foundation for innovation. Ironically, of course, a good many of the hypothetical calamities to which environmentalists draw attention have been recognized only because of the understanding which science has brought. Who would worry about DDT if it had not been for the invention of powerful techniques of chemical analysis, gas chromatography for example? Who would worry about mercury pollution except when there is overt toxicity if it had not been for refinements of chemical analysis? Who would fear the effects of carbon dioxide in the atmosphere if it had not been for the still rudimentary understanding of the atmosphere which the past few decades have brought?

The other side of this coin is that the rational temper of modern science has also helped with the solution of serious social problems. Disease has not of course been eradicated, but it is no longer regarded as a visitation by evil or hostile spirits. City planning, and the amenities which might flow from decent city planning, are now amenable to objective discussion. And society itself can be rationally described. If, as was undoubtedly the case, the habitation of spaceship earth, as it is called, was in part secured in megalithic times by the understanding of the way in which the sun and the moon move across the sky, is it not now foolish to disregard the contributions which scientific understanding and technological development can make to survival and the humane life? One of the most serious errors in the extreme environmentalist position is that by supposing that science and technology have nothing but useless gimmicks to offer the modern world, an indispensable aid to continued survival may be overlooked.

It is too soon to know what permanent effect the more extreme branches of the environmental movement will have had, but they have helped to dramatize a set of problems that might otherwise have been neglected. Public affairs are constantly refreshed by the challenges of those who ask that caricatures of the real world should be taken at their face value and the environmentalists deserve some credit for having pinpricked public administrations into actions that would not otherwise have been natural to them.

The movement as a whole is of course a symptom of the times but especially of the present condition of the United States. Since the beginning of the 1960s, American society seems to have been tormented with self-doubt. Civil rights came first, then Vietnam, now the environment. Outsiders, anxious to please, or at least not to give offense, are bound to ask, while trying to share the sense of doom, whether the fad will last, and for how long. And how much of it stems from an affluence which others cannot yet pretend? What will be the effect of draconian measures against pollution in the United States on imports of less hygienic machinery from elsewhere? Why is it that the people in Washington are so zealous in their public declarations on the environment and so apparently incompetent to persuade municipal authorities to put principle into practice?

One disturbing interpretation is that anxiety about the environment is a convenient lightning rod for diverting anxiety from other topics. And it would be true of course that if everybody was to die tomorrow of mercury in tuna fish, there would be no need to worry now about the problems of next week—foreign aid, relations with mainland China, the social not the technical problems of the urban conurbations and the problem of the young Although an ardent concern for the environment often seems like social prudence, the fact that the environmental movement has managed to unite the extreme left and the extreme right and to bring together some of the very old and the very young (not to mention Mr. Ralph Nader) may sometimes be held to be suspicious. Is it just possible that the environmental movement is a way of expressing anxiety in public, in good and varied company and without the risk of being accused of making mischief?

The way in which the movement's message has been greeted outside the United States, with a mixture of curiosity, politeness and alarm, is less of a feather in anybody's cap. It is true that some governments, that of Sweden for example, have welcomed news of yet another crusade to fight. Governments like the British, more occupied with such goals as economic growth, have made a dutiful acknowledgment of the environmental problem by renaming a government department, appointing a standing Royal Commission and making a number of public speeches. A part of the trouble, unfortunately, is that one of the compromises not far beneath the conscious level in most British minds is how much of a rather well-nurtured environment should now be exchanged for how much economic growth. May it not be a little embarrassing to have to be putting out flags for the environment and still hoping that industry will not merely prosper but expand?

Elsewhere, there has been frank disbelief about the motives for the movement's sudden growth. People in India are as tired of being told by Americans that they should be sterilized so as to damp down the population explosion as were Kenyans tired of being told by the British in the 1950s to stay on their farms so as not to lengthen the unemployment queues in Nairobi. In the past twenty-five years, the governments of the nuclear powers and would-be nuclear powers have become extremely skilled at managing to avoid the threat of war, and indeed they seem somehow to have organized a framework within which nuclear war is less believable than at any time since the Second World War. But the same nations have become much less sensitive in their dealings with the developing countries, or the UDCs as they tend to be called in the literature of the environment. Yet countries such as India and Brazil are sovereign nations: there is no requirement in international law nor precedent in the practice of more prosperous nations to suggest why they should regulate their own affairs in such a way as to conform with standards set by other people.

The political consequences of this tactlessness by people from industrialized societies are serious. For more than a decade it has been clear that the most urgent task in the relationship between advanced societies and the rest is to bridge the economic gulf that separates them. Then there will be a chance to stabilize the world's population. Then there will be time to regulate pollution internationally. So is it not good politics first of all to work for enough foreign aid to put international relations on a basis on which the threat could effectively be dealt with?

The offense which the doomsday men (but not all environmentalists by a long chalk) have given to intellectual life is much more serious. When people say, as Dr. Ehrlich has done, that he will be glad if his somber prediction turns out to be false because then, at least, people will have survived, they are playing a mean trick. The common justification is to say that it is necessary to exaggerate to get people stirred, to get things done. But people are easily anesthetized by repetition and there is a danger that, in spite of its achievements so far, the environmental movement could still find itself falling flat on its face when it is most needed, simply because it has pitched its tale too strongly. The cases of phosphate detergents and cyclamates were close calls.

This is one reason why the environmental movement would be strengthened if it had a wider perspective. It is forever saying that society must learn to balance cost and benefit and then neglecting to point out that economists and others have already done a good deal to show how calculations like these should be performed. The movement's lack of a sense of history is also maiming. Not merely are there lessons to be learned from the past, but some of the conflicts which have been provoked by the movement's attack on the establishment are explicable only historically. Just as it gives offense to tell developing nations not to build blast furnaces that pollute the air, so it gives offense to suggest to relatively poor sections of the community that the time has come for the search for material prosperity to cease. May this explain why black people are conspicuous by their absence among the environmentalists in the United States? In reality, of course, even countries such as the United States, rich as they are, still need a good deal more

The airplane, the atomic bomb and the zipper have cured me of any tendency to state that a thing can't be done.
R. L. DUFFUS

economic growth to be able to afford to pay the teachers who will be needed to teach the disadvantaged children. If there is to be an accommodation between prosperity and pollution, it will have to be worked out in the familiar political arena. Social security, welfare and educational policies will be as prominent as decisions on acceptable standards of air pollution.

Luckily, there is nothing in the record of the past few years that points to an impending cessation of the progressive liberalization and enrichment of society which it was permissible, in Alfred Marshall's time, to call progress. If anything, the process is accelerating. People shake their heads over the way in which the population of the world might be doubling once every thirty-five years, but the population of the universities in Western Europe and North America has for some time been doubling in less than twenty years and the population of secondary schools elsewhere in the world is growing still more quickly. Further, there is no doubt that programs for research and development already begun contain the seeds not merely of further economic growth, much of it free from pollution, but also opportunities for ordinary people to enjoy benefits which have previously been beyond reach.

The environmentalists are fond of using the eloquent metaphor of spaceship earth but this is not the most important point to make about the way in which living things have managed to survive for 3,000 million years and, so far, to evolve. Although everybody seems prepared now to accept that other planets elsewhere in the galaxy are likely to have living things on them, nobody makes light of the evolutionary barriers which the human race has had to surmount. After two million years of near extinction, is it any wonder that instinct should lead to temporary overfecundity? The truth is that the technology of survival has been more successful than could have been imagined in any previous century. It will be of immense importance to discover, in due course, the next important threat to survival, but the short list of doomsday talked of in the past few years contains nothing but paper tigers. Yet in the metaphor of spaceship earth, mere housekeeping needs courage. The most serious worry about the doomsday syndrome is that it will undermine our spirit.

35

The Future Revised: Population of World Outstrips Expectations

Jonathan Spivak

Jonathan Spivak is a staff reporter of *The Wall Street Journal.*

Today—during this one day—the world's population will increase by some 200,000, the size of a Des Moines or a Salt Lake City or a Grand Rapids.

Tomorrow there'll be a similar increase, and the next day. And the next. And the next, for many years to come.

So goes the grim progression that threatens to overcrowd the globe and strain its ability to feed, clothe and shelter its inhabitants. "The magnitude of the new population boom surpasses all earlier expectations," warns a United Nations report on the long-range implications of world population growth. "The longer the high tempo persists . . . the more precarious will be prospects for a healthy life on this planet."

By the year 2000, most experts anticipate the world's population, now 4 billion, will approach 6.5 billion. And if certain assumptions go awry, the 21st century could dawn on a globe with well over 7 billion inhabitants—as many as 2 million more than were expected a decade ago.

Death Rate Falls

"The question is not will the world's population grow, but under what conditions will it stop," says Samuel Baum, assistant chief for international demographic research at the U.S. Census Bureau.

Present attempts at restraint are being overwhelmed. The world's average birth rate is declining as more and more nations emphasize population control. Yet the global death rate has been falling even more rapidly because of improved living conditions and the conquest of disease.

While population growth in the industrial nations of the West is slowing significantly, massive increases are continuing in the developing regions of the world: Latin America, Africa and Asia. In the more primitive lands, birth rates are now three times as high as in the industrialized nations of North America and Europe. Mexico, with less than one-third as many inhabitants as the U.S., adds as many people—1.9 million—to its population each year as this country does.

And most demographers aren't sanguine about the outlook in the less developed countries. "It's difficult for me to see any dramatic reduction in the growth rate in Africa, Asia and Latin America," says Frederick A. Leedy, assistant chief for international program planning evaluation at the Census Bureau.

Reactionary Regimes?

Unquestionably, by the close of this century, the world's population burden will tend to shift from the industrial nations of the North to the more rural, developing nations of the South. North America, Europe and other developed regions, now growing at a rate of less than 1% a year, are expected to increase by only about 200 million people. Expanding more than twice as rapidly, Africa, Asia and South America are expected to add 2 billion inhabitants. By the year 2000, four-fifths of the world's population will be concentrated in the developing countries: the proportion now is less than three-quarters.

Population densities will increase almost everywhere, particularly in South Asia. The pressures will produce unprecedented problems for the less developed regions of the world, even threatening the survival of some nations. A decade ago, this newspaper reported a tentative conclusion that "birth-control programs are making headway in some of the lands whose population problems are most acute." And it appeared then that even the large population growth still anticipated was unlikely "to lead to mass starvation or other catastrophes." Today, demographers are much less optimistic.

The trouble with our times is that the future is not what it used to be.
PAUL VALÉRY

If nothing worse, analysts foresee a spread of reactionary regimes. "Faced with increasingly difficult problems brought on in part by population growth, governments are bound to take a firm hand to bring them under control. The rights of the individual will have to give way to the rights of the community," maintains a State Department population specialist.

But, more than the spread of authoritarian governments, forecasters fear the growth of conflicts within or between densely populated countries. They find omens for the future in the recent past.

Nazli Choucri, an MIT political scientist, has analyzed the causes of 93 conflicts in or between developing nations in the years 1945 to 1969. She concludes that population pressures played a significant role in '66, including the Algerian war for independence, the Nigeria-Biafra conflict and the Arab-Israeli war of 1973. "The higher the rate of growth, the more salient a factor population increase appears to be in the development of conflict and violence," she contends.

More Megalopolises

Particular problems arise from patterns of immigration. Kuwait, for example, has imported 70% of its work force and there's already tension between native citizens and the foreign-born, who have fewer rights. Mexico's large population growth and sparse economic opportunity produce a steady flow of illegal migrants into the Southwestern U.S.; they are becoming a major source of this country's population increase. "It can't get anything but worse," says Justin Blackwelder, president of the Environmental Fund, a population-control group that advocates stricter curbs on immigration.

Urban growth probably will continue undiminished into the next century, creating huge megalopolises. By the year 2000, some 50% of the world's population is expected to live in urban areas, compared with 39% now. The trend will be particularly marked in the developing lands, where urban growth rates are now more than twice as high as in the industrialized countries.

By the turn of the century, demographers calculate, there will be at least 60 cities with 5 million or more people, compared with the current 21. Mexico City, now third-ranking, is expected to swell into the world's largest metropolitan area with 31.5 million inhabitants.

This surge of urbanization will be spurred by better communications, transportation, and industrialization. Rural residents will continue to be pushed from the countryside by poverty and pulled to the cities by hopes of jobs and money. Population planners worry that this vast displacement of the poor will intensify the difficulties of accommodating the huge total increase in mankind.

"What is new about the situation [urbanization] in developing areas today is not poverty per se, but its massiveness, its potentiality for increase, its incongruous association with high technology and its rapidly eroding opportunity for alleviation," declares Kingsley Davis, a demographer at the University of California.

On average, the populations of most developing nations will remain young, because of high birth rates. In the year 2000, it's expected, almost half their inhabitants will still be below the age of 20; the demand for education and jobs will be tremendous.

Moreover, there'll be many women yet to enter their child-bearing years. Thus, even if births dropped rapidly to the replacement level of two children per family, population growth would continue for another 70 to 80 years in developing lands, the experts calculate. (The average family size in those countries now is more than twice the replacement level.) Though population may cease growing by the end of the century in North America and Europe, elsewhere "such an achievement is most unlikely and probably impossible," declares Tomas Frejka, a demographer with the Population Council, a leader in population research.

The momentum for world population growth became grimly clear to demographers more than a decade ago. Experts then noted birth rates were rising and mortality falling in the developing lands. The long-range projections made in the 1960s were not much lower than those made today. But the totals remain similar only because of a happy coincidence of forecasting miscalculations. Some nations grew more slowly than expected; others grew faster.

The Chinese Solution

China, the world's most populous nation, was thought in the early 1960s to be growing as rapidly as the rest of Asia. Many population experts assumed its birth rate was above 35 per 1000 and would drop only a few points in the next decade. (The U.S. rate was then 22.5 and has since dropped to 17.) But now the message carried out from behind the Bamboo Curtain by most visitors is that China's birth rate has dropped by perhaps 10 points because of the Communist regime's emphasis on family planning. Within another 10 years, China's birth rate could be lower than this country's, says Dr. Reimert Ravenholt, who heads the U.S. Agency for International Development's family planning programs abroad.

China embarked on population control in the mid-1950s. While Chairman Mao appeared at first to waiver in his commitment, family planning is now an article of political faith in Peking. "China's is probably the most clearly emphasized and sponsored program in the world," says John Aird, head of the Commerce Department's foreign demographic analysis division.

The two-child family is the Chinese goal except for sparsely settled mountain areas where a larger labor force is needed. Visitors report that local communes set birth quotas and decide which couples can have children each year. Delay in marriage to age 28 for men and age 25 for women—which tends to hold down family size—is emphasized. All forms of contraception are made widely available, including the Chinese version of the pill: a drug-impregnated soluble paper, dissolved in the mouth.

But many China specialists attribute the family planning successes more to the unusual nature of Chinese society. "China's rural population is different from other developing countries. It's better educated. The women are employed and have a more important role in decision-making and society," says Leo Orleans, a Library of Congress expert on the country.

Failure in India

For most other developing nations, demographers now admit they were unduly optimistic that modern medical technology could rapidly reduce birth rates. Only in a few of the smaller countries, such as Taiwan, South Korea and Greenland, have sharp declines occurred. In a span of eight years, Greenland's birth rate was cut by more than half, to 19.5 per 1,000.

But there's little evidence of success in nations with massive population reservoirs, particularly India, Pakistan and Bangladesh. "In the past . . . people felt with the technological breakthrough (in contraceptive methods) they would control world fertility in a decade: now there is a period of re-assessment," notes Shigemi Kono, a United Nations demographer.

More than two decades ago, India embarked on an intensive population control program, and the U.S. aid program made reduction of India's birth rate a prime goal. But the rate has declined only a few points, and India's population growth remains among the highest in the world. At 600 million now, the subcontinent's population could exceed 1.1 billion in the year 2000 and could eventually surpass China's horde in size.

Most demographers attribute the apparent failure in India partly to the government's ineptitude and to prejudice against the pill, based on health

What doth a man gain by a son? . . .
"Food is breath; clothing a protection;
Gold, an ornament. Cattle lead to marriage.
A wife is a comrade, a daughter, a misery . . . ;
And a son, a light in the highest heaven.
A sonless one cannot attain heaven."
AITAREYA BRAHMANA

concerns. But most of the trouble lies in the nature of Indian society, they say. Among the influences which help to sustain the high birth rate are early marriage, the Hindu religion's emphasis on bearing sons, dependence on children for security in old age, and a low level of education among the rural masses.

Contraception No Panacea

There's no doubt that birth control slows population growth. But its impact has been felt mainly in the developed nations. Many experts argue that modern contraception can accelerate a birth-rate decline that's already under way, but won't actually start a downtrend.

So population planners are increasingly abandoning their emphasis on medical solutions to world population growth and focusing on longer-range remedies in social and economic development. Demographers note that birth rates tend to fall when income increases and is more evenly distributed, when education is more widespread, and when more women participate in the work force.

And even though the decline in the world's death rate has contributed to population growth, experts argue that reduction of death rates is essential for lowering birth rates in developing countries. So long as infant mortality is high, they reason, parents will insist on having large numbers of offspring. In India, where average life expectancy is 37 years, parents must have six children to insure that one lives to their old age.

But in fact the death-rate decline has leveled off in the developing nations of Asia and Africa. And during the last three years, death rates have actually risen in India, Bangladesh, Sri Lanka (formerly Ceylon) and the Sahelian regions, south of the Sahara, in Africa, according to Lester Brown, president of Worldwatch Institute, a research organization dealing with population and other issues.

Hunger and the Death Rate

Mr. Brown reasons that the rise in deaths stems largely from huge price increases in wheat, rice and other basic foodstuffs. Nutrition suffered in impoverished lands where people devote 50% to 80% of their income to food. The effect on women and children has been particularly severe; infant mortality rates have risen rapidly, reports Irene Tinker, director of the Office of International Science at the American Association for the Advancement of Science.

Some optimistic planners insist that world growth rates will be cut in half within a decade and the total population by the turn of the century kept well under 5.5 billion. But there are many ifs:

—If India, Pakistan and Bangladesh, after years of futile efforts, curb their rate of population increase.

—If African and Latin American nations, which may double their populations before the year 2000, admit they have growth problems and take action. (At the World Population Conference in Bucharest last year, representatives of many of these nations argued that growth was good and insisted that redistribution of wealth, not population control, would solve their difficulties.)

—If new super-quick methods of sterilizing men and women in out-patient clinics bring the hoped-for results. (AID experts say that more than

100 operations can be performed daily by one doctor, and they say 10,000 clinics at a cost of $100 million would meet the needs of Africa, Latin America and Asia.)

China's Unknown Numbers

Along with these uncertainties, demographers concede they have only a rough knowledge of the size of the world's current population. In some countries of Africa and Asia, a census has never been taken or is badly out of date. In others, like Nigeria, the figures are suspected of being drastically inflated for political reasons. The major uncertainty centers on China. The last full-scale census, taken in 1953, showed its population at 583 million. There's sharp disagreement among China experts on the growth since then.

UN experts estimate the current population at 838 million. But the Commerce Department's John Aird, who doubts birth rates have dropped much, believes that China's population is between 925 million and 1.1 billion. He foresees it rising by the year 2000 to as much as 1.5 billion, far more than other forecasters do.

Thus China remains a demographic puzzle. The Chinese have consistently refused visitors access to their birth and death records in Peking. U.S. experts believe China's rulers themselves don't know how big the population is.

While little can be done to resolve the huge uncertainty on China, the U.S. and the UN are financing a survey to improve data on many other parts of the world. By the time it is completed in 1981, the project will provide better information on fertility in 70 nations, helping to show the impact of family planning programs on birth rates.

But neither this information nor the control efforts now being made can offer assurance of reducing the horde of humanity that may crowd the planet in another quarter-century.

36 Food First!

Frances Moore Lappé and Joseph Collins

Frances Moore Lappé and Joseph Collins are co-directors of the Institute for Food and Development Policy in San Francisco, California, and co-authors of *Food First: Beyond the Myth of Scarcity* (1977). Ms. Lappé is author of the bestselling *Diet for a Small Planet,* which explains how to meet all human nutritional needs from nonmeat sources without skimping on high-grade protein.

- There is no country in the world where people could not feed themselves from their own resources.
- Food security cannot be measured in grain reserve or production figures.
- A nation's per capita food production can double and yet more people can be hungry.
- Increased prices for agricultural exports could lead to increased hunger.

Are you bewildered by any of these statements? We would not be surprised. For the last several years we have struggled to answer the question "Why hunger?" Analyses that call for increasing or improving present development assistance or for reducing our consumption so that the hungry might eat left us with gnawing doubts. We probed and probed. We agonized over the logical consequences of what we were learning that seemed to put us in opposition to groups we previously had supported. But eventually we came to an understanding that feels liberating, that gives us energy instead of paralyzing us with guilt, fear, or despair. The pieces have begun to fit together for the first time.

Here we want to share the six myths that kept us locked into a misunderstanding of the problem as well as the alternative view that emerged once we began to grasp the real issues. Our hope is to help anchor the hunger movement with an unequivocal and cogent analysis. Only then will our collective potential no longer be dissipated.

Myth One: *People are hungry because of scarcity—both of food and of agricultural land.*

Can scarcity seriously be considered the cause of hunger when even in the worst years of famine in the early 70s there was plenty to go around—enough in grain alone—to provide everyone in the world over 3,000 calories a day, not counting all the beans, root crops, fruits, nuts, vegetables, and non-grain-fed meat?

And what of land scarcity?

We looked at the most crowded countries in the world to see if we could find a correlation between land density and hunger. We could not. Bangladesh, for example, has just half the people per cultivated acre that Taiwan has. Yet Taiwan has no starvation while Bangladesh is thought of as the world's worst basketcase. China has twice as many people for each cultivated acre as India. Yet in China people are not hungry.

Finally, when the pattern of *what* is grown sank in, we simply could no longer subscribe to a "scarcity" diagnosis of hunger. In Central America and in the Caribbean, where as much as 70 percent of the children are undernourished, at least half of the agricultural land, and the best land at that, grows crops for export, not food for the local people. In the Sahelian countries of sub-Saharan Africa, exports of cotton and peanuts in the early 1970s actually *increased* as drought and hunger loomed.

Next we asked: What solution emerges when the problem of hunger is defined as scarcity?

Most commonly, people see greater production as the answer. Thus techniques to increase production become the central focus: supplying the "modern" inputs—large-scale irrigation, chemical fertilizers, pesticides, machinery, and the seeds dependent on these other inputs—all to make the land produce more. But when a new agricultural technology enters a system shot through with power inequalities, it brings greater profit only to those who already have some combination of land, money, credit "worthiness," and political influence. This alone has eliminated most of the world's rural population and all the world's hungry.

Once agriculture is viewed as growth industry in which the control of the basic inputs guarantees big money, a catastrophic chain of events is set into motion. Competition for land sends land values soaring (land values have jumped three to five times in the "Green Revolution" areas of India). Higher rents force tenants and sharecroppers into the ranks of the landless. With the new profits the powerful buy out small farmers gone bankrupt in part through having been forced to double or triple their indebtedness trying to partake of the new technology. Moreover, faced with a short planting and harvest time for vast acreages planted uniformly with the most profitable crop, large commercial growers mechanize to avoid the troublesome mobilization of human labor. Those made landless by the production focus, finding ever fewer agricultural jobs, join an equally hopeless search for work in urban slums.

Fewer and fewer people gain control over more and more land. In Sonora, Mexico, before the "Green Revolution," the average farm was 400 acres. After 20 years of publicly funded modernization, the average has climbed to 2,000 acres with some holdings running as large as 25,000 acres.

There are two ways to slice easily through life; to believe everything or to doubt everything. Both ways save us from thinking.
ALFRED KORZYBSKI

We pay the consequences. Total production per capita may be up, yet so are the numbers who face hunger. A strategy to solve hunger by increasing production has led directly to increased inequality, in fact to the absolute decline in the welfare of the majority. A study now being completed by the International Labor Organization documents that in the very South Asian countries—Pakistan, India, Bangladesh, Sri Lanka, Malaysia, the Philippines, and Indonesia—where the focus has been on merely increasing the production and where indeed the Gross National Product per capita has risen,* the rural poor are absolutely worse off than before. The study concludes that *"the increase in poverty has been associated not with a fall but with a rise in cereal production per head, the main component of the diet of the poor."* These seven countries account for 70 percent of the rural population of the non-socialist underdeveloped world.

*Only Bangladesh did not experience a per capita rise in GNP. The pattern, however, is the same, if not more dramatic. While the bottom 80 percent has experienced a decline in income, the top 20 percent has managed to get richer.

But if the scarcity diagnosis, with the implied solution of increasing production by supplying the right technical inputs, has taken us not forward but backward, what is the right diagnosis?

We could answer that question only after our research at IFDP (Institute for Food and Development Policy) led us to conclude that there is *no* country without sufficient agricultural resources for the people to feed themselves and then some. And if they are not doing so, you can be sure there are powerful obstacles in the way. The prime obstacle is not, however, inadequate production to be overcome by technical inputs. The obstacle is that the people do not control the productive resources. When control is in the hands of the producers, people will no longer appear as liabilities—as a drain on resources. People are potentially a country's most underutilized resource and most valuable capital. People who know they are working for themselves will not only make the land produce but through their ingenuity and labor make it ever more productive. Human energy, properly motivated and organized, can transform a desert into a granary.

Myth Two: *A hungry world simply cannot afford the luxury of justice for the small farmer.*

We are made to believe that, if we want to eat, we had better rely on the large landowners. Thus governments, international lending agencies, and foreign assistance programs have passed over the small producers, believing that concentrating on the large holders was the quickest road to production gains. A study of 83 countries that revealed that just over 3 percent of the land holders control about 80 percent of the farmland gave us some idea how many of the world's farmers would be excluded by such a concentration.

In fact, the small farmer is commonly more productive, often many times more productive, than the larger farmer. A study of Argentina, Brazil, Chile, Colombia, Ecuador, and Guatemala found the small farmer to be three to 14 times more productive per acre than the larger farmer. In Thailand plots of two to four acreas yield almost 60 percent more rice per acre than farms of 140 acres or more. Other evidence that justice for the small farmer increases production comes from the experience of countries in which the redistribution of land and other basic agricultural resources like water has resulted in rapid growth in agricultural production: Japan, Taiwan, and China stand out.

We should not romanticize the peasant. He gets more out of the land precisely because he is desperate to survive on the meager resources allowed to him. Studies show that the smaller farmer plants more closely than would a machine, mixes and rotates complementary crops, chooses a combination of cultivation and livestock that is labor intensive, and, above all, works his perceptibly limited resources to the fullest. The control of the land by the large holders for whom land is not the basis of daily sustenance invariably leads to its underutilization. In Colombia, according to a 1960 study, the largest landowners control 70 percent of all agricultural land but actually cultivate only 6 percent. Worldwide studies of the "Green Revolution" have shown that even when the larger farmers are favored with heavy investment in the new seed-fertilizer technology, the net return per acre continues to be less on the large farms than on the small. The large amounts of work put into the small farms more than compensate for the big doses of capital investment on the large.

But where has the grip of the myth that justice and productivity are incompatible led us? As the large holders are reinforced, often with public investment in capital-intensive technologies, the small holders and laborers have been cut out of production through the twin process of increasing land concentration and mechanization. *And to be cut out of production is to be cut out of consumption.*

As fewer and fewer have the wherewithal either to grow food or to buy food, the internal market for food stagnates or even shrinks. But large commercial farmers have not worried. They orient their production to high-paying markets—a few strata of urban dwellers and foreign consumers. Farmers in Sinaloa, Mexico, find they can make 20 times more growing tomatoes for Americans than corn for Mexicans. Development funds have irrigated the desert in Senegal so that multinational firms can grow eggplants and mangoes for air freighting to Europe's best tables. Colombian landholders shift from wheat to carnations that bring 80 times greater return per acre. In Costa Rica the lucrative export beef business expands as the local consumption of meat and dairy products declines. Throughout the non-socialist countries we find a consistent pattern: Agriculture, once the livelihood for millions of self-provisioning farmers, is being turned into the production site of high-value nonessentials for the minority who can pay.

Moreover, entrusting agricultural production to the large farmers means invariably the loss of productive reinvestment in agriculture. Commonly profits of the large holders that might have gone to improve the land are spent instead on conspicuous consumption, investment in urban consumer industries, or job-destroying mechanization. Study after study indicates that small farmers and secure tenants save at rates comparable to or greater than those of large farmers. Indeed it is only rural households with no land to cultivate who do not save.

It is not enough simply to deflate the myth that justice and production are incompatible. We must come to see clearly that the only solution to hunger is a conscious plan to reduce inequality at every level. The reality is that a just redistribution of control over agricultural resources will decrease inequality and increase production; moreover, it is the *only* guarantee that the hungry will eat what is produced.

Myth Three: *We are now faced with a sad trade-off. A needed increase in food production can come only at the expense of the ecological integrity of our food base. Farming must be pushed onto marginal lands at the risk of irreparable erosion. And the use of pesticides will have to be increased even if the risk is great.*

Is the need for food for a growing population the real pressure forcing people to farm lands that are easily destroyed?

Haiti offers a shocking picture of environmental destruction. The majority of the utterly impoverished peasants ravage the once-green mountain slopes in near-futile efforts to grow food to survive. Has food production for Haitians used up every easily cultivated acre so that only the mountain slopes are left? No. These mountain peasants must be seen as exiles from their birthright—some of the world's richest agricultural land. The rich valley lands belong to a handful of elites who seek dollars in order to live an imported lifestyle and to their American partners. These lands are thus made to produce largely low-nutrition and feed crops (sugar, coffee, cacao, alfalfa for cattle) and exclusively for export. Grazing land is export-oriented too. Recently U.S. firms began to fly Texas cattle into Haiti for grazing and re-export to American franchised hamburger restaurants.*

**Editor's note* By constantly increasing their consumption of animal products, the affluent nations are absorbing an immense amount of the increase in agricultural production, for the ever-greater number of animals destined for the slaughterhouse is eating an ever-greater amount of grain and soybeans. Americans consume 120 percent more beef today than they did in 1950; in fact, during his or her lifetime, the average American devours 10,000 pounds of meat. Frances Moore Lappé regards America's excessive production and consumption of meat as totally unsound from an ecological viewpoint, and calls it "the very institutionalization of waste." See her "Fantasies of Famine," *Harper's*, February 1975, pp. 51–54, 87–90.

A World Bank study of Colombia states that "large numbers of farm families . . . try to eke out an existence on too little land, often on slopes of . . . 45 degrees or more. As a result, they exploit the land very severely, adding to erosion and other problems, and even so are not able to make a decent living." Overpopulation? No. Colombia's good level land is in the hands of absentee landlords who use it to graze cattle and to raise animal feed and even flowers for export to the United States ($18 million worth in 1975).

In Africa vast tracts of geologically old sediments perfectly suitable for permanent crops such as grazing grasses or trees have instead been torn up for planting cotton and peanuts for export. In parts of Senegal peanut monoculture has devastated the soils.

The Amazon is being rapidly deforested. Is it the pressure of Brazil's exploding population? Brazil's ratio of cultivatable land to people (and that excludes the Amazon forest) is slightly better than that of the United States. The Amazon forest is being destroyed not because of a shortage of farmland but because the military government refuses to break up the large estates that take up over 43 percent of the country's farmland. Instead the landless are offered the promise of future new frontiers in the Amazon Basin even though most experts feel the tropical forest is not suited to permanent cropping. In addition, multinational corporations like Anderson Clayton, Goodyear, Volkswagen, Nestle, Liquigas, Borden, Mitsubishi, and multibillionaire Daniel Ludwig's Universe Tank Ship Company can get massive government subsidies to turn the Amazon into a major supplier of beef to Europe, the U.S., and Japan.

It is not, then, people's food needs that threaten to destroy the environment but other forces: land monopolizers that export nonfood and luxury crops forcing the rural majority to abuse marginal lands; colonial patterns of cash cropping that continue today; hoarding and speculation of food; and irresponsible profit-seeking by both local and foreign elites. Cutting the number of the hungry in half tomorrow would not stop any of these forces.

Still we found ourselves wondering whether people's legitimate need to grow food might not require injecting even more pesticides into our environment. In the emergency push to grow more food, won't we have to accept some level of damage from deadly chemicals?

First, just how pesticide-dependent is the world's current food production? In the U.S. about 1.2 billion pounds, a whopping six pounds for every American and 30 percent of the world's total, are dumped into the environment every year. Surely, we thought, such a staggering figure means that practically every acre of the nation's farmland is dosed with deadly poisons. U.S. food abundance, therefore, appeared to us as the plus that comes from such a big minus. The facts, however, proved us wrong.

Fact One: Nearly half the pesticides are used not by farmland but by golf courses, parks, and lawns.

Fact Two: Only about 10 percent of the nation's cropland is treated with insecticides, 30 percent with weedkillers, and less than 1 percent with fungicides (the figures are halved if pastureland is counted).

Fact Three: Nonfood crops account for over half of all insecticides used in U.S. agriculture. (Cotton alone received almost half of all insecticides used. Yet half of the total cotton acreage receives no insecticides at all.)

Fact Four: The U.S. Department of Agriculture estimates that, even if all pesticides were eliminated, crop loss due to pests (insects, pathogens, weeds, mammals, and birds) would rise only about seven percentage points, from 33.6 percent to 40.7 percent.

Fact Five: Numerous studies show that where pesticides are used with ever greater intensity crop losses due to pests are frequently *increasing.*

Fact Six: Several recent studies indicate great quantities of pesticides applied annually to croplands are used needlessly.

What about underdeveloped countries? Do pesticides there help produce food for hungry people?

In underdeveloped countries most pesticides are used for export crops, principally cotton, and to a lesser extent fruits and vegetables grown under plantation conditions for export. In effect, then, enclaves of pesticide use in the underdeveloped world function as mere extensions of the agricultural systems of the industrialized countries. The quantities of pesticides injected into the world's environment have little to do with the hungry's food needs.

The alternatives to chemical pesticides—crop rotation, mixed cropping, mulching, hand weeding, hoeing, collection of pest eggs, manipulation of natural predators, and so on—are numerous and proven effective. In China, for example, pesticide use can be minimized because of a nationwide early warning system. In Shao-tung county in Honan Province, 10,000 youths make up watch teams that patrol the fields and report any sign of pathogenic change. Appropriately called "the barefoot doctors of agriculture," they have succeeded in reducing the damage of wheat rust and rice borer to less than 1 percent and have the locust invasions under control. But none of these safe techniques for pest control will be explored as long as the problem is in the hands of profit-oriented corporations. The alternatives require human involvement and the motivation of farmers who have the security of individual or collective tenure over the land they work.

Myth Four: *Hunger is a contest between the "rich world" and the "poor world."*

Terms like "hungry world" and "poor world" make us think of uniformly hungry masses. They hide the reality of vertically stratified societies in which hunger afflicts the lower rungs in both so-called developed and underdeveloped countries. Terms like these turn hunger into a place—and usually a place over there. Rather than being a result of a social process, hunger becomes a static fact, a geographic given.

Worse still, the all-inclusiveness of these labels leads us to assume that everyone living in a "hungry country" has a common interest in eliminating hunger. Thus we look at an underdeveloped country and assume its government officials represent the hungry majority. Well-meaning sympathizers

For a time in the world some force themselves ahead
And some are left behind,
For a time in the world some make a great noise
And some are held silent,
For a time in the world some are puffed fat
And some are kept hungry,
For a time in the world some push aboard
And some are tipped out . . .
LAO TZU

in the industrialized countries then believe that concessions to these governments, e.g., preference schemes or increased foreign investment, represent progress for the hungry when in fact the "progress" may be only for the elites and their partners, the multinational corporations.

Moreover, the "rich world" versus "poor world" scenario makes the hungry appear as a threat to the material well-being of the majority in the metropolitan countries. To average Americans or Europeans the hungry become the enemy who, in the words of Lyndon Johnson, "want what we got." In truth, however, hunger will never be addressed until the average citizens in the metropolitan countries can see that the hungry abroad are their allies, not their enemies.

What are the links between the plight of the average citizen in the metropolitan countries and the poor majority in the underdeveloped countries? There are many. One example is multinational agribusiness shifting production of luxury items—fresh vegetables, fruits, flowers, and meat—out of the industrial countries in search of cheap land and labor in the underdeveloped countries. The result? Farmers and workers in the metropolitan countries lose their jobs while agricultural resources in the underdeveloped countries are increasingly diverted away from food for local people. The food supply of those in the metropolitan countries is being made dependent on the active maintenance of political and economic structures that block hungry people from growing food for themselves.

Nor should we conclude that consumers in the metropolitan countries at least get cheaper food. Do Ralston Purina's and Green Giant's mushrooms grown in Korea and Taiwan sell for less than those produced stateside? Not one cent, according to a U.S. Government study. Del Monte and Dole Philippine pineapples actually cost the U.S. consumers more than those produced by a small company in Hawaii.

The common threat is the worldwide tightening control of wealth and power over the most basic human need, food. Multinational agribusiness firms right now are creating a single world agricultural system in which they exercise integrated control over all stages of production from farm to consumer. Once achieved, they will be able to effectively manipulate supply and prices for the first time on a worldwide basis through well-established monopoly practices. As farmers, workers, and consumers, people everywhere already are beginning to experience the costs in terms of food availability, prices, and quality.

Myth Five: *An underdeveloped country's best hope for development is to export those crops in which it has a natural advantage and use the earnings to import food and industrial goods.*

There is nothing "natural" about the underdeveloped countries' concentration on a few, largely low-nutrition crops. The same land that grows cacao, coffee, rubber, tea, and sugar could grow an incredible diversity of nutritious crops—grains, high-protein legumes, vegetables, and fruits.

Nor is there any advantage. Reliance on a limited number of crops generates economic as well as political vulnerability. Extreme price fluctuations associated with tropical crops combine with the slow-maturing nature of plants themselves (many, for example, take two to ten years before the first harvest) to make development planning impossible.

Often-quoted illustrations showing how much more coffee or bananas it takes to buy one tractor today than 20 years ago have indeed helped us appreciate that the value of agricultural exports has simply not kept pace with the inflating price of imported manufactured goods. But even if one considers only agricultural trade, the underdeveloped countries still come out the clear losers. Between 1961 and 1972 half of the metropolitan countries increased their earnings from agricultural exports by 10 percent each year. By contrast, at least 18 underdeveloped countries are earning *less* from their agricultural exports than they did in 1961.

Another catch in the natural-advantage theory is that the people who need food are not the same people who benefit from foreign exchange earned by agricultural exports. Even when part of the foreign earnings is used to import food, the food is not basic staples but items geared toward the eating habits of the better-off urban classes. In Senegal the choice land is used to grow peanuts and vegetables for export to Europe. Much of the foreign exchange earned is spent to import wheat for foreign-owned bakeries that turn out European-style bread for the urban dwellers. The country's rural majority goes hungry, deprived of land they needed to grow millet and other traditional grains for themselves and local markets.

The very *success* of export agriculture can further *undermine* the position of the poor. When commodity prices go up, small self-provisioning farmers may be pushed off the land by cash-crop producers seeking to profit on the higher commodity prices. Moreover, governments in underdeveloped countries, opting for a development track dependent on promoting agricultural exports, may actively suppress social reform. Minimum wage laws for agricultural laborers are not enacted, for example, because they might make the country's exports "uncompetitive." Governments have been only too willing to exempt plantations from land reform in order to encourage their export production.

Finally, export-oriented agricultural operations invariably import capital-intensive technologies to maximize yields as well as to meet product and processing specifications. Relying on imported technologies then makes it likely that the production will be used to pay the bill—a vicious circle of dependency.

Just as export-oriented agriculture spells the divorce of agriculture and nutrition, food-first policies would make the central question: How can the people best feed themselves with this land? As obvious as it may seem, this policy of basing land use on nutritional output is practiced in only a few countries today; more commonly, commercial farmers and national planners make hit-and-miss calculations of which crop might have a few cents' edge on the world market months or even years hence. With food-first policies industrial crops (like cotton and rubber) and feed crops would be planted only after the people meet their basic needs. Livestock would not compete with people but graze on marginal lands or, like China's 240 million pigs, recycle farm and household wastes while producing fertilizer at the same time.

In most underdeveloped countries the rural population contributes much more to the national income than it receives. With food-first policies agricultural development would be measured in the welfare of the people, not in export income. Priority would go to decentralized industry at the service of labor-intensive agriculture. A commitment to food self-reliance would close the gap between rural and urban well-being, making the countryside a good place to live. Also urban dwellers, like those volunteering to grow vegetables in Cuba's urban "green belts," would move toward self-reliance.

Food self-reliance is not isolationist. But trade would be seen, not as the one desperate hinge on which survival hangs, but as a way to widen choices once the basic needs have been met.

Myth Six: *Hunger should be overcome by redistributing food.*

Over and over again we hear that North America is the world's last remaining breadbasket. Food security is invariably measured in terms of reserves held by the metropolitan countries. We are made to feel the burden of feeding the world is squarely on us. Our overconsumption is tirelessly contrasted with the deprivation elsewhere with the implicit message being that we cause their hunger. No wonder that North Americans and Europeans feel burdened and thus resentful. "What did we do to cause their hunger?" they rightfully ask.

The problem lies in seeing food redistribution as the solution to hunger. We have come to a different understanding. Distribution of food is but a reflection of the control of the resources that produce food. Who controls the land determines who can grow food, what is grown, and where it goes. Who can grow: a few or all who need to? What is grown: luxury nonfood or basic staples? Where does it go: to the hungry or the world's well-fed?

Thus redistribution programs, like food aid, will never solve the problem of hunger. Instead we must face up to the real question: How can people everywhere begin to democratize the control of food resources?

SIX FOOD-FIRST PRINCIPLES

We can now counter these six myths with six positive principles that could ground a coherent and vital movement:

1. There is no country in the world in which the people could not feed themselves from their own resources. But hunger can only be overcome by the transformation of social relationships and only be made worse by the narrow focus on technical inputs to increase production.
2. Inequality is the greatest stumbling block to development.
3. Safeguarding the world's agricultural environment and people feeding themselves are complementary goals.
4. Our food security is not threatened by the hungry masses but by elites that span all market economies profiting by the concentration and internationalization of control of food resources.
5. Agriculture must not be used as the means to export income but as the way for people to produce food first for themselves.
6. Escape from hunger comes not through the redistribution of food but through the redistribution of control over food-producing resources.

What would an international campaign look like that took these truths to be self-evident?

If we begin with the knowledge that people can and will feed themselves if allowed to do so, the question for all of us living in the metropolitan countries is not "What can we do for them?" but "How can we remove the obstacles in the way of people taking control of the production process and feeding themselves?"

Since some of the key obstacles are being built with our taxes, in our name, and by corporations based in our economies, our task is very clear:

Stop any economic aid—government, multilateral or voluntary—that reinforces the use of land for export crops. Stop support for agribusiness penetration into food economies abroad through tax incentives and from governments and multilateral lending agencies. Stop military and counterinsurgency assistance to underdeveloped countries; it is used to oppose the changes necessary for food self-reliance.

Work to build a more self-reliant food economy at home so that we become even less dependent on importing food from hungry people. Work for land reform at home. Support worker-managed producers and distributors to counter the increasing concentration of control over our food resources.

Educate, showing the connections between the way government and corporate power works against the hungry abroad and the way it works against the food interests of the vast majority of people in the industrial countries.

Counter despair. Publicize the fact that 40 percent of all people living in underdeveloped countries live where hunger has been eliminated through common struggle. Learn and communicate the efforts of newly liberated countries in Africa and Asia to reconstruct their agriculture along the principles of food-first self-reliance.

Right now in every country there is a struggle going on over who controls food resources. We must evaluate every one of our actions in light of these struggles. The very existence of these struggles—about which we are kept so ignorant—proves that people are never too oppressed to fight for power over their own lives.

37 Technology with a Human Face

E. F. Schumacher

E. F. Schumacher, a British economist, is the founder of the Intermediate Technology Development Group, which carries out some of the ideas he offers in *Small Is Beautiful.* His forthcoming books are *Good Work* and *Guide for the Perplexed.*

E. F. Schumacher, *Small Is Beautiful: Economics as If People Mattered* (New York: Harper & Row; London: Blond & Briggs, 1973), pp. 138–51.

The modern world has been shaped by its metaphysics, which has shaped its education, which in turn has brought forth its science and technology. So, without going back to metaphysics and education, we can say that the modern world has been shaped by technology. It tumbles from crisis to crisis; on all sides there are prophecies of disaster and, indeed, visible signs of breakdown.

If that which has been shaped by technology, and continues to be so shaped, looks sick, it might be wise to have a look at technology itself. If technology is felt to be becoming more and more inhuman, we might do well to consider whether it is possible to have something better—a technology with a human face.

Strange to say, technology, although of course the product of man, tends to develop by its own laws and principles, and these are very different from those of human nature or of living nature in general. Nature always, so to speak, knows where and when to stop. Greater even than the mystery of natural growth is the mystery of the natural cessation of growth. There is measure in all natural things—in their size, speed, or violence. As a result, the system of nature, of which man is a part, tends to be self-balancing, self-adjusting, self-cleansing. Not so with technology, or perhaps I should say: not so with man dominated by technology and specialisation. Technology

This world, after all our science and sciences, is still a miracle; wonderful, inscrutable, *magical* and more, to whosoever will *think* of it.
THOMAS CARLYLE

recognises no self-limiting principle—in terms, for instance, of size, speed, or violence. It therefore does not possess the virtues of being self-balancing, self-adjusting, and self-cleansing. In the subtle system of nature, technology, and in particular the super-technology of the modern world, acts like a foreign body, and there are now numerous signs of rejection.

Suddenly, if not altogether surprisingly, the modern world, shaped by modern technology, finds itself involved in three crises simultaneously. First, human nature revolts against inhuman technological, organisational, and political patterns, which it experiences as suffocating and debilitating; second, the living environment which supports human life aches and groans and gives signs of partial breakdown; and, third, it is clear to anyone fully knowledgeable in the subject matter that the inroads being made into the world's non-renewable resources, particularly those of fossil fuels, are such that serious bottlenecks and virtual exhaustion loom ahead in the quite foreseeable future.

Any one of these three crises or illnesses can turn out to be deadly. I do not know which of the three is the most likely to be the direct cause of collapse. What is quite clear is that a way of life that bases itself on materialism, i.e., on permanent, limitless expansionism in a finite environment, cannot last long, and that its life expectation is the shorter the more successfully it pursues its expansionist objectives.

If we ask where the tempestuous developments of world industry during the last quarter-century have taken us, the answer is somewhat discouraging. Everywhere the problems seem to be growing faster than the solutions. This seems to apply to the rich countries just as much as to the poor. There is nothing in the experience of the last twenty-five years to suggest that modern technology, as we know it, can really help us to alleviate world poverty, not to mention the problem of unemployment which already reaches levels like thirty per cent in many so-called developing countries, and now threatens to become endemic also in many of the rich countries. In any case, the apparent yet illusory successes of the last twenty-five years cannot be repeated: the threefold crisis of which I have spoken will see to that. So we had better face the question of technology—what does it do and what should it do? Can we develop a technology which really helps us to solve our problems—a technology with a human face?

The primary task of technology, it would seem, is to lighten the burden of work man has to carry in order to stay alive and develop his potential. It is easy enough to see that technology fulfills this purpose when we watch any particular piece of machinery at work—a computer, for instance, can do in seconds what it would take clerks or even mathematicians a very long time, if they can do it at all. It is more difficult to convince oneself of the truth of this simple proposition when one looks at whole societies. When I first began to travel the world, visiting rich and poor countries alike, I was tempted to formulate the first law of economics as follows: "The amount of real leisure a society enjoys tends to be in inverse proportion to the amount of labour-saving machinery it employs." It might be a good idea for the professors of economics to put this proposition into their examination papers and ask their pupils to discuss it. However that may be, the evidence is very strong indeed. If you go from easy-going England to, say, Germany or the United States, you find that people there live under much more strain than here. And if you move to a country like Burma, which is very near to the bottom of the league table of industrial progress, you find that people have an enormous amount of leisure really to enjoy themselves. Of course, as there is so much less labour-saving machinery to help them, they "accomplish" much less than we do; but that is a different point. The fact remains that the burden of living rests much more lightly on their shoulders than on ours.

The question of what technology actually does for us is therefore worthy of investigation. It obviously greatly reduces some kinds of work while it increases other kinds. The type of work which modern technology is most successful in reducing or even eliminating is skilful, productive work of human hands, in touch with real materials of one kind or another. In an advanced industrial society, such work has become exceedingly rare, and to make a decent living by doing such work has become virtually impossible. A great part of the modern neurosis may be due to this very fact; for the human being, defined by Thomas Aquinas as a being with brains and hands, enjoys nothing more than to be creatively, usefully, productively engaged with both his hands and his brains. Today, a person has to be wealthy to be able to afford space and good tools; he has to be lucky enough to find a good teacher and plenty of free time to learn and practise. He really has to be rich enough not to need a job; for the number of jobs that would be satisfactory in these respects is very small indeed.

The extent to which modern technology has taken over the work of human hands may be illustrated as follows. We may ask how much of "total social time"—that is to say, the time all of us have together, twenty-four hours a day each—is actually engaged in real production. Rather less than one-half of the total population of this country is, as they say, gainfully occupied, and about one-third of these are actual producers in agriculture, mining, construction, and industry. I do mean *actual producers,* not people who tell other people what to do, or account for the past, or plan for the future, or distribute what other people have produced. In other words, rather less than one-sixth of the total population is engaged in actual production; on average, each of them supports five others beside himself, of which two are gainfully employed on things other than real production and three are not gainfully employed. Now, a fully employed person, allowing for holidays, sickness, and other absence, spends about one-fifth of his total time on his job. It follows that the proportion of "total social time" spent on actual production—in the narrow sense in which I am using the term—is, roughly, one-fith of one-third of one-half, i.e., 3½ per cent. The other 96½ per cent of "total social time" is spent in other ways, including sleeping, eating, watching television, doing jobs that are not *directly* productive, or just killing time more or less humanely.

Although this bit of figuring work need not be taken too literally, it quite adequately serves to show what technology has enabled us to do: namely, to reduce the amount of time actually spent on production in its most elementary sense to such a tiny percentage of total social time that it pales into insignificance, that it carries no real weight, let alone prestige. When you look at industrial society in this way, you cannot be surprised to find that prestige is carried by those who help fill the other 96½ per cent of total social time, primarily the entertainers but also the executors of Parkinson's Law. In fact, one might put the following proposition to students of sociology: "The prestige carried by people in modern industrial society varies in inverse proportion to their closeness to actual production."

There is a further reason for this. The process of confining productive time to 3½ per cent of total social time has had the inevitable effect of taking all normal human pleasure and satisfaction out of the time spent on this work. Virtually all real production has been turned into an inhuman chore which does not enrich a man but empties him. "From the factory," it has been said, "dead matter goes out improved, whereas men there are corrupted and degraded."

We may say, therefore, that modern technology has deprived man of the kind of work that he enjoys most, creative, useful work with hands and brains, and given him plenty of work of a fragmented kind, most of which he does not enjoy at all. It has multiplied the number of people who are exceedingly busy doing kinds of work which, if it is productive at all, is so only in an indirect or "roundabout" way, and much of which would not be necessary at all if technology were rather less modern. Karl Marx appears to have foreseen much of this when he wrote: "They want production to be limited to useful things, but they forget that the production of too many useful things results in too many useless people," to which we might add: particularly when the processes of production are joyless and boring. All this confirms our suspicion that modern technology, the way it has developed, is developing, and promises further to develop, is showing an increasingly inhuman face, and that we might do well to take stock and reconsider our goals.

Taking stock, we can say that we possess a vast accumulation of new knowledge, splendid scientific techniques to increase it further, and immense experience in its application. All this is truth of a kind. This truthful knowledge, as such, does *not* commit us to a technology of giantism, supersonic speed, violence, and the destruction of human work-enjoyment. The use we have made of our knowledge is only one of its possible uses and, as is now becoming ever more apparent, often an unwise and destructive use.

As I have shown, directly productive time in our society has already been reduced to about 3½ per cent of total social time, and the whole drift of modern technological development is to reduce it further, asymptotically* to zero. Imagine we set ourselves a goal in the opposite direction—to increase it sixfold, to about twenty per cent, so that twenty per cent of total social time would be used for actually producing things, employing hands and brains and, naturally, excellent tools. An incredible thought! Even children would be allowed to make themselves useful, even old people. At one-sixth of present-day productivity, we should be producing as much as at present. There would be six times as much time for any piece of work we chose to undertake—enough to make a really good job of it, to enjoy oneself, to produce real quality, even to make things beautiful. Think of the therapeutic value of real work; think of its educational value. No one would then want to raise the school-leaving age or to lower the retirement age, so as to keep people off the labour market. Everybody would be welcome to lend a hand. Everybody would be admitted to what is now the rarest privilege, the opportunity of working usefully, creatively, with his own hands and brains, in his own time, at his own pace—and with excellent tools. Would this mean an enormous extension of working hours? No, people who work in this way do not know the difference between work and leisure. Unless they sleep or eat or occasionally choose to do nothing at all, they are always agreeably, productively engaged. Many of the "on-cost jobs" would simply disappear; I leave it to the reader's imagination to identify them. There would be little need for mindless entertainment or other drugs, and unquestionably much less illness.

Now, it might be said that this is a romantic, a utopian, vision. True enough. What we have today, in modern industrial society, is not romantic and certainly not utopian, as we have it right here. But it is in very deep trouble and holds no promise of survival. We jolly well have to have the courage to dream if we want to survive and give our children a chance of

---

*Asymptote: A mathematical line continually approaching some curve but never meeting it within a finite distance.

survival. The threefold crisis of which I have spoken will not go away if we simply carry on as before. It will become worse and end in disaster, until or unless we develop a new life-style which is compatible with the real needs of human nature, with the health of living nature around us, and with the resource endowment of the world.

Now, this is indeed a tall order, not because a new life-style to meet these critical requirements and facts is impossible to conceive, but because the present consumer society is like a drug addict who, no matter how miserable he may feel, finds it extremely difficult to get off the hook. The problem children of the world—from this point of view and in spite of many other considerations that could be adduced—are the rich societies and not the poor.

It is almost like a providential blessing that we, the rich countries, have found it in our heart at least to consider the Third World and to try to mitigate its poverty. In spite of the mixture of motives and the persistence of exploitative practices, I think that this fairly recent development in the outlook of the rich is an honourable one. And it could save us; for the poverty

Ideals are like the stars—we never reach them, but like the mariners on the sea, we chart our course by them.
CARL SCHURTZ

of the poor makes it in any case impossible for them successfully to adopt our technology. Of course, they often try to do so, and then have to bear the most dire consequences in terms of mass unemployment, mass migration into cities, rural decay, and intolerable social tensions. They need, in fact, the very thing I am talking about, which we also need: a *different* kind of technology, a technology with a human face, which, instead of making human hands and brains redundant, helps them to become far more productive than they have ever been before.

As Gandhi said, the poor of the world cannot be helped by mass production, only by production by the masses. The system of *mass production,* based on sophisticated, highly capital-intensive, high energy-input dependent, and human labour-saving technology, presupposes that you are already rich, for a great deal of capital investment is needed to establish one single workplace. The system of *production by the masses* mobilises the priceless resources which are possessed by all human beings, their clever brains and skilful hands, *and supports them with first-class tools.* The technology of *mass production* is inherently violent, ecologically damaging, self-defeating in terms of non-renewable resources, and stultifying for the human person. The technology of *production by the masses,* making use of the best of modern knowledge and experience, is conducive to decentralisation, compatible with the laws of ecology, gentle in its use of scarce resources, and designed to serve the human person instead of making him the servant of machines. I have named it *intermediate technology* to signify that it is vastly superior to the primitive technology of bygone ages but at the same time much simpler, cheaper, and freer than the super-technology of the rich. One can also call it self-help technology, or democratic or people's technology—a technology to which everybody can gain admittance and which is not reserved to those already rich and powerful.

Although we are in possession of all requisite knowledge, it still requires a systematic, creative effort to bring this technology into active existence and make it generally visible and available. It is my experience that it is rather more difficult to recapture directness and simplicity than to advance in the direction of ever more sophistication and complexity. Any third-rate engineer or researcher can increase complexity; but it takes a certain flair of real insight to make things simple again. And this insight does not come easily to people who have allowed themselves to become alienated from real, productive work and from the self-balancing system of nature, which never fails to recognise measure and limitation. Any activity which fails to recognise a self-limiting principle is of the devil. In our work with the developing countries we are at least forced to recognise the limitations of poverty, and this work can therefore be a wholesome school for all of us in which, while genuinely trying to help others, we may also gain knowledge and experience how to help ourselves.

I think we can already see the conflict of attitudes which will decide our future. On the one side, I see the people who think they can cope with our threefold crisis by the methods current, only more so; I call them the people of the forward stampede. On the other side, there are people in search of a new life-style, who seek to return to certain basic truths about man and his world; I call them home-comers. Let us admit that the people of the forward stampede, like the devil, have all the best tunes or at least the most popular and familiar tunes. You cannot stand still, they say; standing still means going down; you must go forward; there is nothing wrong with modern technology except that it is as yet incomplete; let us complete it. Dr. Sicco

Mansholt, one of the most prominent chiefs of the European Economic Community, may be quoted as a typical representative of this group. "More, further, quicker, richer," he says, "are the watchwords of present-day society." And he thinks we must help people to adapt "for there is no alternative." This is the authentic voice of the forward stampede, which talks in much the same tone as Dostoyevsky's Grand Inquisitor: "Why have you come to hinder us?" They point to the population explosion and to the possibilities of world hunger. Surely, we must take our flight forward and not be fainthearted. If people start protesting and revolting, we shall have to have more police and have them better equipped. If there is trouble with the environment, we shall need more stringent laws against pollution, and faster economic growth to pay for anti-pollution measures. If there are problems about natural resources, we shall turn to synthetics; if there are problems about fossil fuels, we shall move from slow reactors to fast breeders and from fission to fusion. There *are* no insoluble problems. The slogans of the people of the forward stampede burst into the newspaper headlines every day with the message, "a breakthrough a day keeps the crisis at bay."

And what about the other side? This is made up of people who are deeply convinced that technological development has taken a wrong turn and needs to be redirected. The term "home-comer" has, of course, a religious connotation. For it takes a good deal of courage to say "no" to the fashions and fascinations of the age and to question the presuppositions of a civilisation which appears destined to conquer the whole world; the requisite strength can be derived only from deep convictions. If it were derived from nothing more than fear of the future, it would be likely to disappear at the decisive moment. The genuine "home-comer" does not have the best tunes, but he has the most exalted text, nothing less than the Gospels. For him, there could not be a more concise statement of his situation, of *our* situation, than the parable of the prodigal son. Strange to say, the Sermon on the Mount gives pretty precise instructions on how to construct an outlook that could lead to an Economics of Survival.

—How blessed are those who know that they are poor:
the Kingdom of Heaven is theirs.

—How blessed are the sorrowful;
they shall find consolation.

—How blessed are those of a gentle spirit;
they shall have the earth for their possession.

—How blessed are those who hunger and thirst to see right prevail;
they shall be satisfied;

—How blessed are the peacemakers;
God shall call them his sons.

It may seem daring to connect these beatitudes with matters of technology and economics. But may it not be that we are in trouble precisely because we have failed for so long to make this connection? It is not difficult to discern what these beatitudes may mean for us today:

—We are poor, not demigods.

—We have plenty to be sorrowful about, and are not emerging into a golden age.

—We need a gentle approach, a non-violent spirit, and small is beautiful.

—We must concern ourselves with justice and see right prevail.

—And all this, only this, can enable us to become peacemakers.

The home-comers base themselves upon a different picture of man from that which motivates the people of the forward stampede. It would be very superficial to say that the latter believe in "growth" while the former do not. In a sense, everybody believes in growth, and rightly so, because growth is an essential feature of life. The whole point, however, is to give to the idea of growth a qualitative determination; for there are always many things that ought to be growing and many things that ought to be diminishing.

Equally, it would be very superficial to say that the home-comers do not believe in progress, which also can be said to be an essential feature of all life. The whole point is to determine what constitutes progress. And the home-comers believe that the direction which modern technology has taken and is continuing to pursue—towards ever-greater size, ever-higher speeds, and ever-increased violence, in defiance of all laws of natural harmony—is the opposite of progress. Hence the call for taking stock and finding a new orientation. The stocktaking indicates that we are destroying our very basis of existence, and the reorientation is based on remembering what human life is really about.

In one way or another everybody will have to take sides in this great conflict. To "leave it to the experts" means to side with the people of the forward stampede. It is widely accepted that politics is too important a matter to be left to experts. Today, the main content of politics is economics, and the main content of economics is technology. If politics cannot be left to the experts, neither can economics and technology.

The case for hope rests on the fact that ordinary people are often able to take a wider view, and a more "humanistic" view, than is normally being taken by experts. The power of ordinary people, who today tend to feel utterly powerless, does not lie in starting new lines of action, but in placing their sympathy and support with minority groups which have already started. I shall give two examples, relevant to the subject here under discussion. One relates to agriculture, still the greatest single activity of man on earth, and the other relates to industrial technology.

Modern agriculture relies on applying to soil, plants, and animals ever-increasing quantities of chemical products, the long-term effect of which on soil fertility and health is subject to very grave doubts. People who raise such doubts are generally confronted with the assertion that the choice lies between "poison or hunger." There are highly successful farmers in many

Science cannot solve the ultimate mystery of nature. And that is because, in the last analysis, we ourselves are part of nature and therefore part of the mystery that we are trying to solve.
MAX PLANCK

countries who obtain excellent yields without resort to such chemicals and without raising any doubts about long-term soil fertility and health. For the last twenty-five years, a private, voluntary organisation, the Soil Association, has been engaged in exploring the vital relationships between soil, plant, animal, and man; has undertaken and assisted relevant research; and has attempted to keep the public informed about developments in these fields. Neither the successful farmers nor the Soil Association have been able to attract official support or recognition. They have generally been dismissed as "the muck and mystery people," because they are obviously outside the mainstream of modern technological progress. Their methods bear the mark of non-violence and humility towards the infinitely subtle system of natural harmony, and this stands in opposition to the life-style of the modern world. But if we now realise that the modern life-style is putting us into mortal danger, we may find it in our hearts to support and even join these pioneers rather than to ignore or ridicule them.

On the industrial side, there is the Intermediate Technology Development Group. It is engaged in the systematic study on how to help people to help themselves. While its work is primarily concerned with giving technical assistance to the Third World, the results of its research are attracting increasing attention also from those who are concerned about the future of the rich societies. For they show that an intermediate technology, a technology with a human face, is in fact possible; that it is viable; and that it reintegrates the human being, with his skilful hands and creative brain, into the productive process. It serves *production by the masses* instead of *mass production.* Like the Soil Association, it is a private, voluntary organisation depending on public support.

I have no doubt that it is possible to give a new direction to technological development, a direction that shall lead it back to the real needs of man, and that also means: *to the actual size of man.* Man is small, and, therefore, small is beautiful. To go for giantism is to go for self-destruction. And what is the cost of a reorientation? We might remind ourselves that to calculate the cost of survival is perverse. No doubt, a price has to be paid for anything worth while: to redirect technology so that it serves man instead of destroying him requires primarily an effort of the imagination and an abandonment of fear.

38 Toward an Ecological Solution

Murray Bookchin

Murray Bookchin, professor of community studies at Ramapo State College, New Jersey, has written widely (sometimes under the name Lewis Herber) on environmental problems. His most recent book is *Technology for Peace*.

From *Eco-Catastrophe*, by the Editors of Ramparts (New York: Harper & Row, 1970), pp. 42-53, first published in the May 1970 issue of *Ramparts* Magazine.

Popular alarm over environmental decay and pollution did not emerge for the first time merely in the late '60's, nor for that matter is it the unique response of the present century. Air pollution, water pollution, food adulteration and other environmental problems were public issues as far back as ancient times, when notions of environmental diseases were far more prevalent than they are today. All of these issues came to the surface again with the Industrial Revolution—a period which was marked by burgeoning cities, the growth of the factory system, and an unprecedented befouling and polluting of air and waterways.

Today the situation is changing drastically and at a tempo that portends a catastrophe for the entire world of life. What is not clearly understood in many popular discussions of the present ecological crisis is that the very nature of the issues has changed, that the decay of the environment is directly tied to the decay of the existing social structure. It is not simply malpractices or a given spectrum of poisonous agents that is at stake, but rather the very structure of modern agriculture, industry and the city. Consequently, environmental decay and ecological catastrophe cannot be averted merely by increased programs like "pollution control" which deal with sources rather than systems. To be commensurable to the problem, the solution must entail far-reaching revolutionary changes in society and in man's relation to man.

I

To understand the enormity of the ecological crisis and the sweeping transformation it requires, let us briefly revisit the "pollution problem" as it existed a few decades ago. During the 1930's, pollution was primarily a muckraking issue, a problem of expose journalism typified by Kallet and Schlink's "100 Million Guinea Pigs."

This kind of muckraking literature still exists in abundance and finds an eager market among "consumers," that is to say, a public that seeks personal and legislative solutions to pollution problems. Its supreme pontiff is Ralph Nader, an energetic young man who has shrewdly combined traditional muckraking with a safe form of "New Left" activism. In reality, Nader's emphasis belongs to another historical era, for the magnitude of the pollution problem has expanded beyond the most exaggerated accounts of the '30's. The new pollutants are no longer "poisons" in the popular sense of the term; rather they belong to the problems of ecology, not merely pharmacology, and these do not lend themselves to legislative redress.

What now confronts us is not the predominantly specific, rapidly degradable poisons that alarmed an earlier generation, but long-lived carcinogenic and mutagenic agents, such as radioactive isotopes and chlorinated hydrocarbons. These agents become part of the very anatomy of the individual by entering his bone structure, tissues and fat deposits. Their dispersion is so

This is the first age that's paid much attention to the future, which is a little ironic since we may not have one.
ARTHUR C. CLARKE

global that they become part of the anatomy of the environment itself. They will be within us and around us for years to come, in many cases for generations to come. Their toxic effects are usually chronic rather than acute; the deadly and mutational effects they produce in the individual will not be seen until many years have passed. They are harmful not only in large quantities, but in trace amounts; as such, they are not detectable by human senses or even, in many cases, by conventional methods of analysis. They damage not only specific individuals but the human species as a whole and virtually all other forms of life.

No less alarming is the fact that we must drastically revise our traditional notions of what constitutes an environmental "pollutant." A few decades ago it would have been absurd to describe carbon dioxide and heat as "pollutants" in the customary sense of the term. Yet in both cases they may well rank among the most serious sources of future ecological imbalance and pose major threats to the viability of the planet. As a result of industrial and domestic combustion activities, the quantity of carbon dioxide in the atmosphere has increased by roughly 25 per cent in the past 100 years, a figure that may well double again by the end of the century. The famous "greenhouse effect," which increasing quantities of the gas is expected to produce, has already been widely discussed: eventually, it is supposed, the gas will inhibit the dissipation of the earth's heat into space, causing a rise in overall temperatures which will melt the polar ice caps and result in an inundation of vast coastal areas. Thermal pollution, the result mainly of warm water discharged by nuclear and conventional power plants, has disastrous effects on the ecology of lakes, rivers and estuaries. Increases in water temperature not only damage the physiological and reproductive activities of fish; they also promote the great blooms of algae that have become such formidable problems in waterways.

What is at stake in the ecological crisis we face today is the very capacity of the earth to sustain advanced forms of life. The crisis is being drawn together by massive increases in "typical" forms of air and water pollution; by a mounting accumulation of nondegradable wastes, lead residues, pesticide residues and toxic additives in food; by the expansion of cities into vast urban belts; by increasing stresses due to congestion, noise and mass living; by the wanton scarring of the earth as a result of mining operations, lumbering, and real estate speculation. The result of all this is that the earth within a few decades has been despoiled on a scale that is unprecedented in the entire history of human habitation on the planet.

Finally, the complexity and diversity of life which marked biological evolution over many millions of years is being replaced by a simpler, more synthetic and increasingly homogenized environment. Aside from any esthetic considerations, the elimination of this complexity and diversity may prove to be the most serious loss of all. Modern society is literally undoing the work of organic evolution. If this process continues unabated, the earth may be reduced to a level of biotic simplicity where humanity—whose welfare depends profoundly upon the complex food chains in the soil, on the land surface and in the oceans—will no longer be able to sustain itself as a viable animal species.

II

In recent years a type of biological "cold warrior" has emerged who tends to locate the ecological crisis in technology and population growth, thereby divesting it of its explosive social content. Out of this focus has emerged a new version of "original sin" in which tools and machines, reinforced by sexually irresponsible humans, ravage the earth in concert. Both technology and sexual irresponsibility, so the argument goes, must be curbed—if not voluntarily, then by the divine institution called the state.

The naivete of this approach would be risible were it not for its sinister implications. History has known of many different forms of tools and machines, some of which are patently harmful to human welfare and the natural world, others of which have clearly improved the condition of man and the ecology of an area. It would be absurd to place plows and mutagenic defoliants, weaving machines and automobiles, computers and moon rockets, under a common rubric. Worse, it would be grossly misleading to deal with these technologies in a social vacuum.

Technologies consist of not only the devices humans employ to mediate their relationship with the natural world, but also the attitudes associated with these devices. These attitudes are distinctly social products, the results of the social relationships humans establish with each other. What is clearly needed is not a mindless deprecation of technology as such, but rather a reordering redevelopment of technologies according to ecologically sound principles. We need an eco-technology that will help harmonize society with the natural world.

The same over-simplification is evident in the neo-Malthusian alarm over population growth. The reduction of population growth to a mere ratio between birth rates and death rates obscures the many complex social factors that enter into both statistics. A rising or declining birth rate is not a simple biological datum, any more than is a rising or declining death rate. Both are subject to the influences of the economic status of the individual, the nature of family structure, the values of society, the status of women, the attitude toward children, the culture of the community, and so forth. A change in any single factor interacts with the remainder to produce the statistical data called "birth rate" and "death rate." Culled from such abstract ratios, population growth rates can easily be used to foster authoritarian controls and finally a totalitarian society, especially if neo-Malthusian propaganda and the failure of voluntary birth control are used as an excuse. In arguing that forcible measures of birth control and a calculated policy of indifference to hunger may eventually be necessary to stabilize world populations, the neo-Malthusians are already creating a climate of opinion that will make genocidal policies and authoritarian institutions socially acceptable.

It is supremely ironic that coercion, so clearly implicit in the neo-Malthusian outlook, has acquired a respected place in the public debate on ecology—for the roots of the ecological crisis lie precisely in the coercive basis of modern society. The notion that man must dominate nature emerges directly from the domination of man by man. The patriarchal family may have planted the seed of domination in the nuclear relations of humanity; the classical split between spirit and reality—indeed, mind and labor—may have nourished it; the anti-naturalistic bias of Christianity may have tended to its growth; but it was not until organic community relations, be they tribal, feudal or peasant in form, dissolved into market relationships that the planet itself was reduced to a resource for exploitation.

This centuries-long tendency finds its most exacerbating development in modern capitalism: a social order that is orchestrated entirely by the maxim "Production for the sake of production." Owing to its inherently competitive nature, bourgeois society not only pits humans against each other, but pits the mass of humanity against the natural world. Just as men are converted into commodities, so every aspect of nature is converted into a commodity, a resource to be manufactured and merchandised wantonly. Entire continental areas in turn are converted into factories, and cities into marketplaces. The liberal euphemisms for these unadorned terms are "growth," "industrial society" and "urban blight." By whatever language they are described, the phenomena have their roots in the domination of man by man.

As technology develops, the maxim "Production for the sake of production" finds its complement in "Consumption for the sake of consumption." The phrase "consumer society" completes the description of the present social order as an "industrial society." Needs are tailored by the mass media to create a public demand for utterly useless commodities, each carefully engineered to deteriorate after a predetermined period of time. The plundering of the human spirit by the marketplace is paralleled by the plundering of the earth by capital. The tendency of the liberal to identify the marketplace with human needs, and capital with technology, represents a calculated error that neutralizes the social thrust of the ecological crisis.

The strategic ratios in the ecological crisis are not the population rates of India but the production rates of the United States, a country that produces more than 50 per cent of the world's goods. Here, too, liberal euphemisms like "affluence" conceal the critical thrust of a blunt word like "waste." With a vast section of its industrial capacity committed to war production, the U.S. is literally trampling upon the earth and shredding ecological links that are vital to human survival. If current industrial projections prove to be accurate, the remaining 30 years of the century will witness a five-fold increase in electric power production, based mostly on nuclear fuels and coal. The colossal burden in radioactive wastes and other effluents that this increase will place on the natural ecology of the earth hardly needs description.

In shorter perspective, the problem is no less disquieting. Within the next five years, lumber production may increase an overall 20 per cent; the output of paper, five per cent annually; folding boxes, three per cent annually; metal cans, four to five per cent annually; plastics (which currently form one to two per cent of municipal wastes), seven per cent annually. Collectively, these industries account for the most serious pollutants in the environment. The utterly senseless nature of modern industrial activity is perhaps best illustrated by the decline in returnable (and reusable) beer bottles from 54 billion bottles in 1960 to 26 billion today. Their place has been taken over by "one-way bottles" (a rise from 8 to 21 billion in the same period) and cans (an increase from 38 to 51 billion). The "one-way bottles" and cans, of course, pose tremendous problems in solid waste disposal, but they do sell better.

It may be that the planet, conceived as a lump of minerals, can support these mindless increases in the output of trash. The earth, conceived as a complex web of life, certainly cannot. The only question is, can the earth survive its looting long enough for man to replace the current destructive social system with a humanistic, ecologically oriented society.

The apocalyptic tone that marks so many ecological works over the past decade should not be taken lightly. We are witnessing the end of the world, although whether this world is a long-established social order or the earth as a living organism still remains in question. The ecological crisis, with its threat of human extinction, has developed appositely to the advance of technology, with its promise of abundance, leisure and material security. Both are converging toward a single focus: At a point where the very survival of man is being threatened, the possibility of removing him from the trammels of domination, material scarcity and toil has never been more promising. The very technology that has been used to plunder the planet can now be deployed, artfully and rationally, to make it flourish.

It is necessary to overcome not only bourgeois society but also the long legacy of propertied society: the patriarchal family, the city, the state—indeed, the historic splits that separated mind from sensuousness, individual from society, town from country, work from play, man from nature. The spirit of spontaneity and diversity that permeates the ecological outlook toward the natural world must now be directed toward revolutionary change and utopian reconstruction in the social world. Propertied society, domination, hierarchy and the state, in all their forms, are utterly incompatible with the survival of the biosphere. Either ecology action is revolutionary action or it is nothing at all. Any attempt to reform a social order that by its very nature pits humanity against all the forces of life is a gross deception and serves merely as a safety valve for established institutions.

The application of ecological principles to social reconstruction, on the other hand, opens entirely new opportunities for imagination and creativity. The cities must be decentralized to serve the interests of both natural and social ecology. Urban gigantism is devastating not only to the land, the air, the waterways and the local climate, but to the human spirit. Having

I think we still have some measure of choice in futures. What with all our knowledge and our awareness—maybe our too many treatises—we at least have less excuse for failing than did all the civilizations of the past.
HERBERT J. MULLER

reached its limits in the megalopolis—an urban sprawl that can best be described as the "non-city"—the city must be replaced by a multitude of diversified, well-rounded communities, each scaled to human dimensions and to the carrying capacity of its ecosystem. Technology, in turn, must be placed in the service of meaningful human needs, its output gauged to permit a careful recycling of wastes into the environment.

With the community and its technology sculptured to human scale, it should be possible to establish new, diversified energy patterns: the combined use of solar power, wind power and a judicious use of fossil and nuclear fuels. In this decentralized society, a new sense of tribalism, of face-to-face relations, can be expected to replace the bureaucratic institutions of propertied society and the state. The earth would be shared communally, in a new spirit of harmony between man and man and between man and nature.

In the early years of the 19th century, this image of a new, free and stateless society was at best a distant vision, a humanistic ideal which revolutionaries described as communism or anarchism, and their opponents as utopia. As the one century passed into its successor, the advance of technology increasingly brought this vision into the realm of possibility. The ecological crisis of the late 20th century has now turned the possibility of its early decades into a dire necessity. Not only is humanity more prepared for the realization of this vision than at any time in history—a fact intuited by the tribalism of the youth culture—but upon its realization depends the very existence of humanity in the remaining years ahead.

Perhaps the most important message of Marx a century ago was the concept that humanity must develop the means of survival in order to live. Today, the development of a flexible, open-ended technology has reversed this concept completely. We stand on the brink of a post-scarcity society, a society that can finally remove material want and domination from the human condition. Perhaps the most important message of ecology is the concept that man must master the conditions of life in order to survive.

During the May–June uprising of 1968, the French students sensed the new equation in human affairs when they inscribed the demand: "Be realistic! Do the impossible!" To this demand, the young Americans who face the next century can add the more solemn injunction: "If we don't do the impossible, we shall be faced with the unthinkable."

39 What Has Posterity Ever Done for Me?

Robert L. Heilbroner

Robert L. Heilbroner, Norman Thomas Professor of Economics at the New School for Social Research, is author of *An Inquiry into the Human Prospect.*

From the *New York Times Magazine,* January 19, 1975, pp. 14-15.

Will mankind survive? Who knows? The question I want to put is more searching: Who cares? It is clear that most of us today do not care—or at least do not care enough. How many of us would be willing to give up some minor convenience—say, the use of aerosols—in the hope that this might extend the life of man on earth by a hundred years? Suppose we also knew with a high degree of certainty that humankind could not survive a thousand years unless we gave up our wasteful diet of meat, abandoned all pleasure driving, cut back on every use of energy that was not essential to the maintenance of a bare minimum. Would we care enough for posterity to pay the price of its survival?

I doubt it. A thousand years is unimaginably distant. Even a century far exceeds our powers of empathetic imagination. By the year 2075, I shall probably have been dead for three quarters of a century. My children will also likely be dead, and my grandchildren, if I have any, will be in their dotage. What does it matter to me, then, what life will be like in 2075, much less 3075? Why should I lift a finger to affect events that will have no more meaning for me 75 years after my death than those that happened 75 years before I was born?

There is no rational answer to that terrible question. No argument based on reason will lead me to care for posterity or to lift a finger in its behalf. Indeed, by every rational consideration, precisely the opposite answer is thrust upon us with irresistible force. As a Distinguished Professor of political economy at the University of London has written in the . . . winter [1974–75] issue of *Business and Society Review.*

> Suppose that, as a result of using up all the world's resources, human life did come to an end. So what? What is so desirable about an indefinite continuation of the human species, religious convictions apart? It may well be that nearly everybody who is already here on earth would be reluctant to die, and that everybody has an instinctive fear of death. But one must not confuse this with the notion that, in any meaningful sense, generations who are yet unborn can be said to be better off if they are born than if they are not.

Thus speaks the voice of rationality. It is echoed in the book *The Economic Growth Controversy* by a Distinguished Younger Economist from the Massachusetts Institute of Technology:

> . . . Geological time [has been] made comprehensible to our finite human minds by the statement that the 4.5 billion years of the earth's history [are] equivalent to once around the world in an SST. . . . Man got on eight miles before the end, and industrial man got on six feet before the end. . . . Today we are having a debate about the extent to which man ought to maximize the length of time that he is on the airplane.

> According to what the scientists now think, the sun is gradually expanding and 12 billion years from now the earth will be swallowed up by the sun. This means that our airplane has time to go round three more times. Do we want man to be on it for all three times around the world? Are we interested in man being on for another eight miles? Are we interested in man being on for another six feet? Or are we only interested in man for a fraction of a millimeter—our lifetimes?
>
> That led me to think: Do I care what happens a thousand years from now? . . . Do I care when man gets off the airplane? I think I basically [have come] to the conclusion that I don't care whether man is on the airplane for another eight feet, or if man is on the airplane another three times around the world.

Is this an outrageous position? I must confess it outrages me. But this is not because the economists' arguments are "wrong"—indeed, within their rational framework they are indisputably right. It is because their position reveals the limitations—worse, the suicidal dangers—of what we call "rational

If we believe that mankind should have a future it is time for us to insure that it is a future which we can bequeath to our descendants without feeling shame and guilt. We need in fact to reorient our attitude to time, absorbing more from the past, fussing less about the present and reaching out toward the future in a friendly and helpful way.
MAX NICHOLSON

argument" when we confront questions that can only be decided by an appeal to an entirely different faculty from that of cool reason. More than that, I suspect that if there is cause to fear for man's survival it is because the calculus of logic and reason will be applied to problems where they have as little validity, even as little bearing, as the calculus of feeling or sentiment applied to the solution of a problem in Euclidean geometry.

If the reason cannot give us a compelling argument to care for posterity—and to care desperately and totally—what can? For an answer, I turn to another distinguished economist whose fame originated in his profound examination of moral conduct. In 1759, Adam Smith published *The Theory of Moral Sentiments,* in which he posed a question very much like ours, but to which he gave an answer very different from that of his latter-day descendants.

Suppose, asked Smith, that "a man of humanity" in Europe were to learn of a fearful earthquake in China—an earthquake that swallowed up its millions of inhabitants. How would that man react? He would, Smith mused, "make many melancholy reflections upon the precariousness of human life, and the vanity of all the labors of man, which could thus be annihilated in a moment. He would, too, perhaps, if he was a man of speculation, enter into many reasonings concerning the effects which this disaster might produce upon the commerce of Europe, and the trade and business of the world in general." Yet, when this fine philosophizing was over, would our "man of humanity" care much about the catastrophe in distant China? He would not. As Smith tells us, he would "pursue his business or his pleasure, take his repose or his diversion, with the same ease and tranquillity as if nothing had happened."

But now suppose, Smith says, that our man were told he was to lose his little finger on the morrow. A very different reaction would attend the contemplation of this "frivolous disaster." Our man of humanity would be reduced to a tormented state, tossing all night with fear and dread—whereas "provided he never saw them, he will snore with the most profound security over the ruin of a hundred millions of his brethren."

Next, Smith puts the critical question: Since the hurt to his finger bulks so large and the catastrophe in China so small, does this mean that a man of humanity, given the choice, would prefer the extinction of a hundred million Chinese in order to save his little finger? Smith is unequivocal in his answer. "Human nature startles at the thought," he cries, "and the world in its greatest depravity and corruption never produced such a villain as would be capable of entertaining it."

But what stays our hand? Since we are all such creatures of self-interest (and is not Smith the very patron saint of the motive of self-interest?), what moves us to give precedence to the rights of humanity over those of our own immediate well-being? The answer, says Smith, is the presence within us all of a "man within the breast," an inner creature of conscience whose insistent voice brooks no disobedience: "It is the love of what is honorable and noble, of the grandeur and dignity, and superiority of our own characters."

It does not matter whether Smith's 18th-century view of human nature in general or morality in particular appeals to the modern temper. What matters is that he has put the question that tests us to the quick. For it is one thing to appraise matters of life and death by the principles of rational self-interest and quite another *to take responsibility for our choice.* I cannot imagine the Distinguished Professor from the University of London personally consigning humanity to oblivion with the same equanimity with which he writes off its demise. I am certain that if the Distinguished Younger Economist from M.I.T. were made responsible for determining the precise length of stay of humanity on the SST, he would agonize over the problem and end up by exacting every last possible inch for mankind's journey.

Of course, there are moral dilemmas to be faced even if one takes one's stand on the "survivalist" principle. Mankind cannot expect to continue on earth indefinitely if we do not curb population growth, thereby consigning billions or tens of billions to the oblivion of nonbirth. Yet, in this case, we

The great use of life is to spend it for something that will outlast it.
WILLIAM JAMES

sacrifice some portion of life-to-come in order that life itself may be preserved. This essential commitment to life's continuance gives us the moral authority to take measures, perhaps very harsh measures, whose justification cannot be found in the precepts of rationality, but must be sought in the unbearable anguish we feel if we imagine ourselves as the executioners of mankind.

This anguish may well be those "religious convictions," to use the phrase our London economist so casually tosses away. Perhaps to our secular cast of mind, the anguish can be more easily accepted as the furious power of the biogenetic force we see expressed in every living organism. Whatever its source, when we ask if mankind "should" survive, it is only here we can find a rationale that gives us the affirmation we seek.

This is not to say we will discover a religious affirmation naturally welling up within us as we career toward Armageddon. We know very little about how to convince men by recourse to reason and nothing about how to convert them to religion. A hundred faiths contend for believers today, a few perhaps capable of generating that sense of caring for human salvation on earth. But, in truth, we do not know if "religion" will win out. An appreciation of the magnitude of the sacrifices required to perpetuate life may well tempt us to opt for "rationality"—to enjoy life while it is still to be enjoyed on relatively easy terms, to write mankind a shorter ticket on the SST so that some of us may enjoy the next millimeter of the trip in first-class seats.

Yet I am hopeful that in the end a survivalist ethic will come to the fore—not from the reading of a few books or the passing twinge of a pious lecture, but from an experience that will bring home to us, as Adam Smith brought home to his "man of humanity," the personal responsibility that defies all the homicidal promptings of reasonable calculation. Moreover, I believe that the coming generations, in their encounters with famine, war and the threatened life-carrying capacity of the globe, may be given just such an experience. It is a glimpse into the void of a universe without man. I must rest my ultimate faith on the discovery by these future generations, as the ax of the executioner passes into their hands, of the transcendent importance of posterity for them.

Name Index

Subject Index

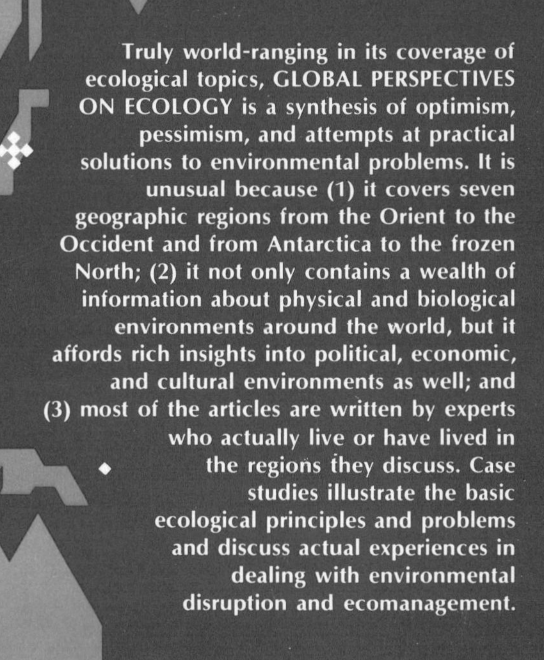

Truly world-ranging in its coverage of ecological topics, GLOBAL PERSPECTIVES ON ECOLOGY is a synthesis of optimism, pessimism, and attempts at practical solutions to environmental problems. It is unusual because (1) it covers seven geographic regions from the Orient to the Occident and from Antarctica to the frozen North; (2) it not only contains a wealth of information about physical and biological environments around the world, but it affords rich insights into political, economic, and cultural environments as well; and (3) most of the articles are written by experts who actually live or have lived in the regions they discuss. Case studies illustrate the basic ecological principles and problems and discuss actual experiences in dealing with environmental disruption and ecomanagement.

Mayfield Publishing Company